FOREWORD

In situ research and investigations have become an integral and essential part of national and international programmes designed to evaluate the feasibility and safety of deep geological repositories for radioactive waste. Accordingly, activities related to in situ research and investigations currently form an important part of the programme of the OECD Nuclear Energy Agency (NEA). In particular, research into the disposal of radioactive waste has been carried out at the former Stripa iron-ore mine in Central Sweden under NEA sponsorship throughout the 1980s. Bilateral co-operation at Stripa began with the joint Swedish-American Co-operative (SAC) programme in 1977, followed in 1980 by three successive international phases within an autonomous OECD International Stripa Project. The Project is managed by the Swedish Nuclear Fuel and Waste Management Company (SKB), under the direction of a Joint Technical Committee with representatives from each of the participating countries. As many as nine countries have participated in the Project.

The principal goals of both phase I and phase II of the Stripa Project were to develop techniques to assess the geology, hydrology, and geochemistry of potential sites for the disposal of radioactive waste, as well as to perform tests to examine groundwater flow within fractured rock and to assess properties of potential backfilling and sealing materials. The experimental work for phase II was completed in 1987, leading to considerable advancement of both investigative techniques and practical knowledge for repository siting and design. Phase III work commenced in 1986 and was completed in 1992. Research focused on three main areas: (1) development and improvement of site assessment methods and concepts, (2) characterisation of the Stripa granite and validation of fracture flow and radionuclide transport concepts, and (3) techniques and materials for the engineered sealing of possible groundwater flow paths through crystalline rock.

The OECD/NEA International Stripa Project has now completed phase III, the final phase of project. The Fourth International Symposium of the Stripa Project was held in Stockholm, Sweden, 14-16 October 1992, to review the accomplishments and developments of the completed Project. These Proceedings reproduce the papers contributed to the Symposium, along with a summary and conclusions prepared by the NEA Secretariat. The opinions, conclusions, and recommendations presented in these Proceedings are those of the authors only, and do not necessarily express the views of any Member country or international organisation. These proceedings are published on the responsibility of the Secretary-General.

OECD DOCUMENTS

of th. .... Inte................. ..... Symposium
St........... ... Sweden, ...tober 1992
.... .D Nuclear Energy Agency (NEA)
and
Swedish Nuclear Fuel and Waste Management Company (SKB)

PUBLISHER'S NOTE

The following texts have been left in their original form to permit faster distribution at lower cost.

ORGANISATION FOR ECONOMIC CO-OPERATION AND DEVELOPMENT

ORGANISATION FOR ECONOMIC CO-OPERATION AND DEVELOPMENT

Pursuant to Article 1 of the Convention signed in Paris on 14th December 1960, and which came into force on 30th September 1961, the Organisation for Economic Co-operation and Development (OECD) shall promote policies designed:

- to achieve the highest sustainable economic growth and employment and a rising standard of living in Member countries, while maintaining financial stability, and thus to contribute to the development of the world economy;
- to contribute to sound economic expansion in Member as well as non-member countries in the process of economic development; and
- to contribute to the expansion of world trade on a multilateral, non-discriminatory basis in accordance with international obligations.

The original Member countries of the OECD are Austria, Belgium, Canada, Denmark, France, Germany, Greece, Iceland, Ireland, Italy, Luxembourg, the Netherlands, Norway, Portugal, Spain, Sweden, Switzerland, Turkey, the United Kingdom and the United States. The following countries became Members subsequently through accession at the dates indicated hereafter: Japan (28th April 1964), Finland (28th January 1969), Australia (7th June 1971), New Zealand (29th May 1973) and Mexico (18th May 1994). The Commission of the European Communities takes part in the work of the OECD (Article 13 of the OECD Convention).

NUCLEAR ENERGY AGENCY

The OECD Nuclear Energy Agency (NEA) was established on 1st February 1958 under the name of the OEEC European Nuclear Energy Agency. It received its present designation on 20th April 1972, when Japan became its first non-European full Member. NEA membership today consists of all European Member countries of OECD as well as Australia, Canada, Japan, Republic of Korea, Mexico and United States. The Commission of the European Communities takes part in the work of the Agency.

The primary objective of NEA is to promote co-operation among the governments of its participating countries in furthering the development of nuclear power as a safe, environmentally acceptable and economic energy source.

This is achieved by:

- *encouraging harmonization of national regulatory policies and practices, with particular reference to the safety of nuclear installations, protection of man against ionising radiation and preservation of the environment, radioactive waste management, and nuclear third party liability and insurance;*
- *assessing the contribution of nuclear power to the overall energy supply by keeping under review the technical and economic aspects of nuclear power growth and forecasting demand and supply for the different phases of the nuclear fuel cycle;*
- *developing exchanges of scientific and technical information particularly through participation in common services;*
- *setting up international research and development programmes and joint undertakings.*

In these and related tasks, NEA works in close collaboration with the International Atomic Energy Agency in Vienna, with which it has concluded a Co-operation Agreement, as well as with other international organisations in the nuclear field.

AVANT-PROPOS

Les recherches et études in situ sont devenus une partie intégrante et essentielle des programmes nationaux et internationaux élaborés pour évaluer la faisabilité et la sûreté des dépôts de déchets radioactifs dans des formations géologiques profondes. De ce fait, les activités de recherche et les études in situ constituent actuellement une part importante du programme du Comité de la gestion des déchets radioactifs de l'AEN. En particulier, des recherches sur le stockage des déchets ont été exécutées, sous le patronage de l'AEN, dans l'ancienne mine de fer de Stripa, en Suède centrale, tout au long des années 80. La coopération bilatérale à Stripa remonte au programme SAC entrepris par la Suède et les Etats-Unis en 1977, qui a été lui-même suivi, en 1980, par les trois phases successives du Projet international autonome de Stripa de l'OCDE. Le projet était géré par SKB (société suédoise de gestion du combustible et des déchets nucléaires) sous la conduite d'un Comité technique commun composé de représentants de chacun des pays participants. Jusqu'à neuf pays ont pris part à ce projet.

Les phases I et II avaient principalement pour objet de mettre au point des techniques permettant d'évaluer les caractéristiques géologiques, hydrologiques et géochimiques de sites envisageables pour le stockage de déchets radioactifs et de procéder à des essais afin d'étudier l'écoulement des eaux souterraines dans des roches fissurées et d'évaluer les propriétés de matériaux de remblayage et de scellement potentiels. Les travaux expérimentaux de la phase II, qui se sont achevés en 1987, ont fait considérablement progresser les techniques d'investigation et les connaissances pratiques sur l'implantation et la conception des dépôts. Les travaux de la phase III, entamés en 1986, se sont achevés en 1991. Les recherches étaient axées essentiellement sur trois domaines : (1) mise au point et amélioration des méthodes et concepts d'évaluation des sites, (2) caractérisation du granit de Stripa et validation des modèles d'écoulement dans les fractures et de transport des radionucléides et (3) techniques et matériaux utilisables pour sceller les voies possibles d'écoulement de l'eau souterraine dans la roche cristalline.

La phase III, à savoir la phase finale du Projet international de Stripa de l'AEN/OCDE est à présent parvenue à son terme. Le quatrième Symposium international du Projet de Stripa a été organisé à Stockholm, Suède, du 14 au 16 octobre 1992 afin de passer en revue les résultats obtenus et les progrès réalisés dans le cadre de ce projet. Ce compte rendu contient les communications présentées à ce symposium ainsi qu'un résumé et des conclusions préparés par le secrétariat de l'AEN. Les opinions, conclusions et recommandations figurant dans ce compte rendu sont ceux des auteurs et ne reflètent pas nécessairement les avis d'un pays Membre ou d'une organisation. Ce compte rendu est publié sous la responsabilité du Secrétaire Général.

TABLE OF CONTENTS
TABLE DES MATIÈRES

OPENING SESSION
SÉANCE D'OUVERTURE

Chairmen - Présidents
E.S. Patera (OECD/NEA, Paris)
H. Carlsson (SGAB, Sweden)

SESSION I - SÉANCE I

NATURAL BARRIERS - CHARACTERISATION
BARRIÈRES NATURELLES - CHARACTÉRISATION DES SITES

Chairmen - Présidents
T. Isaacs (DOE, United States)
H. Sakuma (PNC, Japan)

SESSION I (Cont'd) - SÉANCE I (Suite)

Chairmen - Présidents
*V. Ryhänen (TVO, Finland)
R. Jackson (DOE, United States)*

SESSION II - SÉANCE II

NATURAL BARRIERS - MODELLING
BARRIÈRES NATURELLES - MODÉLISATION

Chairmen - Présidents
*K. Dormuth (AECL, Canada)
B. Levich (DOE, United States)*

SESSION III - SÉANCE III

ENGINEERED BARRIERS
BARRIÈRES OUVRAGÉES

Chairmen - Présidents
R. Lieb (NAGRA, Switzerland)
W. Danker (DOE, United States)

SESSION IV - SÉANCE IV

OVERVIEW REPORTING
RAPPORTS DE SYNTHÈSE

Chairman - Président
P.-E. Ahlström (SKB, Sweden)

SESSION V - SÉANCE V

PANEL DISCUSSION
TABLE RONDE

Chairman - Président
J. Hunter (SNL, United States)

EXECUTIVE SUMMARY

The Fourth International Symposium on the OECD/NEA Stripa Project was held on 14-16 October 1992, in Stockholm. This was the final Project Symposium and it marked the end of the Stripa Project. The Stripa Project began in 1977 as a co-operative effort between the United States and Sweden. The Project was expanded in 1980 to include up to nine countries under the auspices of the OECD Nuclear Energy Agency.

The primary goals of the Stripa Project were to develop techniques to scientifically characterise a granitic rock mass with regards to its hydrological and geochemical properties, to understand the influences of heating a granitic rock mass and develop techniques to seal various parts of a repository system. These issues were of common concern to many countries that had research and development programmes on the disposal of high-level radioactive waste. The Stripa Project provided a place, the old Stripa iron-ore mine in central Sweden, a large body of scientific talent from all the participating countries, a means to share the costs, and a forum for the discussion and the building of confidence regarding these issues. Therefore, the Stripa Project was able to satisfy many of the technical needs of radioactive waste management programmes around the world.

The Fourth Symposium of the Stripa Project was convened to present and review the accomplishments and developments of the complete Project. The Project, over the past twelve years, was divided into three phases.

Phase I of the Stripa Project, carried out between 1980 and 1985, involved: (1) the development of methods for determining the in situ hydraulic conductivity of a fractured rock mass by use of both single and multiple boreholes; (2) the development of procedures for evaluating the geochemical characteristics, including the origin and evolution of groundwaters; (3) studies of the migration of tracers in a single fracture; and (4) evaluations of the behaviour of Na-bentonite-based buffer materials under simulated repository conditions.

Phase II, carried out from 1983 until 1988, was concerned with: (1) the development of crosshole geophysical techniques, including radar, seismic and hydraulic methods, for the detection and characterisation of fracture zones; (2) an evaluation of migration of tracers in a large volume of a fractured rock mass; (3) an evaluation of the sealing ability of highly compacted Na-bentonite when used to plug drilled boreholes and excavations; (4) additional studies of the hydrologic and geochemical characteristics of the Stripa granite in order to enhance methods of data interpretation.

Phase III, carried out between 1986 and 1992, had two main objectives involving: (1) predictions of groundwater flow and nuclide transport within a previously undisturbed and uncharacterised large volume of rock in the Stripa mine and the comparison of those predictions with data collected by the use of improved methods for site characterisation; and (2) the selection and verification of the suitability of materials for the long-term sealing of fractures and fracture zones in crystalline rock. Phase III of the Stripa Project was different from the first two phases in that the Site Characterisation and Validation (SCV) programme was designed as a staged prediction-validation exercise utilising multi-disciplinary teams of site characterisers and modellers.

The Symposium was organised into five sessions: (1) Natural Barrier-characterisation; (2) Natural Barrier-modelling; (3) Engineered-barriers; (4) Overview Reporting; and (5) Panel Discussion. In the first session the Site Characterisation and Validation Programme was described for the large-scale volume of rock (150 x 150 x 50 m) within the Stripa mine used in the study. A staged multi-disciplinary investigation using boreholes and a drift was conducted by first characterising the site, making predictions, conducting experiments, and then analysing the results with regards to the predictions. This process was repeated on a more detailed level at the next stage. The data gathered during the site characterisation included: (1) fracture mapping; (2) stress measurement and analysis of the nature of stress-dependent joints; (3) geophysical logging, including borehole seismics and borehole radar, both techniques developed throughout the phases of the Stripa Project; (4) hydraulic properties of the rock mass and the fracture zones; (5) groundwater chemistry data; and (6) tracer transport experiments were conducted based on the knowledge gained during phases I and II from the single fracture tracer test and the 3-D migration test. All of this information was used to develop a conceptual model of the large volume characterisation block.

With the information obtained, the four modelling groups were able to model this site. Predictions were made on fracture occurrences, distribution of groundwater inflows and tracer transport into the validation drift. Both porous media and discrete fracture modelling were conducted. In order for the modellers to make predictions, a conceptual model of the site needed to be developed. This was a crucial area of integration between the scientists responsible for characterising the site and those scientists involved in the modelling. Also a process of validation was established based on a comparison between the predictions and the measured values of inflow. The predicted inflows of groundwater into the boreholes were in good agreement with the measured values. However, the predictions of groundwater inflow into the validation drift were overpredicted relative to the measured amounts. This overprediction is assumed to be accounted for by drift excavation effects, which decreased the inflows into the drift.

The session on engineered barriers reported the accomplishments and developments achieved in the area of sealing technology. During phase I, the emphasis was to demonstrate the use of a Na-bentonite buffer material under simulated repository conditions. Highly-compacted bentonite blocks were emplaced between a heater, simulating a waste package, and the granite walls of the emplacement borehole. Temperature and moisture content were monitored. Another experiment performed was the backfilling of a drift with highly-compacted bentonite and a bentonite/sand mixture. Both experiments demonstrated the usefulness of bentonite as a buffer and backfill material.

The phase II sealing programme was designed to develop and demonstrate the technology for sealing boreholes, shafts, and tunnels using highly-compacted Na-bentonite clay. The studies were designed to develop emplacement techniques, monitor the physical characteristics of the clay as it matured, and measure the sealing effectiveness of the designs. These experiments were successful by demonstrating that boreholes, shafts, and tunnels can be effectively sealed using bentonite clay. Much of this technology has already been put to use in the Swedish Final Repository for low- and intermediate-waste (SFR). The exploratory boreholes used to characterise the site were sealed with bentonite and the large concrete silo will be totally surrounded by a bentonite/sand mixture to provide a large scale buffer between the silo and the crystalline rock.

The sealing programme in phase III investigated techniques for sealing of fractures and fracture zones with both Na-bentonite and cement-based grouts. In addition to the techniques of emplacing grout, a study was made towards developing emplacement techniques of grout into very small aperture fractures and towards understanding the nature of long-term chemical changes of these materials. The work conducted on sealing of fractures in the Stripa Project provides a sound basis for further studies on the abilities of these materials to seal fractured rock.

The last two sessions of the Symposium were devoted to presentations of the summary overview reports being drafted for the Project and a panel discussion. The panel consisted of representatives from each of the participating countries. The discussion leader asked each member to give a brief account of what the Stripa Project meant for this country. All panel members agreed that the Stripa Project was a success from both a technical and an international perspective. The technical successes of the Stripa Project are many and are highlighted by:

- the development of geophysical characterisation tools, borehole seismics and borehole radar, which are now being used in Member countries site characterisation activities;

- the understanding of tracer transport properties in a fractured medium, including matrix diffusion and channelling;

- the development and use of both porous media and discrete fracture hydrological modelling for understanding the hydrology and making predictions;

- the integration of different disciplines of geology, geophysics, hydrology, geochemistry, and computer modelling to achieve a sensible conceptual model and the use of this information to make predictions and validate these techniques according to present criteria;

- the demonstration of sealing technologies utilising Na-bentonite and cement-based grouts.

The non-technical successes of the Stripa Project are equally as important and all members agreed the Stripa Project:

- provided a place and an opportunity to conduct a pure research and development programme that might not have been possible on a national scale;

- provided public credibility of the science and technology developed for radioactive waste management within an open international forum;

- spread the costs of conducting a research and development programme amongst many nations.

In conclusion, the Stripa Project was a well-conceived, well-managed and a highly successful international Project. The Stripa Project could be used as a model for other large scale international projects, not only concerned with radioactive waste management, but also for other environmental problems that face all nations.

EXPOSÉ DE SYNTHÈSE

Le quatrième Symposium international sur le Projet de Stripa de l'AEN/OCDE s'est tenu à Stockholm du 14 au 16 octobre 1992. Ce Symposium, qui était le dernier, marquait la fin du Projet de Stripa. Lancé en 1977 sous forme d'une coopération entre les Etats-Unis et la Suède, ce projet a pris une plus grande envergure à partir de 1980 avec la participation de plusieurs pays, jusqu'à neuf, sous les auspices de l'Agence de l'OCDE pour l'énergie nucléaire.

Le projet de Stripa avait pour objectifs premiers de mettre au point des techniques de caractérisation pour définir de manière scientifique les propriétés hydrologiques et géochimiques d'une masse rocheuse granitique, de comprendre les effets d'une élévation de température sur cette masse et d'élaborer des techniques permettant de sceller diverses parties d'un dépôt. Ces questions intéressaient de nombreux pays qui consacraient des programmes d'études et de recherches à l'évacuation de déchets de haute activité. Le Projet de Stripa avait l'avantage d'offrir un site, en l'occurrence l'ancienne mine de fer de Stripa en Suède centrale, de rassembler de nombreux experts scientifiques venant de tous les pays participants, de permettre de partager les coûts, de favoriser les échanges de vue et de développer la confiance dans ce domaine. C'est pourquoi ce projet était susceptible sur de nombreux aspects techniques d'apporter des éléments de réponses aux programmes de gestion des déchets radioactifs exécutés dans le monde.

Le quatrième Symposium international du Projet de Stripa a été organisé pour présenter et passer en revue les résultats obtenus et les progrès accomplis à l'issue de ce projet. Le projet, pendant les douze années de son existence, a comporté trois phases.

La phase I, conduite entre 1980 et 1985 portait sur : (1) la mise au point de méthodes de mesure de la conductivité hydraulique in situ d'une masse rocheuse fissurée à l'aide de forages isolés ou multiples, (2) la mise au point de procédures d'évaluation des caractéristiques géochimiques, et notamment étude de l'origine et de l'évolution des eaux souterraines, (3) les études de la migration de traceurs dans une fissure isolée et (4) l'évaluation du comportement des matériaux tampons à base de bentonite sodique dans des conditions de stockage simulées.

La phase II, qui s'est poursuivie de 1983 à 1988, a été consacrée à (1) la mise au point de techniques d'investigation géophysiques entre forages, y compris de méthodes d'investigation sismique, hydraulique et par radar, afin de détecter et de caractériser les zones fissurées; (2) la mesure de la migration de traceurs dans un grand volume de roche fissurée; (3) l'estimation de la capacité de scellement de la bentonite sodique fortement compactée utilisée pour colmater des forages et des excavations; (4) des études complémentaires des caractéristiques hydrologiques et géochimiques du granit de Stripa dans le but de perfectionner les méthodes d'interprétation des données.

La phase III, qui a duré de 1986 à 1992, visait deux objectifs principaux : (1) obtenir des prévisions de l'écoulement des eaux souterraines et du transport des nucléides dans une grande masse rocheuse encore non perturbée et non caractérisée de la mine de Stripa et comparer ces prévisions avec des données obtenues avec des méthodes perfectionnées de caractérisation des sites et (2) choisir des matériaux appropriés pour sceller des fissures et des zones fissurées dans des roches cristallines et vérifier leur adéquation. La phase III du projet de Stripa se distinguait des deux premières phases par le fait que le programme de caractérisation et de validation du site consistait en un exercice de prévision-validation en plusieurs étapes auquel ont participé des équipes pluridisciplinaires de spécialistes de la caractérisation et de la modélisation des sites.

Le Symposium comprenait cinq sessions : (1) caractérisation des barrières naturelles, (2) modélisation des barrières naturelles, (3) barrières ouvragées, (4) bilan des travaux, et (5) table ronde. La première session a été consacrée à la description du programme de caractérisation et de validation du site appliqué à une grande masse rocheuse de la mine de Stripa (150 x 150 x 50 m). Une investigation pluridisciplinaire en plusieurs étapes, réalisée sur des forages et une galerie, a commencé par une caractérisation du site, des prévisions et des expériences, suivies d'une comparaison des résultats aux prévisions. L'étape suivante a consisté à répéter les mêmes opérations mais plus en détail. Parmi les données recueillies pendant la caractérisation du site, on peut citer : (1) cartographie des fissures, (2) mesures des contraintes et analyse de la nature des joints sous contrainte, (3) diagraphie géophysique, y compris les techniques d'investigation sismique et par radar en forages qui ont été mises au point au cours des différentes phases du projet de Stripa, (4) propriétés hydrauliques de la masse rocheuse et des zones fissurées, (5) données chimiques sur les eaux souterraines; et (6) des expériences sur le transport de traceurs ont été réalisées en s'appuyant sur les connaissances fournies, au cours des phases I et II, par l'essai de migration de traceurs dans une fissure isolée et par un essai de migration tridimensionnel. Toutes ces informations ont été utilisées pour élaborer un modèle théorique du grand bloc de granit ayant fait l'objet de la caractérisation.

Les quatre groupes de modélisation ont pu, grâce aux informations obtenues, élaborer un modèle du site. Des prévisions ont été faites sur l'emplacement des fissures, la distribution des apports d'eau souterraine et le transport de traceurs dans la galerie de validation. Des milieux poreux et des fissures discontinues ont été modélisés. Il fallait qu'un modèle théorique du site soit élaboré pour que les responsables de la modélisation puissent faire des prévisons. Pour créer ce modèle, les experts de la caractérisation et de la modélisation ont dû faire un effort d'intégration. D'autre part, la validation a été basée sur la comparaison des prévisions et des valeurs mesurées de l'apport. On a ainsi observé une bonne concordance entre les valeurs prévues de l'apport d'eau souterraine dans les forages et les valeurs mesurées. Par contre, pour l'apport dans la galerie de validation, les prévisions dépassaient les valeurs mesurées. Cette surestimation est attribuée aux effets de l'excavation de la galerie qui ont réduit les apports dans la galerie.

La session sur les barrières ouvragées a été consacrée à la description des progrès et des innovations dans les techniques de scellement. Au cours de la première phase, il fallait avant tout démontrer la possibilité d'utiliser de la bentonite sodique comme matériau tampon dans des conditions simulées de stockage en dépôt. Des blocs de bentonite fortement compactée ont été placés entre un élément chauffant utilisé pour simuler un conteneur de déchets et les parois en granit du forage de mise en place des déchets. La température et la teneur en humidité étaient surveillées. Une autre expérience a consisté à remblayer une galerie avec de la bentonite fortement compactée et un mélange de bentonite et de sable. Les deux expériences ont démontré que la bentonite était un bon matériau tampon et de remblayage.

La phase II du programme de scellement devait permettre d'élaborer et de démontrer une technique de scellement des forages, des puits et des tunnels avec de l'argile de type bentonite sodique. Les études avaient pour objet de mettre au point des techniques de mise en place, de surveiller l'évolution des caractéristiques physiques de l'argile dans le temps et d'évaluer l'efficacité de scellement de diverses méthodes. Ces expériences ont permis de démontrer que les forages, les puits et les tunnels pouvaient être scellés efficacement avec de la bentonite. Une grande partie de cette technologie a déjà été utilisée dans l'installation suédoise de stockage définitif de déchets de faible et moyenne activités (SFR). Les forages de reconnaissance effectués pour caractériser le site ont été scellés avec de la bentonite, et un mélange de bentonite et de sable sera installé tout autour du grand silo de stockage en béton afin de faire tampon entre ce silo et la roche cristalline.

La phase III du programme de scellement avait pour but d'étudier les techniques de scellement de fissures et de zones fissurées avec de la bentonite sodique et du coulis à base de ciment. En plus des études consacrées aux techniques de mise en place du coulis, une étude a été réalisée pour mettre au point des techniques d'injection de coulis dans des fissures à très petites ouvertures et pour tenter de comprendre la nature des changements chimiques qui se produisent à long terme dans ces matériaux. Les travaux sur le scellement des fissures dans le projet de Stripa constituent une bonne base pour des études complémentaires sur l'aptitude de ces matériaux à sceller une roche fissurée.

Les deux dernières sessions du symposium ont été consacrées à la présentation de rapports de synthèse rédigés sur le projet et à une table ronde. Ont pris part à la table ronde des représentants de chacun des pays participants. L'animateur a demandé à chacun des membres d'exposer brièvement le point de vue de leur pays sur le projet de Stripa. Ceux-ci ont déclaré que le projet avait été une réussite tant du point de vue technique que sur le plan de la collaboration internationale. Les succès techniques sont nombreux comme en attestent :

- la mise au point d'outils de caractérisation géophysique, et de méthodes d'investigation sismique et par radar en forages, qui sont utilisés à présent dans les pays Membres pour la caractérisation des sites;

- les connaissances acquises sur les propriétés de transport de traceurs dans un milieu fissuré, et notamment sur la diffusion dans la matrice et les voies d'écoulement préférentielles;

- la mise au point et l'utilisation de modèles hydrologiques pour des milieux poreux et des fissures discontinues pour comprendre l'hydrologie et faire des prévisions;

- l'intégration de diverses disciplines, géologie, géophysique, hydrologie, géochimie, et de la modélisation informatique afin d'élaborer un bon modèle théorique et l'utilisation de celui-ci pour faire des prévisions et valider ces techniques en fonction des critères actuels;

- la démonstration des techniques de scellement avec de la bentonite sodique et des coulis à base de ciment.

Les succès du projet de Stripa dans des domaines non techniques sont tout aussi importants, et l'ensemble des membres reconnaissent que le projet de Stripa

- a non seulement fourni un site mais aussi l'occasion de réaliser un programme de recherches pures qui n'aurait peut-être pu être mené à bien à l'échelle nationale;

- a donné à la science et aux techniques de gestion des déchets radioactifs élaborées dans ce cadre international ouvert une crédibilité aux yeux du public;

- a permis de partager les coûts du programme de recherche et développement entre les pays Membres.

En conclusion, on peut dire que le projet de Stripa a été un projet international bien conçu et très fructueux. Il peut servir de modèle à d'autres projets internationaux de grande envergure qui seraient entrepris non pas uniquement dans le domaine de la gestion des déchets radioactifs mais aussi pour étudier d'autres problèmes de l'environnement rencontrés par tous les pays.

OPENING SESSION

SÉANCE D'OUVERTURE

Chairmen - Présidents
E.S. Patera (OECD/NEA)
H. Carlsson (Sweden)

WELCOMING ADDRESS

Jean-Pierre Olivier
OECD Nuclear Energy Agency

On behalf of the OECD Nuclear Energy Agency, I have the pleasure to welcome you to the Fourth and final joint NEA/SKB Symposium on the Stripa Project. After some 12 years of international co-operation within the Stripa Project, where many of you have been involved from the beginning, we can measure the high interest that the Project has raised by the attendance of a large number of managers and experts at this symposium. The Stripa Project has enhanced considerably the level of international co-operation on geological disposal and in situ research. It has also served as a signal to many countries, inviting them to launch their own field work shortly after. The symposium will give us a last opportunity to review the historical and scientific developments within the Stripa Programme, as well as the main achievements and lessons from this co-operation. However, rather than to elaborating further on the various topics which will be covered later at this meeting, I would propose to make some personal remarks as to where we seem to stand today in the field of radioactive waste management in general. In doing so, we may realise that we have finally made a big step forward during the life of the Stripa Project.

In the area of high-level waste management, many countries have today relatively clear objectives and plans which have been translated into specific criteria, standards, laws and regulations. These objectives and plans indicate that, in practice, geologic disposal is considered as the only realistic solution available. In this respect, there are currently no credible alternatives to the long-term isolation of high-level waste and spent fuel into geological systems. The two possible options which are sometimes suggested - indefinite storage with surveillance and separation and transmutation of long-lived radioactive nuclides - are by no means real alternatives. The first one, indefinite storage, cannot be regarded as a final solution in the sense that it does not meet a number of widely accepted ethical criteria, such as the need to protect populations and the environment in the far future and the need to avoid undue burden for our descendants. Therefore, indefinite storage should only be regarded as an interim solution which may help to extend the time until final and demonstrably safe disposal measures can be adopted. The second option, separation and transmutation of long-lived radionuclides, does not seem to have the potential to provide a complete satisfactory solution. Even if such processes prove cost-effective in the future and could be introduced in waste management schemes, they are likely to leave behind residual waste for which geologic disposal will almost certainly be required.

So we are left with geologic disposal and it is interesting to make a quick comparison with what we knew in the late 70's when Stripa was set up, and what we know now. The major difference is that we obviously know much more about every aspect of our problem. In particular now we have long-term performance assessment techniques which are becoming extremely sophisticated to the point that we run the risk of losing the sense of realism and forget that a certain degree of uncertainty will always remain. My feeling is that for each practical case, both sophisticated simulations and simple and robust assessments are useful and not mutually exclusive. The possibility of assessing sufficiently well the long-term behaviour of high-level waste disposal systems and the availability of the relevant techniques have been widely recognised following the publication, at the beginning of 1991, of a joint NEA/IAEA/CEC Collective Opinion on this subject.

This Collective Opinion, on the other hand, stressed the need to collect data from actual sites, in such a way that real data are incorporated into appropriate models and full site-specific performance assessments are done. Thanks to Stripa and similar national activities, we know how to characterise potential disposal sites, at least to a certain extent. We know also how to design and construct multibarrier systems which could be adapted to specific site conditions and satisfy long-term isolation requirements. The common objective of national programmes during the next 10 to 20 years is clearly to identify and characterise sites for disposal and develop complete disposal systems for implementation.

Looking at the state-of-the-art today in geologic disposal of long-lived radioactive waste, I have the impression that additional scientific progress as such is probably going to remain marginal. R&D activities are likely to become more and more selective. The emphasis of national plans seems to be in the field of technical applications rather than in the area of research. The remaining issues seem to be mainly connected to the detailed knowledge of potential disposal sites and to which degree of certainty their specific features can be assessed. This requires, in particular, close co-operation and interface between experimentalists and modellers and between performance assessment specialists and geoscientists in charge of site characterisation activities. The key words are going to be interface and integration. It is a worthwhile exercise to put together, in a coherent way, a great deal of information from many disciplines. The careful and rational integration of this information is absolutely necessary to build up or to support a safety case. I would like to mention, in this respect, the initiative of the regulatory authorities in Sweden with SKI Project-90. For SKI, Project-90 was a first step in this direction and we already know that the next version called

SITE-94 will go further, using site data from the Hard Rock Laboratory at Äspö. We also look forward to reading the new safety assessment, SKB-91 by our SKB colleagues and to hearing about progress elsewhere, notably in Canada, Finland, Japan, Switzerland, United Kingdom and the United States.

The integrated results of performance assessment and site characterisation efforts will indeed need to be scientifically and conceptually advanced and transparent. They should be scientifically and conceptually advanced because the multidisciplinary teams involved will have to satisfy themselves that they have gone far enough to prove their sense of responsibility and professional competence. The calculations should also be traceable and transparent in order to help pass a convincing message to others outside the scientific and technical community, such as the decision-makers. We have no option, but to make a first-class job on these two aspects of the same issue.

Of course, to be successful in the licensing and implementation of radioactive waste disposal systems requires some additional skills having to do with the art of communication and decision-making. But based on the considerable progress made in the low-level waste management area with three new disposal sites open this year, in France, Finland and Spain respectively in that order, I see reasons to be optimistic. The interface between environmental protection in general and radioactive waste management is increasing and our long-term isolation concepts and safety analysis approach are getting better and better recognised. Another type of interface concerns the one involving implementing agencies, such as SKB, and the various authorities, locally and nationally, which need to be part of the site selection and licensing process. An open and gradual approach (which Sten Bjurström is probably going to describe to you), may have the potential to promote a constructive dialogue and a more favourable attitude towards disposal plans. In particular, the type of conventional licensing which was inherited from past nuclear reactor practices with strict milestones and formal procedures leading to a final operating licence, may be less suitable in our field than a softer, less rigid step by step procedure, allowing gradual confidence build-up. In this regard, the function of "nuclear waste negotiator" which has been proposed and even implemented in several countries, may become an important element of such a procedure.

In this context, the role of international co-operation within NEA will probably continue to be at two main levels:

- globally, to contribute to an improved presentation and perception of the broad issues involved and of the solutions acceptable through independent assessments; and

-   topically, to stimulate discussion and progress in specific areas where there is an obvious common interest, such as the handling of human intrusion situations, the use of expert judgements in a technical and regulatory context, the definition of typical reference biospheres to be used in very long-term performance assessment calculations, etc.

In conclusion, as I have mentioned earlier, we have to be sophisticated and transparent at the scientific and technical level in order to present a convincing case for geologic disposal, in spite of the limitations which we will always face. We have also to be sophisticated and transparent for the sake of communication with the outside world. I am confident that this symposium will confirm these trends and usefully contribute to the ongoing debate. I am also confident that the legacy of the Stripa Project, through the efforts of many of you, will continue to inspire national programmes in many positive ways.

I wish you a very successful meeting and thank you for your attention.

The Stripa Project
in a Swedish Waste Management Perspective

Sten Bjurström
Swedish Nuclear Fuel and Waste Management Co.

The Stripa Project has for a long time been a very important part of the Swedish nuclear waste management programme. It has not been the only part of our programme but it has during the years played a very important key role of our work

During the late 60s when nuclear started in Sweden, we had an attitude toward waste programmes as in most other countries with nuclear power. From engineering point of view one founds that the potential for safe disposal solutions was great and the problem could be solved later when necessary.

At that time one could not anticipate the later extremely strict requirements where society should ask us to prove the safety in details for many thousands of years ahead. Never before has society asked industry to show - on beforehand that its activities should never be of any danger for humans and environment at any time.

The waste became, however, relatively early a political question here in Sweden and played an important role in the debate from the 70s. This, in turn, led to a view from the Swedish society to put on very strict requirements and in particular the Stipulation Act of 1977 marked this. In order to allow start-up of reactors one had to show that the waste could be disposed of in an absolutely safe way.

Since then comprehensive and intensive research work has been going on to show first feasibility and then more concrete solution for the nuclear waste problem. During the 70s we planned for the system we built during the 80s, which today takes care of all radioactive residues in the country.

As a result of this the low and medium level operational waste is now disposed of in an industrial scale in this country and could be looked on as a solved problem.

The target of the research work during the 80s was the final disposal of the spent fuel we today have for interim storage CLAB.

We have just some weeks ago submitted a programme to the authorities and the Government where we propose to go from the research and development period to use gained knowledge in its practical application in order to get encapsulated spent fuel and dispose encapsulated fuel as soon as possible. An initial phase to demonstrate all steps involved is planned to be operational 5-10 years after the turn of the century.

The Stripa Project has played a very important role in all this development work for us here in Sweden.

In the early 70s the Stripa Project gave us what could be called "a flying start" of the research programme. Within SKB and the KBS project one had at that time decided to work on a - for that time - very large programme on disposal in rock. Early in that research work the Stripa mine came into the picture as a possibility for research development.

The early participation by the United States at that time certainly became a booster for the whole research programme.

The collaboration helped us to better understand that the waste disposal could not be treated as a normal underground engineering problem. One needed in all aspects, a State of the Art-approach.

The early American proposals about very basic research together with a more down to the earth Swedish research programme on buffer tests etc. was the start of a very fruitful period which set the standard and ambitions for this kind of work.

The next important step was taken around 1980 and the NEA initiative to organize an international participation in the Stripa Project. By this the KBS project and its people and also Sweden got the necessary international network established within a very short time and got access to a number of large laboratories and brilliant brains around the world.

The work done at Stripa was internationally discussed and scrutinized which in turn - certainly contributed to quality and confidence in the work, not least from the Swedish authorities.

From this time on the Stripa Project also dealt with many important vital questions for a disposal in geological medium.

So, the investigation methods developed and tested have, as we all know, been very important for our work today. Lessons learned from rock characterization are much used at Äspö and all the tests, function of buffers and paths for migrations increased our fundamental knowledge.

It is not my task this morning to comment on this - there will be a number of presentations on this later on during this seminar.

I would rather come back to our new programme to achieve encapsulated spent fuel and to get this disposed of in a deep repository.

After the many years of research we now feel ready to use the gained knowledge to define as well the technical system we believe most in as demands and possibilities for siting of a repository in Sweden.

Although we have a system where all radioactive residues in this country are well taken care of and we could wait - the general opinion is that we should use all possibilities available to achieve a safe disposal as soon as practical. Regarding the question - if the waste disposal is solved problem or not - many may accept that we have the general knowledge available. At the same time quite a large number doubt if we in practioe build a facility. They do not see our job finished before encapsulated wastes are disposed of in a deep repository.

In the programme, now to be reviewed by authorities and expertise in Sweden and then submitted to the Government next year for final consideration, we present a programme and a procedure to decide and build the necessary facilities in steps.

In the new programme we define the canister type we believe most in and the SKB Board have also decided to build the necessary facilities for encapsulation at CLAB.

Regarding the deep repository we plan to start with the disposal of a small quantity of around 6-800 tonnes of spent fuel as a kind of demonstration. After a short time of evaluation the remaining spent fuel should be disposed of.

The whole process of the siting and for the disposal is intended to be very open process that should clearly demonstrate the various steps to outside groups, interested and concerned.

By this strategy we will be able to demonstrate encapsulation, handling, licensing, decision processes, disposal technology, and the quality of the work.

We will of course not be able to demonstrate the long-term safety - this has to be modelled.

By the existence of our interim storage we are enabled to take back the fuel if for some reason future generations would like to do so.

Along with this we will carry out quite a large programme of supporting research in areas of particularly importance to safety related matters. Our ambition to keep quality by focusing on the most promising lines of the programme we might be a little one-eyed. To avoid that we will also carry out an alternative but smaller programme on alternative methods and research.

This new strategy will in its first step really close the remaining part to demonstrate a safe disposal and at the same time be very open for as well demonstration of all processes involved as to new facts or alternative knowledge that will be gained during the relatively long period to carry out the demonstration. It is also - we understand an advantage not to ask for further going decisions than necessary. The important decision - to complete the disposal - will be taken around 2020 by those who have to carry out the job. They can decide whether they would like to carry on or do something else.

I would like to emphasize that we are not demonstrating the safety to ourselves. We believe we have a solid scientific ground for a safe way to take care of the waste. At the same time we must however work on an as flexible method as possible for the future and certainly be open for new facts, alternatives or changed conditions.

Acknowledgement I would like to end my talk by expressing our thanks to our international colleagues who have shown SKB the confidence to administrate the Stripa Project. We have certainly tried to do our best and I know that our work has been very much appreciated.

I also would like to emphasize that the Stripa Project can be seen as an example of a most fruitful international research project in it absolutely best sense. It is very seldom that research people around the world can use facts and data measured by others for their own work; in the Stripa Project many have used data and exchanged information openly and with very few commercial and other limitations.

We are very pleased that this has set the trend of behaviour in later international development. I believe that much of our co-operation, for instance with Canada and certainly within our new laboratory at Äspö, can carry on this tradition.

Relevance of the International Stripa Project to the National Nuclear Waste Management Programme

K.W. Dormuth and S.H. Whitaker

AECL Research
Whiteshell Laboratories
Pinawa, Manitoba, ROE 1LO Canada

Abstract

For over a decade the experiments at the Stripa Mine have been a major aspect of the worlds research into the geological disposal of nuclear fuel waste. The multi-national involvement in the Project has made the program cost-effective for each nation, and has enhanced the pool of world-class expertise available for the planning and execution of experiments. The impact on the Canadian program is an example of the impact on national programs. Information from the experiments enters our conceptual engineering, performance assessment, and experimental programs. Perhaps the greatest benefit of the Project has been the ground-breaking work in establishing the feasibility and value of international cooperation for in situ experiments on waste disposal.

Introduction

The experimental program at the Stripa Mine has made a major contribution to the global technology for underground disposal of high level radioactive waste. Insofar as the information from the program is published, all nations can benefit from the knowledge developed. However, the countries that have contributed directly to the program benefitted not only from the information generated, but from the experience gained in the planning and execution of the experiments.

The Stripa Project evolved from an experimental program performed, between 1977 and 1981, by the Swedish agency responsible for Nuclear Waste Management (now known as SKB) and Lawrence Berkeley Laboratories, under a cooperative agreement between SKB and the United States Department of Energy. This initial program clearly established the feasibility and value of international cooperation to make use of the versatile underground facility and attracted the interest of other countries involved in research on geological disposal of nuclear waste. This led to the first four-year phase of the International Stripa Project, established under the auspices of the OECD/NEA in January of 1981. A second phase overlapped the first, spanning the years 1983 to 1986. A third and final phase covers the period 1986-1992.

Canada has been involved with the Stripa Project since 1981, as an associate participant in Phase 1 and as a full participant in Phase 2 and Phase 3. Participation in the Project has been a valuable element of the Canadian Nuclear Fuel Waste Management Program, complementing research at Canadian laboratories and geological research areas. The impacts of the Stripa Project on the Canadian program are discussed here as an example of the Project's influence on national waste management programs.

Background on the Canadian Program

In 1978, the governments of Canada and Ontario established the Canadian Nuclear Fuel Waste Management Program to investigate the safety, security, and desirability of a concept for the long-term management of nuclear fuel waste: permanent disposal in a deep underground repository in intrusive igneous (plutonic) rock (Gov 1978). Subsequently, the governments of Canada and Ontario announced that no disposal site selection would be undertaken until after the concept had been reviewed and accepted (Gov 1981). The review of the disposal concept is now being undertaken by a federal Environmental Assessment Review Panel, which will also examine a broad range of issues related to nuclear fuel waste management. As the proponent, AECL must present its evidence for the acceptability of the concept in an Environmental Impact Statement, which will form the basis for the review of the concept.

The characteristic features of the disposal concept being investigated are:

1. The waste is either used CANDU fuel or a solid incorporating the highly radioactive waste products of a reprocessing operation. (CANDU fuel is not currently reprocessed for recycle, but the option remains available to do so at some time in the future.)

2. The waste is enclosed in containers designed to last at least 500 years.

3. The containers of waste are emplaced in excavated rooms or in the rock surrounding excavated rooms, which are nominally 500 to 1000 metres deep in well-characterized plutonic rock of the Canadian Shield.

4. The containers are surrounded by a buffer material, separating them from the rock.

5. Eventually all rooms, tunnels, shafts, and boreholes are filled with sealing materials, after which institutional controls are not required to maintain safety.

Within the scope of the concept, many options for siting, materials, and repository design are possible. Final choices among the options will not be made until the concept is implemented.

Sealing Technology Development

The technology for sealing a disposal vault has been investigated as part of the assessment of the disposal concept, and the Stripa Project has played a key role in this technology development. The program of sealing experiments at Stripa has been timely, allowing Canadian experiments on shaft and borehole sealing, for example, to be scheduled for a later time, following the review of the disposal concept.

Concrete bulkheads would most likely be employed to seal emplacement rooms. Tests in Stripa have shown that gaskets of highly compacted bentonite placed next to concrete bulkheads can be used to virtually eliminate water flow along the plug-rock interface (Pusch et al. 1987b). As well, methods for plugging both horizontal and vertical boreholes up to 100 m long with highly compacted bentonite have been demonstrated at Stripa (Pusch et al. 1987a).

The Stripa Buffer Mass Test (Pusch et al. 1985) was designed to determine the moisture transients in a highly compacted bentonite buffer and their effects on the mechanical and thermal performance of the buffer. The test showed, for example, that the saturation front advancing from the buffer-rock interface toward the inner portions of the buffer is geometrically regular and does not reflect the non-uniform distribution of visible inflow channels at the interface. Similarly, observations on the backfill above the boreholes showed that, while non-uniform inflow distribution was visible along the floor, walls, and back of the test room, wetting of the backfill was spatially uniform, controlled more by the water uptake properties of the backfill than by the non-uniformity of flow conditions at the backfill-rock interface.

The Stripa Project complemented AECL's program to assess the potential of cement-based sealing materials. A high performance grout was developed and successfully field-tested in 1987 at the URL. Later this and other similar grouts were studied at Stripa. When the Stripa Task Force on Sealing Materials proposed a program of experiments and modelling to assess the longevity of grouts in a repository environment, AECL participated directly by determining key physical and chemical characteristics of the grouts (Onofrei et al. 1992). The resultant performance modelling approach has helped AECL to establish a methodology for assessing grout longevity and to further refine the grout to be used in future field experiments.

Near-Field Conditions and Mass Transport

Experiments at Stripa have complemented those performed at the Canadian Underground Research Laboratory (URL) to investigate hydrogeological conditions in the near field. When conventional drilling and blasting techniques are used for excavation, as done at Stripa, measurements and analyses indicate (Pusch and Gray 1989) that the hydraulic conductivity of the rock has been significantly increased to distances of 0.5 m to 1 m into the rock from the excavation surfaces. Whereas the hydraulic conductivity of the undisturbed rock is likely to be less than 10^{-11} m/s, that in the disturbed zone may be as high as 10^{-7} m/s.

Data from Stripa (Pusch et al. 1985) and at the URL (Kozak 1990) show that hydraulic pressures in granite rocks can increase to values corresponding to those expected from assessments of the regional groundwater flow field within one opening diameter of the excavation. Results from the Stripa Buffer Mass Test show that, despite groundwater pressures of up to 1.5 MPa only ten metres away from the face of the excavation, the groundwater pressures at the excavation face had only risen to about 45 kPa 2.5 years after backfilling.

Development of Equipment and Methods

During both Phase 2 and Phase 3 of the Stripa Project (1983-1991) the development of borehole radar and borehole seismic equipment, survey methods and data analysis methods has been a significant activity. The Canadian program was particularly interested in these methods because of their potential for noninvasive investigation of interborehole volumes of rock. We had begun supporting the development of cross-hole seismic equipment and methods in 1980 and had conducted successful tests of a prototype system over interborehole distances of up to 175 m at our URL in 1982 (Wong et al. 1983).

Our participation in the Stripa Project enabled us to compare an alternative approach to the one we were following for the development of borehole seismic equipment and to evaluate borehole radar logging technology without having to split our own effort between the two technologies. We tested the radar system developed in the Stripa Project at the URL during 1987 and subsequently purchased a system (including data analysis software) for use in our program. Examples of results from both single borehole and crosshole surveys in a pair of boreholes are given by Holloway et al. 1992). For these boreholes there is good correspondence between reflectors in the single borehole radargrams, changes in velocity in the crosshole tomogram, and hydraulic connectivity determined from hydrogeological testing between the boreholes.

We are also using the software developed in the Stripa Project for the analysis of seismic survey data in our program.

Demonstration of Site Characterization Methods

Hydrogeochemical investigations were an important component of Phases 1 and 2 of the Stripa Project (1981-1986). The comprehensive nature of the analytical program undertaken (eg., Nordstrom et al. 1985) provided a good model for the development of hydrogeochemical investigations in our program. In addition, results from the Stripa Project raised issues that we have subsequently investigated in our program as well. A particular example was the conclusion that fluid inclusions in the Stripa granite could account for the salinity in the Stripa groundwater (Nordstrom 1983, Lindblom 1984). This led us to undertake a variety of leaching investigations both in the laboratory and in situ at the URL. Some of these investigations have shown leachable Cl concentrations in the granite of the Lac du Bonnet batholith comparable to those reported for the unfractured Stripa granite (Gascoyne et al. 1989). Studies of the 36 Cl/Cl ratio support derivation of the bulk of the Cl content in the saline groundwater of the Lac du Bonnet batholith from the rock matrix (Gascoyne et al. 1992).

In all three Phases of the Stripa Project, investigation of the migration of tracer substances through the rock has been an important component. We have been particularly interested in this aspect of the project, because tracer migration experiments in two domains of the rock at the URL (migration in intensely fractured zones and migration in moderately fractured rock) are a substantial part of the experimental program at the URL and are critical to developing our approach to the assessment of long term safety.

Throughout the Stripa Project we have emphasized the importance of establishing both the three-dimensional distribution of hydraulically conductive features in the vicinity of a tracer experiment and the hydraulic boundary conditions for the experiment. During Phase 3, monitoring of heads was begun in several boreholes in the mine to provide boundary conditions for the SCV block (Carlsten et al. 1988). The head distribution suggested that transport from some of the injection points for the 3-D Migration Experiment could have been southwestwards towards the old mine workings rather than to the collection points in the drift. This was confirmed when Eosin Y was discovered in a gallery about 150 m southwest of the injection point (Neretnieks et al. 1989). The improved characterization of the boundary conditions for the tracer experiments conducted in the SCV block led to significantly better recoveries of tracers. The recoveries from the SCV tracer tests are more consistent with those we have observed in our tracer experiments at the URL.

Hydraulic boundary conditions are monitored in an array of isolated intervals in boreholes in the volume of rock surrounding the URL. Initially, migration experiments in highly fractured rock were conducted in a radial flow field towards the URL Shaft through a major fracture zone. However, with time the monitoring array detected changes in the gradients that would be sufficiently rapid to change the boundary conditions during experiments, so the location of additional experiments was changed to a deeper fracture zone. Because there is no radial flow field developed in the deeper zone, the experiments are run as continuously recirculating injection-withdrawal tests (Frost et al. 1992). Because of the extensive characterization of the flow field prior to migration experiments, we have been able to account for greater proportions of the tracer mass (40-90%), thus reducing a major source of uncertainty in the interpretation of the results of the experiments.

Modelling of groundwater flow has been an important component of Phase 3 of the Stripa Project. We have used equivalent porous medium models of groundwater flow in our program, and, so far, our experience has been that they are appropriate for the analysis of flow. Therefore, we have been interested in the comparison of a variety of fracture flow modelling approaches with the equivalent porous medium model in the fracture modelling activity during Phase 3. The results appear to have substantiated both the applicability of the equivalent porous medium modelling approach for larger volumes and the applicability of discrete fracture modelling approaches at small size scales (Hodgkinson and Cooper 1992a, b). The results from our drawdown experiment conducted in association with the excavation of the shaft for the URL support the conclusions from Stripa regarding the porous medium modelling approach. Observed drawdowns corresponded well with those predicted using an equivalent porous medium modelling approach in which fracture zones were treated as planar finite elements embedded in a three dimensional array of volume elements (Davison 1986).

The Canadian Underground Research Laboratory

The Swedish-American experiments in Stripa were influential in the planning stages of the URL in that they illustrated the types of underground experiments that could be proposed to address technical issues relevant to waste disposal (Simmons et al. 1992). Subsequent phases of the Project contributed significantly to the development of the experiments for the operating phase of the URL. The successes and the shortcomings recognized in the Stripa experiments provide a valuable resource in developing the scope of our experimental program. The principal issues, experiment plans and project organization developed for Stripa have been useful in establishing the organization, objectives, rationale and scope for the URL operating phase experiments.

The experiments being conducted at Stripa were a significant factor in establishing the schedule for the URL experiments. For example, as previously noted, because of the sealing studies being performed at Stripa, it was possible to schedule our own sealing experiments for a later time, when there will be a greater accumulation of relevant knowledge that can feed into the design and execution of the experiments.

Conclusion

Information from the International Stripa Project is integral to the assessment of the Canadian Nuclear Fuel Waste Disposal Concept. It enters our conceptual engineering, performance assessment, and experimental programs.

The multi-national involvement in the Project has made the program cost-effective for each nation, and has enhanced the pool of world-class . expertise available for the planning and execution of experiments.

Perhaps the greatest benefit of the Project has been the ground breaking work in establishing the feasibility and value of international cooperation for in situ experiments on waste disposal. Much of the international cooperation in nuclear waste management today can trace its origins to the scientific contacts, cooperative principles, and technical effectiveness of the Stripa Project.

Acknowledgments

We are grateful to Mr. L.H. Johnson, Mr. G.R. Simmons, and Dr. M. Onofrei for providing valuable information.

The Canadian Nuclear Fuel Waste Management Program is jointly funded by AECL and Ontario Hydro under the auspices of the CANDU Owners Group.

REFERENCES

Carlsten, S., O. Olsson, O. Persson and M. Sehlstedt. 1988. Site characterization and validation - monitoring of head in the Stripa mine during 1987. Stripa Project TR 88-02, KBS, Stockholm, Sweden.

Davison, C.C. 1986. URL drawdown experiment and comparison with model predictions. Proceedings of the Twentieth Information Meeting of the Canadian Nuclear Fuel Waste Management Program. Atomic Energy of Canada Limited Technical Record TR-375, volume I, pp. 103-124.

EARP. 1991. Environmental Assessment Panel Reviewing the Nuclear Fuel Waste Management and Disposal Concept, "Guidelines for the Preparation of an Environmental Impact Statement," Federal Environmental Assessment Review Office, Fontaine Building, 13th Floor, 200 Sacre-Coueur Boulevard, Hull, Quebec, Canada K1A 0H3 (June).

Frost, L.H., N.W. Scheier, E.T. Kozak and C.C. Davison. 1992. Solute transport properties of a major fracture zone in granite. Proceedings of the 6th International Symposium on Water Tracing, Karlsruhe, Germany, September 21-26.

Gascoyne, M., J.D. Ross, R.L. Watson and D.C. Kamineni. 1989. Soluble salts in a Canadian Shield granite as contributors to groundwater salinity. Proceedings of the 6th International Symposium on Water Rock Interaction. Balkema, Rotterdam, pp. 247-249.

Gascoyne, M., D.C. Kamineni and J. Fabryka-Martin. 1992. Chlorine-36 in groundwaters in the Lac du Bonnet granite, southeastern Manitoba, Canada. Proceedings of the 7th International Symposium on Water Rock Interaction. Balkema, Rotterdam, pp. 931-933.

Gov. 1978. The Federal Minister of Energy, Mines and Resources and the Ontario Energy Minister, "Canada/Ontario Radioactive Waste Management Program," Printing and Publishing, Supply and Services Canada, Ottawa, Canada K1A 0S9 (June 5).

Gov. 1981. The Federal Minister of Energy, Mines and Resources and the Ontario Energy Minister, "Canada/Ontario Joint Statement on the Nuclear Fuel Waste Management Program," Printing and Publishing, Supply and Services Canada, Ottawa, Canada K1A 0S9 (August 4).

Hodgkinson, D. and N. Cooper. 1992a. A comparison of measurements and calculations for the Stripa validation drift inflow experiment, Stripa Project TR 92-07, KBS, Stockholm, Sweden.

Hodgkinson, D. and N. Cooper. 1992b. A comparison of measurements and calculations for the Stripa tracer experiments, Stripa Project TR 92-20, KBS, Stockholm, Sweden.

Holloway, A.L., K.M. Stevens and G.S. Lodha. 1992. The results of surface and borehole radar profiling from Permit Area B of the Whiteshell Research Area, Manitoba, Canada. Proceedings of the Fourth International Conference on Ground Penetrating Radar June 8-13, 1992. Rovaniemi, Finland, Geological Survey of Finland, Special Paper 16, pp. 329-337.

Kozak, E.T. 1990. Personal communication.

Lindblom, S. 1984. Hydrogeological and hydrogeochemical investigations in boreholes - fluid inclusion studies in the Stripa granite. Stripa Project TR 84-07, KBS, Stockholm, Sweden.

Neretnieks, I., L. Moreno, H. Abelin, L. Birgersson, H. Widen and T. Agren. 1990. A large scale flow and tracer experiment in granite. In Situ Experiments Associated with the Disposal of Radioactive Waste. Proceedings of the 3rd NEA/SKB Symposium on the International Stripa Project. OECD/NEA, Paris, pp. 123-134.

Nordstrom, D.K. 1983. Preliminary data on the geochemical characteristics of groundwater at Stripa. Proceedings of the OECD/NEA Workshop on Geological Disposal of Radioactive Waste, In Situ Experiments in Granite. OECD/NEA, Paris, pp. 143-153.

Nordstrom, D.K., J.N. Andrews, L. Carlsson, J.-C. Fontes, P. Fritz, H. Moser and T. Olsson. 1985. Hydrogeological and hydrogeochemical investigations in boreholes - final report of the Phase I geochemical investigations of the Stripa groundwaters. Stripa Project TR 85-06, KBS, Stockholm, Sweden.

Onofrei, M., M.N. Gray, W.E. Coons and S.R. Alcorn. 1992. High performance cement-based grouts for use in a nuclear waste disposal facility. Waste Management Vol. 12, pp. 133-154.

Pusch, R.L., L. Borgesson and G. Ramquist. 1985. Final report of the buffer mass test - Vol. 2: test results. Stripa Project TR 85-12, KBS, Stockholm, Sweden.

Pusch, R., L. Borgesson and G. Ramqvist. 1987a. Final report of the Borehole, shaft, and tunnel sealing test - volume I: borehole plugging. Stripa Project TR 87-01, KBS, Stockholm, Sweden.

Pusch, R., L. Borgesson and G. Ramqvist. 1987b. Final report of the Borehole, shaft, and tunnel sealing test - volume III: tunnel plugging. Stripa Project TR 87-03, KBS, Stockholm, Sweden.

Pusch, R.L. and M.N. Gray. 1989. Sealing of radioactive waste repositories in crystalline rock. OECD Workshop, Braunschweig, FDR.

Simmons, G.R., D.M. Bilinsky, C.C. Davison, M.N. Gray, B.H. Kjartanson, C.C. Martin, D.A. Peters, L.D. Keil and P.A. Lang. 1992. Program of experiments for the operating phase of the underground research laboratory. Atomic Energy of Canada Limited Technical Record, AECL-10554.

Wong, J., P. Hurley and G.F. West. 1983. Crosshole seismology and seismic imaging in crystalline rocks, Geophysical Research Letters, vol. 10, no. 8, pp. 686-689.

SESSION I

NATURAL BARRIERS - CHARACTERISATION

SÉANCE I

BARRIÈRES NATURELLES - CHARACTÉRISATION DES SITES

Chairmen - Présidents
T. Isaacs (United States)
H. Sakuma (Japan)

The Site Characterization and Validation Project

Olle Olsson

Conterra AB, Uppsala, Sweden

Abstract

The Site Characterization and Validation (SCV) Project was set up to test our ability to characterize the structural features of a site and to quantify groundwater flow and transport through the site. The size of the target volume was about 150x150x50 m and it was located within a granite pluton about 400 m below ground. A staged multi-disciplinary investigation program was applied from boreholes and drifts. A Fracture Zone Index (FZI) was used to define the occurrence and width of fracture zones where they intersected the boreholes. The orientation and extent of the zones was obtained through borehole radar and seismic techniques. The hydraulic significance of the zones was then tested by cross-hole hydrologic measurements which also yielded values of the average hydraulic properties of the zones.

The conceptual model of the site was used as input to four different numerical models used to predict the inflow and tracer transport to a series of boreholes and a drift in the central portions of the site. A process of validation was set up based on comparison of model predictions and measurements against a set of predefined criteria. General agreement was obtained between predicted and measured inflows to boreholes. Predicted inflows to the drift were generally larger than measured due to unaccounted-for drift excavation effects.

Introduction

The Site Characterization and Validation (SCV) Project formed one of the main components of Phase 3 of the Stripa Project. The overall objective of the SCV Project was to determine how well the techniques and approaches used in site characterization can be used to predict groundwater flow and radio-nuclide transport in a fractured rock medium. The central aims of the project were:

- to develop and apply an advanced site characterization methodology which integrates different tools and methods in order to predict groundwater flow and solute transport

- to develop and apply a methodology to validate that the models (both conceptual and numerical) are appropriate to the processes under examination in a fractured rock mass

The basic experiment within the SCV Project was to predict the distribution of water flow and tracer transport through a volume of granitic rock, before and after excavation of a sub-horizontal drift, and to compare these predictions with actual field measurements. To test different approaches to numerical prediction of groundwater flow and transport, four different research groups were engaged for the numerical modelling.

The concept underlying the SCV Project was that model-based predictions should be checked against experimental results on an iterative basis. Hence, the SCV Project consisted of two complete cycles of data-gathering, prediction, and validation. The first cycle consisted of Stages I to III and the second cycle of Stages III to V.

Stage I comprised the drilling and investigation of 5 boreholes from existing drifts in the Stripa Mine for the preliminary characterization of an unexplored volume of rock, which was to be the site for the SCV Project. Three 200 m long semi-horizontal boreholes (N2-N4) were drilled 60 m apart towards the north. Two 150 m long boreholes were drilled towards west 70 m apart (W1 and W2). The volume of rock investigated was situated around 380 m below ground to the north of the mined-out region of the mine (Figure 1).

During **Stage II** the data were analyzed and a conceptual model of the site devised. This model was the basis for preliminary numerical predictions of the groundwater inflow to six 100 m long parallel boreholes (the D-boreholes) that outlined a cylinder (diameter 2.4 m) centrally located within the SCV site. Four different types of numerical groundwater flow models were used for the inflow predictions.

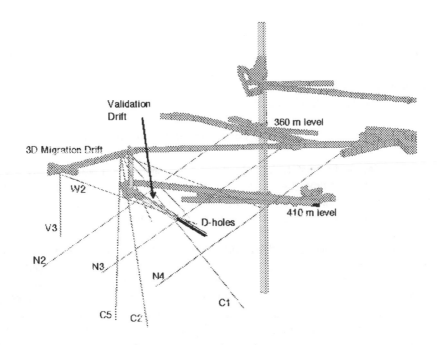

Figure 1   The SCV site is located north of existing mine workings. The
            site was investigated from boreholes drilled from the 360 m
            level. The Validation Drift and the D boreholes at the 385 m
            level were used to check the predictions.

In **Stage III** five boreholes (the C-boreholes) were drilled from essentially
the same point at the 360 m level towards the central portion of the site and
investigations made in them to provide data for detailed predictions of inflow to
the Validation Drift (to be excavated in Stage V, Figure 1) and to check on the
first conceptual model. In this stage the newly developed directional radar and
high resolution borehole seismic techniques were applied. An access drift was
excavated from the 410 m level to the 385 m level. The six 100 m long D-
boreholes were drilled from the end of the access drift, the inflow measured,
and compared to predictions. This constituted the first attempt at validating the
models.

In **Stage IV** the conceptual model was updated based on the additional
data available from Stage III. This model was used as input to the upgraded
numerical models which were used to make predictions on fracture
occurrences, distribution of groundwater inflows, and tracer transport to the
Validation Drift.

At the beginning of **Stage V** the Validation Drift was excavated in place of the first 50 m of the cylinder outlined by the D-boreholes. This was followed by fracture mapping, measurements of groundwater inflow, and tracer transport to the drift. A comparison was made of the measured data and the predictions provided by the four research groups on numerical flow and transport modelling in order to validate the models.

The characterization program

The detailed characterization of the SCV site, which was located several hundred meters below the ground surface, required all investigations to be performed from drifts, and boreholes drilled from drifts. The program for characterization of the SCV site included the following items:

Fracture mapping in drifts and boreholes: The fractures in the drifts adjacent to the SCV site were mapped along scanlines. Areal mapping of the drift walls was made in selected locations. Detailed areal mapping was also made to study the variability in fracturing within a fracture zone intersected by several drifts. The cores from all boreholes drilled were mapped and oriented by identifying reference fractures by TV-logging. The fracture mapping program provided data on fracture orientations, trace lengths, termination modes, and spacing.

Borehole radar: Cross-hole and single hole radar measurements were made to find the orientation and extent of fracture zones at the site. The directional borehole radar system developed within this phase of the Stripa Project proved particularly useful as it provided data directly on the orientation of fracture zones based on measurements in a single borehole. Radar tomography was also used to show how saline tracer injected in a borehole became distributed in the rock mass as it traversed three survey planes.

Borehole seismics: Seismic techniques were also used successfully to find the orientation and extent of fracture zones. The seismic program included both cross-hole reflection and tomography measurements. In this case the reflection measurements provided the best data for characterization of fracture zones. The success of the seismic method was largely due to a novel processing technique developed within the project.

Single borehole geophysics: To obtain *in situ* data on the physical properties of the rock in the vicinity of the boreholes the following logs were run; borehole deviation, sonic velocity, single point resistance, normal resistivity, caliper, temperature, borehole fluid conductivity, natural gamma radiation, and neutron porosity. The sonic velocity, single point resistance, and normal resistivity were found to be useful in identifying fractures and fracture zones.

Hydrogeological characterization: Initially, single borehole testing was made to provide data on transmissivity and head along the boreholes. New equipment was developed to ensure that reliable information could be collected within the mine environment in reasonable time scales. The system was built around a multiple packer probe which allowed rapid testing of permeable features with high spatial resolution. The single borehole testing was followed by cross-hole testing to define hydraulic properties of the fracture zones on the scale of the site ($\approx$100 m). Another important aspect of the cross-hole testing was that it provided a check on the hydraulic significance of the geophysically identified fracture zones. The hydraulic program also included monitoring of head in more than 50 locations across the site. This yielded information on the head distribution across the site as well as hydraulic responses to various activities in the mine which could be used to characterize hydraulic connections across the site.

Hydrochemical characterization: Groundwater samples were taken during hydraulic testing and analyzed for major constituents. The analysis showed that there were three types of groundwater present. These were classified as "shallow", "mixed", and "deep". The groundwater was also found to contain about 3% of dissolved gas (by volume), mainly nitrogen.

Rock mechanical characterization: An important aspect of groundwater flow through fractures is the effect of stress on fracture transmissivity. To study this, flow through fractures under different stress loads was studied on several samples and in one *in situ* test. This yielded stress-permeability relationships which were used in the modelling. Stress measurements were made to get data on the in situ stresses. At the level of the Validation Drift the maximum principal stress was $\approx$24 MPa, oriented parallel to the drift (i.e. NNW-ESE).

Structural model of the SCV site

A structured approach was developed to combine site characterization data into a geological and hydrological conceptual model of a site. The conceptual model of the site was based on a binary representation of the rock mass in terms of "fracture zones" and "averagely fractured rock". A "Fracture Zone Index" (FZI) was defined based on principal component analysis of data from the single borehole measurements. The following parameters were included in the analysis; normal resistivity, sonic velocity, hydraulic conductivity, fracture frequency, and occurrence of single hole radar reflectors. The frequency distribution of the FZI showed that a binary representation of the rock mass in "fracture zones" and "averagely fractured rock" was motivated.

Based on this concept of a binary representation of the rock mass a structured procedure was defined for constructing the structural conceptual model of the site. The procedure is based on identification of fracture zone locations in the boreholes (using the FZI) and finding the extent of the zones through the use of remote sensing geophysical techniques (i.e. radar and seismics). The hydrogeologic significance of the geometric model thus obtained was then examined by means of cross-hole hydraulic testing, which also yielded data on the hydraulic properties of the zones. Further checking of the consistency of the conceptual model was made by comparison with geologic and geochemical data. The procedure is by nature iterative and should as an end result produce lists of the geometry and properties of identified features, as well as lists of inconsistencies and unexplained anomalies. The procedure is outlined graphically in Figure 2.

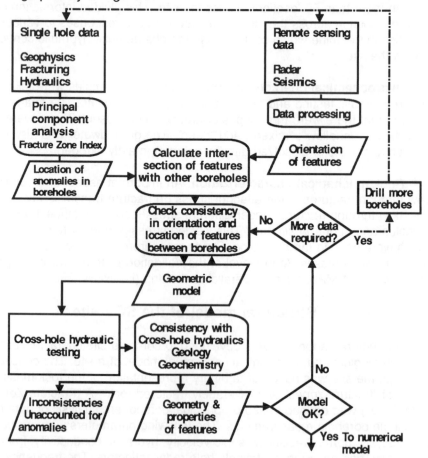

Figure 2   Outline of procedure used for construction of the structural conceptual model of the SCV-site.

The conceptual model of the SCV-site contains three major features or fracture zones named A, B, and H (Figure 3). These features were considered to extend beyond the limits of the SCV-site to the ground surface. The connection between the SCV-site and the surface provided by these features was thought to cause the high heads observed at the site. The properties and width of these major features are highly variable where they are observed intersecting the boreholes. The thickness of these features varies from 2 m to 12 m. At the borehole intersections the features generally exhibit anomalous properties compared to "averagely fractured rock". These major features are important for the groundwater flow system across the SCV-site in that they account for 75 % of the hydraulic transmissivity as measured by single hole hydraulic tests.

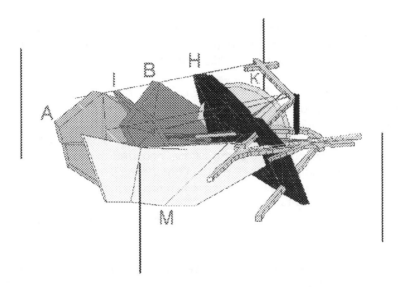

Figure 3   Perspective view of the SCV site and the features contained in the conceptual model.

Three minor features named I, K, and M with an extent of 50-100 m were also identified. These features are observed in the remote sensing data (radar and seismics) and it has thus been possible to determine their orientation and extent. The cross-hole hydraulic testing showed that these features provided hydraulic connections between the major features A, B, and H. These minor features account for 4 % of the single hole hydraulic transmissivity measured in the boreholes.

From the Figure 3 it is evident that A, B, and H intersect just north of W2. Feature I connects A and B with each other and borehole W1. K connects borehole N2 with zone H. The consistency of the model and its relevance for groundwater flow through the site was confirmed by the crosshole hydraulic testing.

Validation experiments

A set of validation experiments was designed in order to check the predictions made by the groundwater modelling teams. These experiments were performed in the D boreholes and the Validation Drift which was excavated in place of the first 50 m of the D boreholes (Figure 4). The modelling groups were first asked to predict the inflow to the D boreholes and the distribution of flow within the boreholes. This was followed by predictions of inflow and tracer transport to the Validation Drift.

In the Simulated Drift Experiment (SDE), the inflow distribution to the D boreholes was measured under three different heads in order to study if there were significant non-linear effects (e.g. due to changes in effective stress). The D boreholes had been arranged in a circle in order to simulate a drift without introducing excavation effects. Large inflows occurred at two locations where fracture zones H and B intersected the D boreholes. The inflows were essentially localized to a few fractures within the fracture zones. For a large portion of the borehole length the inflow was below the measurement limit. The fracture zones intersected the D boreholes within a couple of meters of the predicted locations.

The inflows to the Validation Drift were measured by collecting the flows to the upper part of the drift in plastic sheets while flows to the lower part of the drift were collected in "sumps" which were drilled into "wet" fractures. To account for evaporation from unsheeted areas a bulkhead sealed the end of the drift and the net outflow of water through ventilation air was measured. Among the many hundreds of fractures seen in the drift, two fractures in zone H contributed 90% of the inflow to the drift. These were the same fractures through which water flowed into the D boreholes.

The total flow to the Validation Drift was only about 1/8th of the flow to the corresponding part of the D boreholes. The relative reduction in flow to the drift was greater for the "averagely fractured rock" which was reduced by roughly a factor of 40 while the flow through the fracture zone H was reduced by a factor of 8. A reduction of inflow by a factor of two was also observed when the inflow to the D-boreholes were remeasured at a lower pressure than used the first time. Also in this case, the relative reduction in flow to the averagely fractured rock was significantly greater than for the fracture zone. The excavation of the

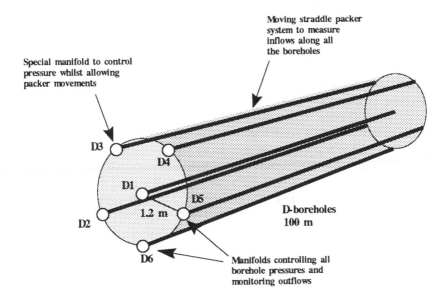

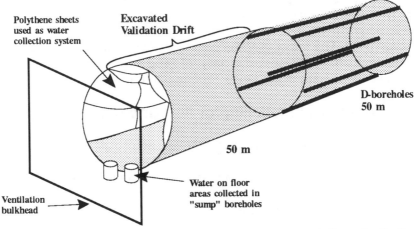

Figure 4   In the Simulated Drift Experiment (SDE) the inflow distribution
to six 100 m long boreholes held at the same head was
measured. In Stage V a 50 m long drift was excavated along
the boreholes.

Validation Drift caused heads in the surrounding rock to increase relative to the
heads prevailing when the D-boreholes were open. The head increase during
excavation was consistent with the reduction in inflow caused by the
excavation. This shows that there was no significant diversion of flow locally,
e.g. axially along the drift. The total inflow to the drift remained nearly constant

during the first year after excavation. However, significant redistributions of inflow location were observed on time scales less than 100 hours.

A series of tracer experiments was also performed, with tracer collection in the D boreholes and the Validation Drift. In the first experiment, tracer was injected with a rate of 200 ml/min in a borehole in zone H approximately 28 m away from the D boreholes which were used as a sink. Transport was dominated by two pathways within zone H with mean residence times of approximately 33 and 220 h, respectively. The experiment was repeated after excavation and in this case the drift was used as a sink. The tracer transport times to the drift were 2-3 times longer than the transport times to the boreholes. The relative delay was basically because the drift constituted a significantly weaker sink than the boreholes. In these experiments, a saline (electrically conductive) tracer combined with radar difference tomography were used to map the spreading of the tracer through three planes surrounding the injection point. It was shown that tracer was largely confined to zone H but that a minor fraction of the tracer was transported out of zone H through minor intersecting fractures.

A series of tracer experiments was also performed where metal complexes and dyes where injected in seven different locations with very low injection rates (2-30 ml/hour) in order to avoid disturbance of the flow field at the injection point. The tracer injection points were located at distances of 10 to 28 m from the drift. The tracer essentially followed the H zone to the drift. Mean residence times were in the range 1200 to 5000 h.

Numerical modelling of flow and solute transport

Numerical modelling of groundwater flow and solute transport was performed by four different research groups. Each group had their own approach to modelling and interpretation of the data provided by the experimental groups. The diversity in approach proved very useful in evaluating the capabilities of the new modelling techniques.

Equivalent porous medium modelling was performed by Fracflow Consultants, St. John's, Newfoundland, Canada. This approach was used in order to provide a reference case for the new codes developed during the course of the project. The finite element code CFEST was used to model groundwater flow in a sequence of four models from the regional scale to a detailed model of the SCV site. In the detailed model, finite element sizes down to 2.5 m were used.

Discrete fracture flow modelling was performed with the NAPSAC code developed by AEA Decommissioning & Radwaste, Harwell, United Kingdom. In this approach a random fracture network is generated based on fracture statistics obtained from fracture mapping of boreholes and drifts. Different fracture networks were generated for the "averagely fractured rock" and the "fracture zones" to represent correctly their respective fracture intensities. The fractures were given transmissivities consistent with the fracture transmissivity distribution derived from the single hole hydraulic testing. The flow through the network was then calculated for a number of different network realizations. The model can handle more than 50,000 fractures. The size of the volume that can be modelled depends on the fracture intensity.

Another discrete fracture flow code, FracMan, was developed by Golder Associates, Seattle, USA. This code calculates the flow through a fracture network in much the same way as NAPSAC but the approach used in interpreting the experimental data was different. In the approach used by Golder only the fractures that were assumed to be hydraulically conductive on the basis of the single hole testing were included in the model. This resulted in a relatively sparse fracture network which facilitated modelling of a volume encompassing the entire site.

Another type of model, referred to as an equivalent discontinuum model, was developed by Lawrence Berkeley Laboratories, Berkeley, USA. In this model a template (rectangular grid) of linear conductors was used to represent the heterogeneity of the flow system within the fracture zones. An inversion of cross-hole hydraulic test data was used to find the distribution of hydraulic conductors within the fracture zones. Once the inversion has been accomplished the model can be used to predict behavior in the fracture system under different flow conditions. Both two and three-dimensional models were constructed to model the SCV site.

All models explicitly included the data from the structural model on the location and extent of the fracture zones. This implies that the models assumed the location of major permeable features to be known. This was to account for the experimental data which showed that 80-90% of the flow was through the fracture zones and that the location of the fracture zones could be reliably defined through the remote sensing methods. In fact, the LBL model only included flow through fracture zones. However, an important aspect of the discrete fracture and equivalent discontinuum models is that they can represent the observed variability in properties within fracture zones.

A process of validation was set up in order assess the predictive capabilities and usefulness of these models. A number of criteria was specified for comparison of predictions and measurements. The modelers were first asked to predict the inflow to the D boreholes during the SDE. This was done half way through the project and predictions were based on incomplete site characterization, and on flow models still under active development. The models, being stochastic, produced predictions within a relatively wide range, but the mean values of predicted total inflow to the D boreholes were quite close to the measured value.

The inflow distribution to the Validation Drift was then predicted. As the drift was in the same place as the boreholes and the boundary conditions were essentially the same, the observed reduction in inflow to the drift must have been due to drift excavation effects. Based on empirical evidence of reduced fracture transmissivity in the disturbed zone around drifts, Golder obtained inflows in close agreement with the observed factor of eight reduction. LBL made a similar assumption of reduced transmissivity, but obtained a considerably smaller reduction in inflow. However, in calculating the inflows to the drift none of the modelling teams were able to predict the mechanism causing the observed reduction in inflow. AEA tested the assumption of a power law dependence between fracture transmissivity and normal stress in their simulations which resulted in a small increase in inflow to the drift contrary to observations.

Finally predictions were made of the experiments where tracer was injected in several locations at distances of 10-28 m from the drift. In general the predictions of tracer concentration and recovery were in reasonable agreement to observed values, but the predicted arrival times were generally shorter, and the breakthrough curves steeper, than observed.

To improve the understanding of drift excavation effects, several analyses were performed of the stress redistribution around the drift. These included a three dimensional continuum model, two-dimensional (UDEC-BB) and three-dimensional (3-DEC) discrete fracture models. A fully coupled two-dimensional hydromechanical model (UDEC-BB) was used to simulate the hydraulic effects of drift excavation. The model simulations showed a decrease in inflow to the drift due to closure of radial fractures and increased axial permeability. However, the coupled model could not incorporate the most frequent fracture set which was perpendicular to the drift. According to the three-dimensional modelling performed, the stress changes due to drift excavation would be small on this fracture set.

Conclusions

A structured approach to combining the data from a site characterization program into a geological and hydrological conceptual model of a site has been devised. This approach can be transferred to repository site characterization exercises in similar hard, fractured rock environments. The approach is based on a "Fracture Zone Index" which is used objectively to define location and width of fracture zones in the boreholes. The geometric model essentially provided by remote sensing methods is verified through cross-hole hydraulic testing and geochemical data.

The geological and hydrogeological conceptual model was in good agreement with observations. Hence, the combination of single-hole and cross-hole investigations which were applied can be used to give a reliable and sufficiently detailed description of a site at the 100 m scale. A prerequisite to obtain a reliable geometric model, is the use of remote sensing techniques such as radar and seismics, which provide data on rock structure at large distances from boreholes and drifts.

The Simulated Drift Experiment quantified the magnitude of drift excavation effects. The flow to the drift was found to be 1/8th of the flow to the corresponding parts of the boreholes. The reduction in flow through the "averagely fractured rock" was found to be significantly greater than the reduction in flow through the "fracture zone". Two-phase flow conditions due to degassing of the groundwater and drying is considered to be the principal cause for the flow reduction. Dynamic loading effects during excavation and shear displacements are also foreseen as significant mechanisms. However, it has not been possible to quantify the relative significance of these mechanisms based on available data. For the Validation Drift case, changes in normal stress on fracture planes seem to be of limited significance. Further research is warranted in order to adequately understand the hydrology of the disturbed zone.

Groundwater flow at the SCV-site was concentrated within the fracture zones. Within the fracture zones a large variability (one to two orders of magnitude) in hydraulic transmissivity over small distances (meter scale) was found. In spite of the fracture zones at the SCV site being relatively minor from a construction point of view, significant hydraulic connections were found over large distances (several hundred meters). Occasionally, highly transmissive fractures were found outside fracture zones.

The discrete fracture flow codes have evolved during the course of the project from being research tools to practical tools capable of representing flow through physically realistic fracture systems. The models have been shown to provide realistic predictions of groundwater flow and transport of nonsorbing solutes through undisturbed rock. The attempts to model the inflow to the drift were not successful due to inadequate understanding of the hydrology of the disturbed zone around drifts. It has also been demonstrated that the discrete fracture flow models can be constructed from data obtained during a site characterization program containing the items described above. Hence, data for these models can be collected with a reasonable effort.

The staged approach where data collection has been followed by blind predictions and subsequent validation in several cycles has proved to be very useful. This forced an interaction between modelers and experimentalists which helped to focus the investigation program towards parameters required for modelling. In its turn this guided model development towards an adequate representation of flow through fractured rock.

Acknowledgement

The SCV project represents the combined effort of many research groups. The Project was managed by the Stripa Project Manager, Bengt Stillborg and the Scientific Coordinators John Black and Olle Olsson. The major project tasks were assigned to the following Principal Investigators:

- Nick Barton, NGI, Norway. Rock mechanics testing and modelling.
- John Black, Golder Associates, United Kingdom. Hydrogelogical testing and analysis.
- Calin Cosma, Vibrometric OY, Finland. Borehole seismic investigations.
- Bill Dershowitz, Golder Associates, USA. Discrete fracture network modelling.
- Tom Doe, Golder Associates, USA. Analysis of single fracture packer tests.
- John Gale, Fracflow Consultants, Canada. Fracture characterization and equivalent porous media modelling.
- Alan Herbert, AEA Decommissioning & Radwaste, United Kingdom. Discrete fracture network modelling.
- David Holmes, British Geological Survey, United Kingdom. Hydrogeological equipment and testing.
- Marcus Laaksoharju, Geovision, Sweden. Hydrochemical investigations.
- Jane Long, Lawrence Berkeley Laboratories, USA. Discrete fracture network modelling.
- Ivars Neretnieks, Royal Institute of Technology and Lars Birgersson, Chemflow, Sweden. Tracer migration experiments to the Validation Drift.
- Olle Olsson, Conterra AB, Sweden. Borehole radar investigations.

References

Olsson, O. (ed), 1992. Site Characterization and Validation - Final Report. Stripa Project TR 92-22, SKB, Stockholm, Sweden.

Characterizing the Geometry
of the Stripa Fracture System for
Flow and Transport Studies

Gale[1], J., MacLeod[1], R., Bursey[1], G.,
Strahle[2], A., Carlsten[2], S. and S. Tiren[2].

[1]Fracflow Consultants Inc., St.John's, Nfld, Canada.

[2]SG AB, Uppsalla, Sweden.

Abstract

The large and small scale geometry of the fracture system at Stripa have been determined using regional scale remote imaging techniques and detailed data from drillcore logging and drift mapping. These data identified two sub-vertical dipping sets, a strong north-south striking set and a weaker set with a northwest-southeast strike, and two sub-horizontal fracture sets. Fracture mapping showed a clear difference between fracture geometry in the average rock and that in the largest fracture zone, the H-zone, that was identified in the test site. The small scale fractures in the averagely fractured granitic rock at the Stripa site appear to mimic the large scale features, both in orientation and density.

Introduction

The regional geology of the Stripa area has been summarized in the earlier work of Geijer (1938) and the more recent work by Olkiewicz, et al. (1979), Wollenburg et al. (1982), and Lundström (1983). The work of the above authors, on the geology and structure of the Stripa area, is briefly reviewed here to define the structural framework of the Stripa research site. The primary focus of this paper is on the geometry of the large scale fractures within both the Stripa area and the Stripa site and the statistical description of the orientations, trace lengths and spacings, for each set of the small scale fractures in both the averagely fractured rock as well as in the major fracture zone, the H-zone, that cuts through the SCV site (Olsson et al., 1992). Data on the different scales of fracturing were collected as part of the fracture characterization program during the different parts of the overall Stripa Project as well as during the three stages of the final phase, Phase III, of the Stripa Project.

Regional Geology

The Fennoscandian Shield bedrock in the Stripa area was deformed about 1,800 to 2,000 Ma ago (Lundstrom, 1983). Locally, the bedrock consists primarily of calc-alkaline metavolcanic and metasedimentary (primarily, mica gneiss) supracrustal rocks (Figure 1). The Stripa mine is completely surrounded by the metavolcanics, locally called leptites, which in turn are partly enclosed to the south, southeast, and southwest by the metasedimentary rocks.

The bedrock in the Stripa area shows at least two important periods of deformation. The major fold structures are thought to be due to E-W compression with the resulting fold axes being mainly horizontal with a roughly N-S trend (Lundstrom, 1983). This inference is supported by the Stripa mine maps (Geijer, 1938) which show small scale, N-S trending, folds with horizontal fold axes. The second phase of deformation (Lundstrom, 1983), is thought to be related to a N-S compression. The resulting refolding has produced northeast to east plunging fold axes such as the truncated syncline (Figure 1) in which the Stripa mine was developed.

Major Fracture Zones

In order to determine the orientation of the major features in the Stripa area, the fracture zones or lineaments in a 20 km by 20 km area, centred on Stripa, were analyzed using relief illumination techniques. This analysis (Gale et al., 1992) shows that the major fracture zones, those with continuous traces and the strongest topographical expression, have NE, NNE, or NW orientations. Most of these major fracture zones are curved or sinuous.

The most abundant lineaments or fracture zones in this 20 km by 20 km area have orientations between N10E and N20W. However, it should be noted that the strong topographic expression of these fracture zones has been enhanced by the N-S direction of glacial movement and hence these features are preferentially high-lighted by the relief illumination technique. Much smaller numbers of fracture zones or lineaments have NE or E-W orientations. The E-W trending zones do not have a strong topographic expression and are more irregular and discontinuous. The abundant N-S to NNW trending fracture zones are narrow and have short trace lengths compared to the other fracture zones. However, abundance may not be a good measure of the structural or hydraulic importance of fracture zones.

Local Mine Geology

The Stripa mine is situated north of an extensive NW trending fracture zone and is bordered by minor N-S zones. An irregular ENE trending, relatively wide, fracture zone is located north of the mine. Locally, the metavolcanic rocks (leptites), that were intruded by the Stripa Pluton have a NE-SW strike. However, the local geological structure consists of a eastward plunging syncline, or synclinorium, that changes bearing at about the 120 m level to N60E. The syncline, which contained the ore zones (Geijer, 1938) is asymmetric with the southern limb partly truncated (Figure 1).

The Stripa pluton intrudes under the north limb of the syncline, where the SCV site is located, and outcrops over an area of about 0.3 km^2, immediately north of the Stripa mine. The true size of the pluton at the surface is unknown due to the extensive cover of glacial debris. The Stripa granite is a grey, fine to medium grained, relatively uraniferous granitic rock. Several stages of fracturing are evident in the granite, including fractures that have been welded or bonded together by fracture mineralization.

The old Stripa mine maps (Geijer, 1938) show the location and orientation of the large scale features, faults and fracture zones, that intersect the mine workings. Apparent stratigraphic offsets along fault and fracture zones range up to 20 to 25 m on the northern limb of the syncline and up to 50 m at the junction of the base of the syncline with the southern limb. In general, both the mapped and inferred vertical offsets along major fracture zones appear to increase, in a scissors-like motion, to the east and northeast in the direction of the SCV site. The east to northeast trending faults cutting the northern limb of the syncline are spaced about 50 m apart and have trace lengths that are generally less than 200 m. However, the east to northeast trending faults that are located in the southwestern part of the mine cutting the southern limb have similar spacings but have trace lengths of 300 to 800 m (Gale et al., 1991)

In contrast to the regional lineaments, the mine maps show very few faults with northerly trends. The few that are shown on the old mine maps have trace lengths that are generally less than 70 to 80 m. However, the regional fracture zone/lineament analysis indicates a greater likelihood that the largest fracture zones in the SCV site will have a N-S trend with a smaller likelihood that major E-W to NE trending major fracture zones will be encountered. Given the long trace lengths and higher density of the east-northeast trending faults or fracture zones shown on the mine maps, one would expect that these structures would be more abundant in the SCV site than the northerly trending features shown on the old mine maps.

Geometric Characteristics of Zone H

Application of the crosshole and single hole geophysical tools (Olsson et al., 1992) shows that the SCV site is cut by three large features or fracture zones. Based on the geophysical data, the strike of fracture zone H was calculated to be N5W with a dip of about 70 to 80 degrees to the east. With this orientation, the projected intersections of the H zone with the existing mine drifts were confirmed at the 310, 360, 385, 410 m levels and the Access drift and Validation drift. Where possible, the H zone was mapped on both sides of the drift, creating both scanline and areal maps, for each of these drift intersections for a total of 11 drift sections, not including the Validation drift (Gale et al., 1991).

The H zone is characterized by a red colour or alteration both where it is exposed in the drifts as well as where it intersects the drill core. In addition, selected fracture planes within the main fracture zone are filled with breccias or mylonites, up to several cm's wide, that consist of fragments of granitic rock. Combining the locations of the H zone from the SCV drift and borehole intersections (Figure 2) shows that the overall geometry of the H zone (Gale et al., 1991) is planar (Figure 3). However, while the zone maintains its overall dip its southward extension appears to be offset or to change strike where it crosses the drifts at the 360 m level (Figure 3).

The width of the H zone appears to vary along both strike and dip, based on the apparent thicknesses from both the drift mapping and borehole data as well as the geophysical cross-hole data. If the thickness of the H zone is limited to the zone of most intense fracturing, then the apparent width of the H zone in the drifts ranges from 2.15 to 4.25 m with a mean of 3.62 m (Gale et al., 1991). The geophysical borehole program suggests that the H zone splits into two segments where it is intersected by borehole W2 (Olsson et al., 1989) and has an apparent width of more than 10 m. This contrasts with the well defined H zone that was mapped at the 310 m level. However, it is not inconsistent with the more diffuse zone mapped in the 2D access drift at the 360 m level. Given its strong definition at the 310 m level and its impact on the hydrology of the SCV site, it is safe to assume that the H zone extends to the surface as a structural feature and should form a well defined surface lineament at least at the local mine scale.

The general impression obtained from the composite areal maps (Gale et al., 1991) of the north side of the drifts is one of decreasing fracture strength with increasing depth. The 390 and 410 m levels appear to be characterized by shorter fracture traces than those present at the 310 m level. However, well defined zones of mylonite or crushed rock have been mapped at the 390 m level. This suggests that significant deformation still persists to this level, and deeper, but that it is restricted to these small lenses of intense deformation.

The areal and scanline mapping showed that, in addition to the mylonites and breccias, the H zone is characterized by anastomosing structures, duplexes and splays (Gale et al., 1991). During the H zone fracture mapping program, fractures that showed evidence of shear displacement were identified and the orientations of the striations were mapped. The poles to these fracture planes are contoured in Figure 4 (left side) and the mean orientations of the three clusters for this data set are also shown on the same figure. Lineations on these fracture planes, plotted as bearing and plunge, are contoured in Figure 4 (right side). The pattern of lineation orientations suggests an oblique dip-slip fault movement similar to that described in Petit, 1987. Dip slip movement is also consistent with what appears to be an abrupt change in the strike or offset of the H zone at the 360 m access drift.

The pattern produced by the mean orientation of the fracture planes making up each of the three clusters (Figure 4) can be interpreted as representing a set of Reidel shears and P shears (Tchalenko, 1970; Scholz, 1990). The fracture pattern formed by the fracture planes on which the striations where mapped, clearly extends at least ten metres into the rock adjacent to the fracture zone (Gale et al., 1991). It is also obvious that it is difficult to rely on only borehole data to clearly define the effective width of a fracture zone and the zone or area in which the fracture geometry has been influenced by the forces that created the large scale feature or fracture zone.

Small scale fracture system

Mapping the small scale fractures in the Stripa Site included (a) drift mapping on the boundaries of the SCV site, (b) coreholes through the SCV site, (c) drift mapping of the major fracture zone (H-zone) cutting through the test block, with (d) final detailed drift mapping and core logging in the centre of the test block that was completed using a three-stage approach (Olsson, et al., 1992). These data were used to develop a statistical description of the orientation (Figures 5 and 6), trace length and spacing, for each fracture grouping or set, in both the averagely fractured rock as well as in the H-zone.

The initial set of borehole and scanline fracture data from the Stage I program (Gale and Strahle, 1988), identified three main fracture clusters or groups. The three clusters included two sub-vertical dipping sets, one strong north-south striking set and one weaker set with a northwest to southeast strike, and a weaker, or poorly defined, sub-horizontal dipping fracture set. Given the horizontal bias of both the scanlines and the N and W boreholes, it was not possible to determine if a strong sub-horizontal fracture set existed at the SCV site or if it did exist that it was merely poorly sampled. Also, the initial set of data suggested that the north-south striking fracture set had a much higher density than the other two clusters or sets.

The Stage I data were corrected for the orientation bias using the procedure outlined by Terzaghi (1965) and this corrected data set was used to predict the fracture patterns for the C and D boreholes. Comparison of the Stage III measured fracture orientations in the C and D boreholes with the predicted patterns (Gale et al., 1990) shows good agreement for borehole C3 and the D boreholes which had orientations similar to the Stage I boreholes on which the predictions were based. However, C1 and C2 which had steep plunges, about 40 degrees, intersected more sub-horizontal fractures than predicted. Thus, it is apparent that the procedure outlined in Terzaghi (1965) does not properly correct for the borehole orientation bias and is not a good procedure for predicting fracture patterns when the sampling directions are not represented in the original data base.

Additional data from boreholes C4 and C5, which also had steep plunges, confirmed the presence of a sub-horizontal fracture system in the area of the proposed Validation Drift. Thus, the predicted pattern for the Validation Drift was assumed to be the same as that obtained from the D boreholes, with the addition of the sub-horizontal set sampled by the C boreholes and the northwest-southeast set that parallels the D boreholes, both of which were not sampled by the D boreholes. Also, the D boreholes all showed similar overall patterns, indicating that there is no variation in the fracture orientations within the rock mass defined by and perpendicular to the ring of six D boreholes. Similarly, comparison of the fracture orientations, along the length of all of the D holes, showed no significant variation in the orientation of the main cluster even when the fracture zone sections were removed from the data set. Hence, we concluded that the main north-south trending fracture set did not show any obvious variation in orientation in the cylinder of rock defined by the D boreholes.

Comparison of the fracture data from the Validation Drift mapping (Figure 7) with the D borehole pattern showed the expected borehole orientation bias (Bursey et al., 1991). Where the D borehole pattern was dominated by a single group of subvertical fractures, the Validation Drift data show the same dominant subvertical fracture group plus the weaker sub-vertical, northwest-southeast trending, set that was present in the Stage I data. In addition, the Validation Drift data clearly show the presence of two sub-horizontal fracture groups that were not present in the D borehole data. However, these sub-horizontal sets appeared in the Stage I and III data, but were usually lost in the tails of the distributions for the larger groups.

Fracture data from the Stage III H zone mapping (Gale et al., 1992) showed a distinct difference between the geometry of the fractures in the average rock (Figure 5) and that in the H zone (Figure 6). The H zone drift and borehole data predicted that the northwest-southeast trending sub-vertical fracture set would not be present in the H zone in the Validation Drift and this was confirmed by the Validation Drift mapping. In addition, the Validation Drift mapping showed that the mean trace lengths of the fractures making up the H zone were shorter than those in the average rock as predicted from the H zone mapping (Gale et al., 1991). The corrected statistics (Bursey et al., 1991) for both sets of trace length data for each cluster in the average rock and H zone are given in Tables 1 and 2. (n equals the number of fractures.)

The equations that we have used to correct for shape bias assumes a log-normal distribution and a square fracture shape (Herbert, A., Pers. Comm.), and calculate an approximate mean (u_l) and standard deviation (s_l) of the fracture lengths:

$$u_l = u_{tl} - (2s_{tl}^2) + 0.215$$

$$s_l = s_{tl}^2 - 0.064$$

where u_{tl} and s_{tl} are the mean trace length and standard deviation, respectively, for raw data. In each case, the estimates of the mean ($\tilde{u}$) and standard deviation ($\tilde{s}$) for original distributions are calculated from the log-normal values using the following relations (Bury, 1975):

$$\tilde{u} = EXP[u_{ln} + s^2_{ln}/2]$$

$$\tilde{s} = \{EXP(2u_{ln} + s^2_{ln})[EXP(s^2_{ln})-1]\}^{1/2}$$

where $u_{ln} = u_{tl}, \hat{u}_{tl},$ or u_l and $s_{ln} = s_{tl}, \hat{s}_{tl},$ or s_l.

Tables 1 and 2 provide the detailed trace length and length statistics for the average rock and the H zone, respectively, based on the Validation Drift map data. The statistics of orientation and spacing for each cluster are given in Bursey et al., 1991, and Gale et al., 1991. These statistics confirm the decision to use a separate set of fracture statistics to characterize the H zone and a separate set to characterize the average rock for flow and transport studies.

Comparison of the fracture orientation predictions (Gale et al., 1991) with the measurements in the Validation Drift, shows that a more rigorous method is required to assign the individual fractures to the correct cluster or set. However, it is clear that the cluster analysis approach provides a reasonable interim approach to the problem. In addition, calculating the statistics of trace length and spacing provides a means of properly weighting each fracture set when sample line or borehole orientation bias results in the oversampling of one fracture set with respect to the other sets.

Summary

The initial set of borehole and scanline data from Stage I identified three main fracture clusters or groups. The three clusters included two sub-vertical dipping sets, one strong north-south striking set and one weaker set with a northwest to southeast strike, and a weaker, or poorly defined, sub-horizontal dipping fracture set. The Stage III data showed that the Stage I data did not adequately sample the sub-horizontal fracture system. Comparison of the fracture data from the Validation Drift mapping with the Stage I and III data showed the same dominant subvertical fracture group plus the weaker sub-vertical, northwest-southeast trending, set that was present in the Stage I data. In addition, the Validation Drift data clearly show the presence of two sub-horizontal fracture groups that were not identified in the Stage III borehole data. Fracture data from the Stage III H-zone mapping showed a distinct difference between the geometry of the fractures in the average rock and that in the H-zone. The small scale fractures at the SCV site appear to mimic the large scale features, both in orientation and density.

References

Bursey, G. G., Gale, J.E., MacLeod, R., Strahle, A., Tiren, S., 1991. Site characterization and validation - Validation Drift data, stage IV. Stripa Project TR 91-19, SKB, Stockholm, Sweden.

Bury, K. V., 1975. Statistical models in applied sciences. John Wiley, 625 p.

Chung, C-J., 1988. Statistical analysis of truncated data in geosciences. Sciences de la Terre-Serie Informatique Geologic.

Gale, J. E., Strahle, A., 1988. Site characterization and validation -Drift and fracture data, stage 1. Stripa Project IR 88-10, SKB, Stockholm, Sweden.

Gale, J. E., MacLeod, R, Strahle, A. and Carlsten, S., 1990. Site characterization and validation - Drift and borehole fracture data, stage 3. Stripa Project IR 90 02, SKB, Stockholm, Sweden.

Gale, J. E., MacLeod, R, Bursey, G. G., Strahle, A. and Tiren, S., 1991. Characterization of the structure and geometry of the H fracture zone at the SCV site. Stripa Project TR 91-37, SKB, Stockholm, Sweden.

Geijer, P., 1938. Stripa odalfalts geologi (Geology of the Stripa mining field). SGU Report No. 28., Stockholm, Sweden.

Lundstrom, I., 1983. Beskrivning till berggrundskartan Lindesberg SV. SGU Af 126, Geological Survey of Sweden, Uppsala, Sweden.

Olkiewicz, A., Gale, J.E., Thorpe, R. and Paulsson, B., 1979. Geology and fracture system at Stripa Lawrence Berkeley Lab. Rep. LBL-8907, SAC-21.

Petit, J. P., 1987. Criteria for sense of movement on fault surfaces in brittle rocks. Journal of Structural Geology, 9, 597-608.

Scholz, C.H., 1990. The mechanics of earthquakes and faulting. Cambridge University Press, Cambridge, UK, 439 pp.

Tchalenko, J. S., 1970. Similarities between shear zones of different magnitudes. Geol. Sur. Am. Bull, 81, 1625-1640.

Terzaghi R. D., 1965. Sources of error in joint surveys. Geotechnique, 15, 287304.

Wollenberg, H., Flexser, S. and Andersson, L., 1980. Petrology and radiogeology of the Stripa pluton. Lawrence Berkeley Lab. Rep. LBL-11654, SAC-36.

Table 1. Trace length statistics for fractures in the good rock sections of the Validation drift.

CLUSTER	1	2	3	4	5	6	7
Raw data, not corrected for truncation or censoring:							
n	14	280	44	86	24	41	101
$u)_{tl}$ (m)	1.08	1.20	1.06	0.96	1.13	1.33	1.21
s_{tl} (m)	0.85	0.92	1.01	0.70	0.76	1.36	0.93
u_{tl} (ln,m)	-0.199	-0.084	-0.264	-0.262	-0.089	-0.043	-0.08
s_{tl} (ln,m)	0.810	0.729	0.790	0.666	0.663	0.794	0.76
$\tilde{u}_{tl}$ (m)	1.14	1.20	1.05	0.96	1.14	1.31	1.23
$\tilde{s}_{tl}$ (m)	1.10	1.00	0.98	0.72	0.85	1.23	1.08
Corrected for truncation and censoring (MULTI; Chung, 1988):							
$\hat{u}_{tl}$ (ln,m)	-0.200	-0.082	-0.350	-0.262	-0.089	-0.043	-0.07
$\hat{s}_{tl}$ (ln,m)	0.609	0.532	0.626	0.438	0.421	0.616	0.58
$\tilde{\hat{u}}_{tl}$ (m)	0.99	1.06	0.86	0.85	1.00	1.16	1.10
$\tilde{\hat{s}}_{tl}$ (m)	0.66	0.61	0.59	0.39	0.44	0.79	0.70
Corrected for length (Herbert; Pers. Comm.):							
u_l (ln,m)	-0.726	-0.433	-0.919	-0.430	-0.228	-0.587	-0.53
s_l (ln,m)	0.554	0.468	0.573	0.357	0.336	0.562	0.52
$\tilde{u}_l$ (m)	0.56	0.72	0.47	0.69	0.84	0.65	0.68
$\tilde{s}_l$ (m)	0.34	0.36	0.29	0.26	0.29	0.40	0.38

Table 2. Trace length statistics for fractures in the H zone section of the Validation drift.

CLUSTER	1	2	3	4
Raw data, not corrected for truncation or censoring:				
n	213	56	32	22
u_{tl} (m)	0.91	0.80	1.08	0.71
s_{tl} (m)	0.68	0.63	0.75	0.50
u_{tl} (ln,m)	-0.336	-0.445	-0.103	-0.535
s_{tl} (ln,m)	0.686	0.664	0.573	0.585
$\tilde{u}_{tl}$ (m)	0.90	0.80	1.06	0.69
$\tilde{s}_{tl}$ (m)	0.70	0.59	0.66	0.44
Corrected for truncation and censoring (MUTLI; Chung, 1988):				
$\hat{u}_{tl}$ (ln,m)	-0.336	-0.444	-0.103	-0.534
$\hat{s}_{tl}$ (ln,m)	0.468	0.433	0.318	0.327
$\tilde{\hat{u}}_{tl}$ (m)	0.80	0.70	0.95	0.62
$\tilde{\hat{s}}_{tl}$ (m)	0.39	0.32	0.31	0.21
Corrected for length (Herbert; Pers. Comm.):				
u_l (ln,m)	-0.559	-0.604	-0.090	-0.533
s_l (ln,m)	0.394	0.351	0.193	0.207
$\tilde{u}_l$ (m)	0.62	0.58	0.93	0.60
$\tilde{s}_l$ (m)	0.25	0.21	0.18	0.13

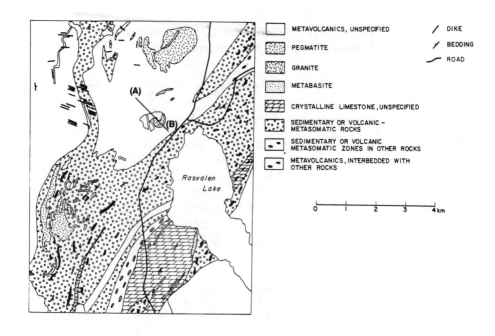

Legend:

METAVOLCANICS, UNSPECIFIED

PEGMATITE

GRANITE

METABASITE

CRYSTALLINE LIMESTONE, UNSPECIFIED

SEDIMENTARY OR VOLCANIC – METASOMATIC ROCKS

SEDIMENTARY OR VOLCANIC METASOMATIC ZONES IN OTHER ROCKS

METAVOLCANICS, INTERBEDDED WITH OTHER ROCKS

/ DIKE

/ BEDDING

— ROAD

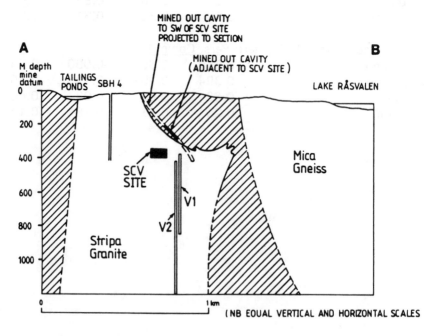

Figure 1. (A) Regional geology of the area around Stripa (after Lundstrom, 1983) and (B) vertical cross-section through the Stripa mine.

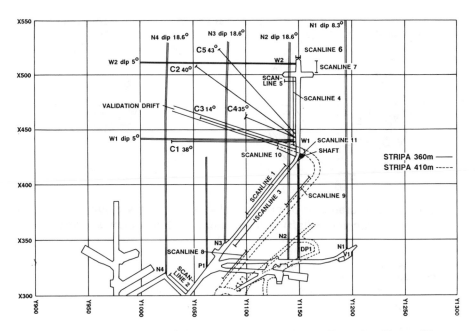

Figure 2. **Location of drifts, boreholes and scanlines for Phase III.**

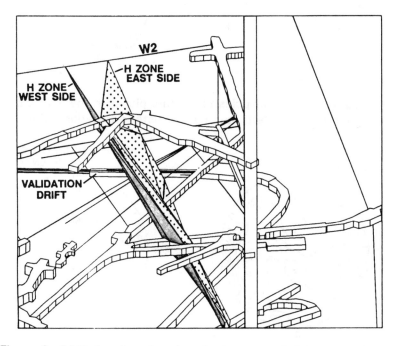

Figure 3. **CAD drawing showing the location of the east and west sides of Zone H at the SCV site.**

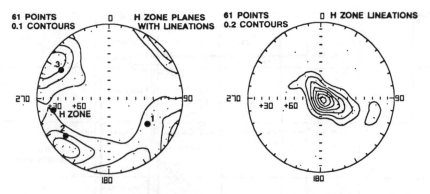

Figure 4. Contoured plot of poles to planes that show evidence of shear displacement (left) and lineations on these planes (right).

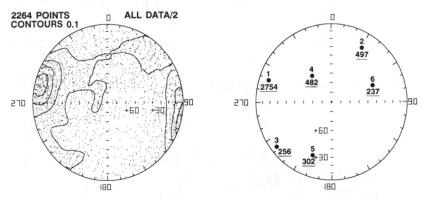

Figure 5. Contoured plot of poles to fracture planes in the SCV block minus fracture zones (left), and mean cluster orientations (right).

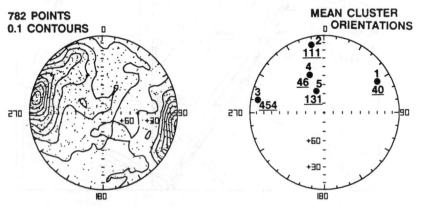

Figure 6. Contour plot of poles to fracture planes (left) and mean cluster orientations (right) for fractures in the H zone.

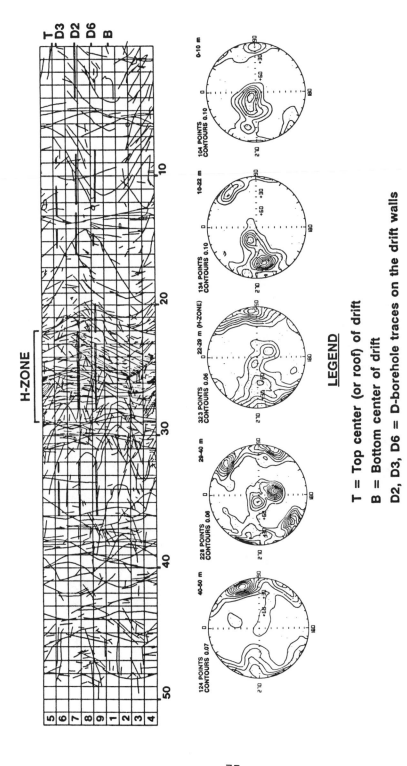

T = Top center (or roof) of drift

B = Bottom center of drift

D2, D3, D6 = D-borehole traces on the drift walls

Figure 7. Fracture trace map of the Validation drift and fracture orientations by section.

75

Geotechnical Predictions of the Excavation Disturbed Zone at Stripa

N. Barton
A. Makurat
K. Monsen
G. Vik
L. Tunbridge
Norwegian Geotechnical Institute
(Oslo, Norway)

Abstract

The 3m wide by 2m high by 50m long validation drift at Stripa surprised the site characterisation and validation project investigators, by the limited amount of water inflow compared to borehole measurements and predictions. Rock mechanics characterisation and laboratory and field tests performed by NGI are reviewed in an attempt to explain the reduced inflows. Discrete element UDEC-BB modelling was used to predict the effects of excavation induced disturbance in two-dimensional models. Full coupling of hydro-mechanical effects was incorporated in some models and shown to be important compared to mechanical modelling.

Prédictions géotechniques de la zone perturbée autour d'une excavation à Stripa

Résumé

La galerie d'accés, de trois mètres de large, deux mètres de haut et cinquante mètres de long, à Stripa, utilisée pour les validations, a surpris les enquêteurs chargés de la charactérisation du site et du projet de validation, par le faible débit d'eau comparé aux prédictions et mesures de sondage. La charactérisation de mécanique des roches et les tests in situ et au laboratoire effectués par NGI sont repassés en revue afin d'expliquer les débits réduits. Une modélisation par la méthode des éléments discrets UDEC-BB a été effectuée pour prédire les effets de perturbation induite par l'excavation dans des modèles à deux dimensions. Le couplage des effets hydro-mécaniques a été incorporé dans certains modèles et se révèle important comparé aux cas où la modélisation est purement mécanique.

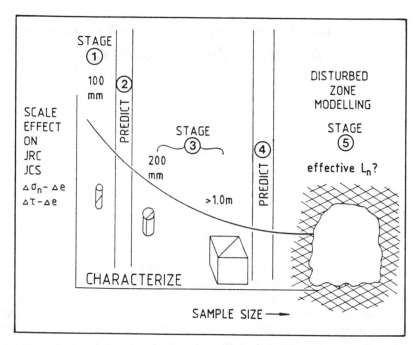

Fig. 1.   Five stages of joint characterisation and prediction.

INTRODUCTION

The relatively small 3m wide by 2m high validation drift at Stripa has surprised modellers by the limited water inflow recorded. Suggested reasons for the reduced inflow compared to that recorded in peripheral boreholes have included blast vibration and blast gas effects, stress redistribution causing reduced radial permeability, shear displacements on joints, and two phase flow conditions due to degassing of the groundwater.

Rock mechanics tests and numerical modelling were included in the Site Characterisation and Validation project, in the hope of gaining some insight into the mechanical processes caused by tunnel excavation. However, the rock mechanics modelling was not incorporated in the Fracture Flow Task Force modelling nor in the planning of the validation drift inflow experiment.

STAGE BY STAGE CHARACTERISATION

A schematic presentation of the various stages of the rock mechanics test programme is given in Fig. 1. Joints were characterised at successively larger scales and in smaller numbers, as Stage 1 passed into Stage 3.

- Stage 1   201 joints   100mm diameter core (NGI)
- Stage 3   5 joints   200mm core (NGI, MUN, LULEÅ)
- Stage 3   1 joint 1000×1000mm block test (NGI)

Coupled stress-flow tests were performed at 200mm scale in NGI (Makurat et al., 1990a) and in the Memorial University of Newfoundland by Gale et al.,

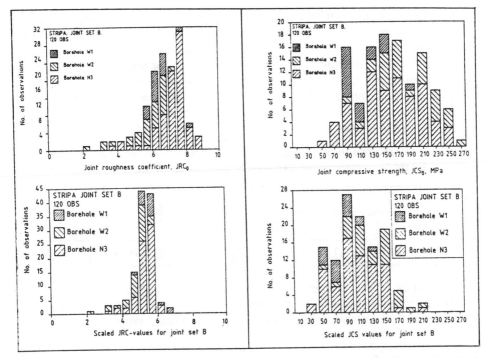

Fig. 2.　Examples of JRC and JCS statistics for Stripa SCV joint set B, showing
block size scaling from 100mm core to assumed 0.5m block in situ.
(Vik and Johansen, 1990)

1990a (TR 90-11), and at the Technical University of Luleå by Hakami, 1989
(TR 89-08).

Stage 1: Characterisation

Joints recovered in drill core from some of the first exploratory holes W1,
W2 and B3 were selected for characterisation. One hundred and twenty-two
joints represented the N-S trending Set A, while fifty-two joints represented
the NW-SE trending Set B. A further twenty-seven joints were tested that
were not matched with one of the sets identified by Gale and Stråhle, 1988
(TR 88-10).

The Stage 1 tests were purely mechanical; utilising tilt (very low gravity-
induced shear and normal stress) index tests to characterise joint roughness
(JRC), and Schmidt hammer rebound testing to characterise joint wall
strength (JCS). As indicated in Fig. 1, these parameters are sample size
dependent, declining in value as block size or joint length is increased. The
peak friction angle (ϕ) can be predicted as follows (Barton and Choubey,
1977):

$$\phi = JRC \log\left(\frac{JCS}{\sigma_n}\right) + \phi_r$$

79

where JRC = joint roughness coefficient, JCS = joint wall compression strength, and ϕ_r = residual friction angle. Stress-closure behaviour is also strongly governed by JRC and JCS values (Bandis et al., 1983).

Fig. 2 shows histograms of the measured data for joint roughness (JRC) and wall strength (JCS), and conversion to values appropriate to *in situ* block sizes (Barton and Bandis, 1982). Table 1 gives median values of the key joint strength parameters with example values of predicted peak friction angles for assumed normal stress levels of 5 and 25 MPa ($\phi_5°$ and $\phi_{25}°$).

Table 1. **Predicted median values of joint strength parameters for stripa joints (approximately 100mm scale). JCS in units of MPa.**

Joint Set	JRC	JCS	$\phi_r°$	$\phi_5°$	$\phi_{25}°$
B (N-S strike)	6.4	140	25.5°	35.5°	30.7°
A (NW-SE strike)	7.1	120	24.3°	34.1°	29.1°

The values of JCS ranged from 60 to 220 MPa, and ϕ_r ranged from 20° to 30°, reflecting the effect of several mineral fillings and coatings in the various sets. In general, joint roughness showed more uniformity.

Stage 2: Prediction

With the assumed *in situ* block size of 0.5m, corrected "full-scale" values for JRC_n were 5.3 and 5.4, while "full-scale" values of JCS_n were 95 and 80 MPa for sets B and A respectively (Fig. 2 shows examples).

The above input data was used in the Barton-Bandis joint sub-routine (which is part of the discrete element model UDEC-BB) to generate sets of joint behaviour curves for the following variables:
A) shear strength - displacement   B) dilation - displacement
C) conductivity - displacement   D) normal stress - closure
               E) normal stress - conductivity.

The predicted behaviour for joints with the median strength parameters given in Table 1 is reported in Vik and Barton, 1988 (TR 88-08). Sketches of the general form of these five sets of behaviour curves are given in Fig. 3 (A to E).

Stage 3a: CSFT Coupled Stress-Flow Testing of 200mm Cores

Coupled stress-flow testing was performed in NGI's biaxial CSFT apparatus, on two joints recovered from 200mm drillcore. Test I was performed on a planar, mineralized joint representing Set B (N-S strike) from the 2D drift. Test II was performed on a nearly planar, mineralized joint representing Set A (NW-SE strike) from the 3D drift. This core was taken from the same joint that formed the diagonal to the 1m × 1m × 2m *in situ* block test. Table 2 gives the mean values obtained from tilt and Schmidt

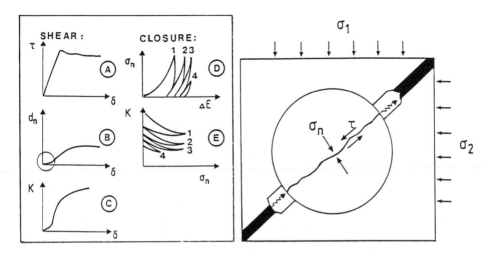

Fig. 3.   Shear-dilation-permeability and stress-closure-permeability coupling and
their measurement in CSFT tests.

hammer tests on these larger 200mm joint samples (from parts of the same
joint, immediately adjacent to the CSFT sample).

Table 2. **Approximate values of joint strength parameters for the 200mm
CSFT test (JCS$_n$ in units of MPa).**

Joint Set	Sample	JRC$_n$	JCS$_n$	ϕ_r
B (N-S strike)	No. 1	1.9	150	26.5°
A (NW-SE strike)	No. 2	3.8	125	25.1°

Table 2 indicates that the two large core samples had fairly typical JCS
and ϕ_r values, representative of the mineralized nature of many of the joints
at Stripa.   However, the planarity (and probably the persistence) were
approaching extreme values.

i) Stress - Closure - Flow testing
The effect of the three normal stress cycles to 25 MPa on conducting
apertures is illustrated in Fig. 4. Initial (unloaded) apertures reduced from
175μ to below 15μm for sample No. 1 (JRC$_n$ $\approx$ 1.9) and from 285μm to
about 30μm for sample No. 2 (JRC$_n$ $\approx$ 3.8). The measurement of smaller
initial and final apertures for smoother joints was consistent with previous
experience and with predictions (Barton et al., 1985).

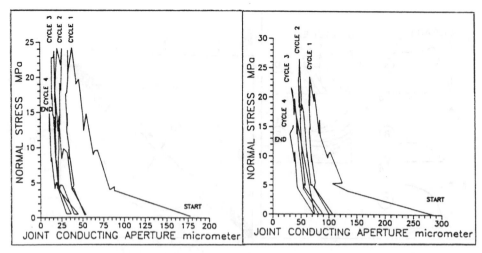

Fig. 4. Normal loading effect on conducting apertures measured in joint samples 1 and 2; 200mm cores. (Left: Set B, N-S, JRC$_n$ = 1.9, and Right: Set A, NW-SE, JRC$_n$ = 3.8.)

ii) Shear - Dilation - Flow testing

Three shearing increments were subsequently applied to give shear displacements of approximately 0.3mm, 1.0 - 1.2mm and 2.0 - 2.3mm. The first shear increments appeared to cause no increase in conducting aperture, and may, as in the case of joint No. 2, cause a reduction if the normal stress increases during the shearing event. In the next shear increment from approximately 0.3mm to 1.0 or 1.2mm, a consistent increase in conducting aperture was measured in both tests. This increment caused conducting apertures to increase from approximately 8 to 16μm in joint No. 1, and from approximately 15 to 30μm in joint No. 2. Normal stress was held at a high level (23 to 28 MPa) throughout this increment.

Conductivity reduced in the static period following several of the shearing increments under high normal stress, possibly due to the action of creep in the chlorite coatings. *This type of behaviour is also likely to have occurred in the validation drift.*

Stage 3b: CSFT Coupled Stress Flow Testing of 2m³ Block

The final step in the rock mechanics test programme was the performance of an *in situ* block test. The persistent, chlorite-coated joint that was sampled in test No. 2 at 200mm scale, was selected as a diagonal in the 1m × 1m × 2m block test. It was therefore sampled at 1400mm scale. The test was performed in the 3D drift, north of the cross and close to the west wall of the drift. The block was loaded with pairs of 1m² flatjacks placed within steel boxes that were cemented into core-drilled slots on the four sub-vertical sides of the block. Fig. 5 shows the test set-up.

The undisturbed joint (prior to flatjack slot drilling) was permeability tested

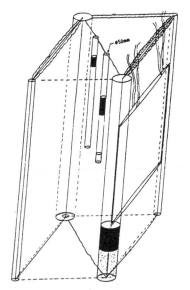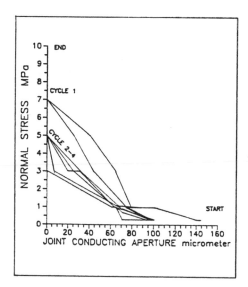

Fig. 5. *In situ block test for coupled stress, shear, flow testing of joint Set A*
(as CSFT No. 2). NGI performed this 1 × 1 × 2 × m test in the N end of
the 3D Migration drift.

by a small scale crosshole procedure, and indicated conducting apertures (e)
in the range of 5 to 7μm. Following the unloading caused by the drilling of
the flatjack slots (2m in depth), the conducting aperture increased to 145μm.

During four load cycles to 10 MPa normal stress (Fig. 5), the conducting
aperture reduced to zero (non-measurable) on each cycle. On Cycles 1 and
2, normal stresses of 5 to 7 MPa were sufficient; on Cycles 3 and 4, normal
stresses of 3 to 5 MPa were sufficient.

The measured physical closures (ΔE) on each cycle were 98μm, 43μm and
8μm, respectively. Just prior to commencement of shearing on the fourth
cycle, the joint had physically closed a total of 95μm (= ΔE). However, the
original (unloaded) conducting aperture of 145μm had reduced to zero
(unmeasurable flow) (= Δe). This suggests some physical changes in the
flow path, perhaps caused by the deformable chlorite fill. *Such behaviour is*
likely to have occurred in the validation drift, with its numerous mineralized
joints.

Shearing of up to 1.8mm was achieved under a normal stress of 10 to 11
MPa. No flow was registered during any of the shearing events. A measured
joint dilation of only 8μm occurred during the first 360μm of shear.
Thereafter, a continuous reduction of physical aperture was measured during
shear, amounting to some 10μm. The joint, as measured at this 1400mm
scale, was clearly too planar to dilate significantly, and was effectively sealed
to water flow at a moderate normal stress level, perhaps due to "damage" to
pre-existing flow channels in the chlorite infill.

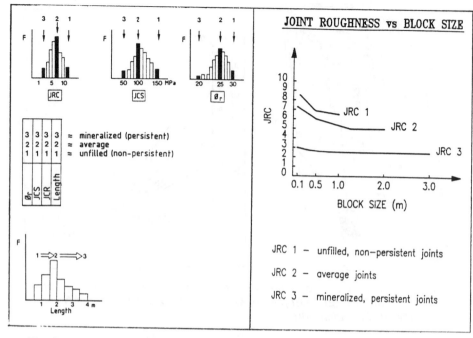

Fig. 6.   The relevant values of JRC, JCS and ϕ_r were chosen with respect to
joint length, with lower values for the persistent mineralized joints.

Stage 4: Predicting Input Data for Numerical Modelling

As shown in Fig. 2, the characterisation data for the joints was organised in the form of histograms, so that the variability of the parameters JRC (Joint Roughness Coefficient), JCS, (Joint Wall Compressive Strength) and ϕ_r (Residual Friction Angle) was accounted for. The first four UDEC-BB models utilised mean values of these input parameters (material model 2 "average") in the histograms sketched in Fig. 6, while models 5, 6, 7 and 8 (with unchanged joint structures) utilised three sets of length-dependent values of the joint strength parameters.

The shortest joints were therefore given high values to represent rough, interlocked, higher stiffness behaviour. Joints of intermediate length were given intermediate properties. The more persistent mineralized joints were given the lowest values of JRC, JCS and ϕ_r.

Table 3 gives the final form of the input data for the joints as required for the Stage 5 UDEC-BB studies. Definition of terms are given on the right-hand side of the table.

The one-dimensional joint behaviour was calculated by use of a Lotus spreadsheet version of the Barton-Bandis joint model (see Makurat et al., 1990b). The joint conductivity vs shear displacement behaviour predicted for the three joint models is shown in Fig. 7. Note that the physical aperture (E)

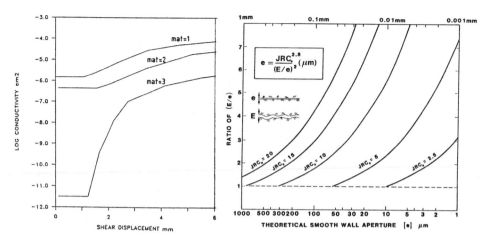

Fig. 7. *(Left) Prediction of shear-dilation-conductivity coupling for the three joint material models. (Right) Method of converting physical joint apertures (E) to conducting apertures (e).*

is converted to the conducting aperture (e) by the empirical model shown in Fig. 7 (Barton, 1982). The intact rock was modelled as an elastic isotropic medium acting under plane strain conditions, with elastic properties as listed in Table 4.

Table 3. **Joint material input parameters.**

parameter	unit	mat = 3	mat = 2	mat = 1	explanation
L	m	L > 2	1 < L < 2	L < 1	joint length
JRC_0		3.0	7.25	8.5	joint roughness coefficient, lab. scale
JCS_0	MPa	70	140	200	joint wall compressive strength, lab. scale
L_0	m	0.1	0.1	0.1	block size, lab. scale
L_n	m	1.0	1.0	0.5	block size, *in situ*
ϕ_r	°	21	25	29	residual angle of friction
σ_c	MPa	240	240	240	uniaxial compressive strength
JKN	$\frac{GPa}{mm}$	1.3e7	7.2e7	7.1e10	joint normal stiffness limit
JKS	$\frac{GPa}{mm}$	8.4e3	8.2e3	14.3e3	joint shear stiffness limit
E_0	mm	0.100	0.100	0.100	zero stress aperture

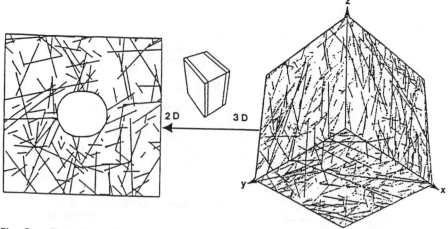

Fig. 8. Two-dimensional jointed (UDEC-BB) models were created from Harwell's 8m × 8m × 8m stochastically generated cubes. (Herbert, 1989)

Table 4. **Intact rock material input parameters.**

parameter	parameter	No.	units
density of rock	γ	2600	kg/m³
bulk modulus	K	38.0	GPa
shear modulus	G	22.8	GPa

Bulk and shear moduli of the intact rock have been calculated based on classical theory of elasticity. Young's modulus E = 57 GPa and Poisson's ratio v = 0.25 gives:

$$Bulk\ modulus : K = \frac{E}{3(1-2v)} = 38.0\ GPa$$

$$Shear\ modulus : G = \frac{E}{2(1+2v)} = 22.8\ GPa$$

Stage 5: Discrete Element (UDEC-BB) Modelling

UDEC-BB is a two-dimensional version of the distinct element method which is specifically designed to simulate the predominant features of naturally jointed or of fractured rock masses, including:
* variable rock deformability
* complex joint structures
* non-linear, inelastic joint behaviour
* fluid flow in joints
* far-field static or dynamic boundary conditions
* transient heat flow and thermally-induced stresses

The BB subroutine simulating the behaviour of joints has the following major features:

1) hyperbolic normal closure function (joint stiffness increases with closure) (Bandis et al., 1983);
2) closure limit (joints cannot continue to close at very high normal stress);
3) normal reversal logic (an unloaded joint will experience hysteresis, and will stiffen each time it is reloaded);
4) joint opening (if the joint is pulled apart, stiffness is reduced to that of disturbed joint conditions upon reloading);
5) shear reversal and damage (shear stiffness of joint depends on number of reversals and extent of shear displacement) (Barton et al., 1985);
6) dilation with shear; joint dilation is a function of shear displacement and normal stress, joint normal stiffness decreases with shear and dilation.
7) The calculated physical joint apertures (E) resulting from joint closure, opening, shear and dilation are converted to conducting apertures (e) for flow calculations as shown in Fig. 7.

The UDEC code was developed by Cundall (1980), and its application as UDEC-BB with the Barton-Bandis joint logic is described by Makurat et al. (1990b). The BB joint logic is described by Barton et al. (1985), and Bandis et al. (1981; 1983).

Modelled joint geometries

Systematic joint mapping performed by Gale et al., 1990b (IR 90-02) provided the statistical data for Harwell's initial 8m × 8m × 8m stochastically generated jointed cubes (Herbert, 1989). The four appropriate end faces of these two cubes were used by NGI to define possible discrete two-dimensional joint geometries for rock mechanics modelling. One of the stochastic models and an end face are illustrated in Figure 8 (Herbert, 1989).

The modelling, all of which was predictive in nature, was divided into four stages:

I   Modelling with average joint material parameters, Models 1, 2, 3 and 4

II   Modelling with length-dependent joint material parameters, Models 5, 6, 7 and 8 (see Fig. 9)

III   Comparison with continuum analyses

IV   Hydromechanical (coupled) D-hole and validation drift inflow modelling using Models 5 and 8.

For the most part, only two joint sets (A and C) were represented in the eight UDEC-BB models. Set B crossing the future drift direction nearly at right angles could not be modelled by UDEC, but was represented in Itasca's 3DEC modelling (see later). Note that the discontinuous nature of the jointing shown in Fig. 9 was achieved by having numerically glued sections of the joints.

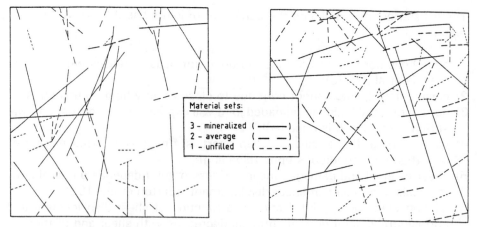

Fig. 9. *Joint geometries assumed for models 1 and 4 (with average No. 2 material properties). The same geometries were used for models 5 and 8 (with length dependent material properties Nos 1 to 3, Fig. 6).*

The modelled *in situ* stresses were based on McKinnon and Carr, 1990 (TR 90-09). Horizontal and vertical boundary stresses of 18 and 10 MPa were applied in the UDEC-BB models. The lower boundary of the model was a roller boundary. Simulations were performed under plane strain conditions (i.e., no deformation in the third dimension).

In the case of the coupled hydro-mechanical modelling, the *in situ* joint pressure was initially set to 2.27 MPa at the model boundaries. After the drift excavation, this pressure was reduced to 0.4 MPa at the model boundaries to approximately satisfy measured gradients (Herbert, 1989).

MECHANICAL RESPONSE

Key features of the mechanical (non-coupled) UDEC-BB modelling have been described by Barton et al., 1992 (TR 92-12) and can be summarised as follows:

1) Initial apertures of joint types No. 1, 2 and 3 were approximately 40, 25 and 1μm or less prior to excavation.
2) Excavation caused some joint shearing (generally 0.3 to 0.9mm) and some channel formation at block corners, mostly in the first 0.5m. Maximum channel dimensions ranged from almost 0.6mm at 0.4m, to 0.15mm some 1.5m inside the walls.
3) Channels can possibly provide local increases in axial permeability in the first half diameter (0 to 1.5m).
4) Peak tangential stresses of 55, 74, 64 and 62 MPa were registered in Models 5, 6, 7 and 8, while a continuum model registered a maximum tangential stress of 43 MPa (compared to virgin stresses of σ_v = 10 MPa and σ_h = 18 MPa).

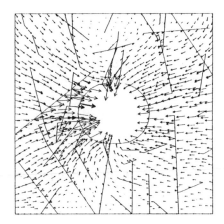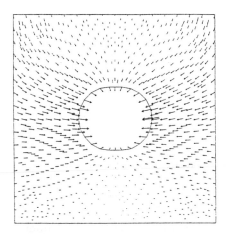

*Fig. 10. Displacement vectors in jointed model No. 5 compared with the continu-
um model. Maxima were 1.55 and 0.75mm.*

5) Joint planes intersecting the drift tend to be closed by the increased tangential stress levels (or do not shear enough to dilate) and will therefore tend to cause reduced radial permeability.
6) Excavation induced deformations showed maxima of 1.1 to 1.6mm in Models 1, 2, 3 and 4, and maxima also of 1.0 to 1.6mm in Models 5, 6, 7 and 8. Drift closures were therefore generally limited to 2 to 3mm (see Figure 10).
7) Deformations were greatest in the more jointed zones of the models, and the level of tangential stress was lower in these same jointed zones.
8) Significantly larger magnitudes of joint shearing were seen in the three-material models (5 to 8) at some 2 to 3 metres (2R) from the drift, due to the reduced shear strength of the persistent mineralized joints.
9) In these small drift simulations (3m span × 2m height) most of the disturbed zone (zone of changed apertures) was of nearly the same thickness as the assumed blast-damaged zone. This will not be the case when excavation dimensions are increased.

HYDRO-MECHANICAL (H-M) RESPONSE
Key features of the coupled hydro-mechanical UDEC-BB modelling have been described by Monsen et al., 1992 (TR 92-11) and Barton et al., 1992 (TR 92-12) and can be summarised as below:

A. D-hole Simulations (H-M model)
1. The modelling of water flow to the D-holes that were drilled around the future periphery of the future drift was limited to one model (Model 8 geometry).
2. Poorly connected parts of the model joint network showed high residual pore pressures.

3. Due to the presence of some extremely tight (material 3) joints with conducting apertures less than $1\mu m$, time steps had to be very small and calculation times correspondingly larger. Analysis of results of D-hole excavation in Model 8 indicated that even at 300,000 cycles flow transients still existed, but mechanical equilibrium appeared to have been reached.
4. At 60,000 calculation cycles, inflow to the D-holes from two joints with connection to the boundary showed approximately $4 \times 10^{-12} m^3/sec$ (0.004 litres/sec, or 0.24 litres/min. per 1m length of D-holes).
5. At 300,000 cycles, the inflow to the D-holes from the same two joints had reduced to half the above values, i.e., to 0.12 litres/min. per 1m length of D-hole.

B. Drift Excavation (H-M Models)
Figure 11 shows results from one of the H-M drift excavation models. In general:
1. H-M models (5 and 8) showed stress rotations and moderate reductions in *stress* in the neighbourhood of the drifts, but occasional extra high values of stress compared to the M (uncoupled behaviour) models. (Model 8: maximum stresses M 61.9 MPa, H-M 82.2 MPa.)
2. H-M models (5 and 8) showed some general 10 to 30% reductions of *displacements* and some rotations in the walls, but individual maximum displacements in the arch and invert were between 120 and 140% of those in the uncoupled models. (Model 8: maximum displacements M 1.32mm, H-M 1.61mm.)
3. H-M models (5 and 8) showed some local 40 to 60% reduction and some local 20 to 40% increases in *joint shearing*, compared to the uncoupled models. (Model 5: maximum joint shearing M 0.97mm, H-M 1.26mm.)
4. H-M models (5 and 8) showed few changes to "far-field" *apertures*, but close to the drift some major apertures were 20 to 100% larger than in the uncoupled models. (Model 8: maximum apertures M 0.29mm, H-M 0.54mm.)
5. In general, maximum deformations, stresses, joint shearing and apertures tended to be somewhat larger in the H-M models. There were, however, exceptions.

C. Comparison of D-hole and Drift Excavation H-M Models
1. The whole D-hole model showed *stresses* of almost original magnitude (approx. 18 MPa horizontally, and 10 MPa vertically) while the drift excavation showed much greater anisotropy (ratio of tangential to radial stress in the range 5 to 10 or more in the first 0.5 to 0.75m from the drift walls). Maximum Stresses Model 8 D-hole (H-M) = 24.0 MPa, Model 8 Drift (H-M) = 82.2 MPa.
2. The *displacements* in the D-hole simulations were generally 1 to 2 orders of magnitude less than in the drift simulations, in the future drift region.

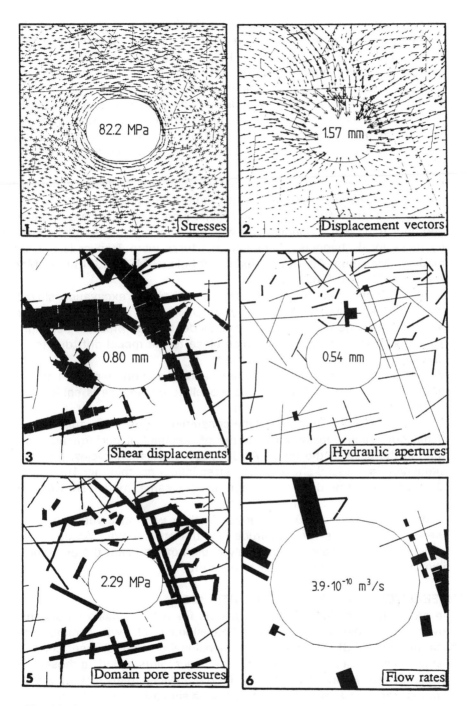

Fig. 11. Validation drift excavation with Model 8 jointing, using fully coupled
H-M version of UDEC-BB.

However, the upper boundary in the D-hole simulation was displaced about 0.4mm as a result of the pore pressure reduction. Maximum Displacements Model 8 D-hole (H-M) = 0.4mm, Model 8 Drift (H-M) = 1.6mm.

3. Large differences in *joint shearing* were evident between the D-hole and drift simulations. Maximum shearing in the D-hole model occurred out at the boundary between the jointed discrete element zone and the surrounding boundary element (continuum) zone. Maximum shearing (at least an order of magnitude larger) occurred closest to the drift. Maximum Joint Shearing Model 8 D-hole (H-M) = 0.096mm, Model 8 Drift (H-M) = 0.80mm.

4. Out at the boundaries of both models, joint *apertures* appeared to be identical. However, intersections close to the drift were of course affected by the excavation and showed local major increases, especially in the arch and right-hand wall. Connectivity was however not necessarily improved by these local channels. Minimum apertures were $1\mu m$ or less in both models. Maximum Apertures Model 8 D-hole (H-M) = 0.043mm, Model 8 Drift (H-M) = 0.54mm.

5. The high tangential stresses caused by the tunnel excavation slow the dissipation of some of the high *pore pressures*. More efficient pore pressure drainage to the D-holes was evident. Pore pressures as high as 1.6 MPa were registered in a region close to the model drift due to poor drainage through extremely tight joints.

6. With much longer running times, these extremely tight joints will drain the high pore pressures. Perhaps this modelling experience is equivalent to transients just after tunnel blasting.

7. There is evidence that the time-consuming runs with fully coupled behaviour reach equilibrium in terms of mechanical performance earlier than in the case of the flow. At 300,000 calculation cycles, the D-hole simulation in Model 8 showed 0.12 litres inflow per min. per 1m length of D-holes while the drift excavation in the same model showed 0.03 litres/min. per 1m of drift.

8. In practice, an 80 to 90% reduction in flow was observed, but inflows were about an order of magnitude less than the above. As pointed out earlier, the 2D models unfortunately could not represent the predominant jointing that was perpendicular to the drift.

THREE-DIMENSIONAL MODELLING WITH 3DEC

In an attempt to find additional reasons for the surprising reductions in inflow to the validation drift, a three-dimensional model was commissioned. Itasca Inc. utilised 3DEC, their three-dimensional discrete element code, and managed to model a selection of the more major joints including those of Set B, which in the H-zone caused most inflow. (More than 99% of the flow emanated from this zone, 57% coming from a single $2m^2$ area. Flow from the "average rock" was so small that it could only be detected by evaporation/ventilation measurements; Harding and Black, 1992.)

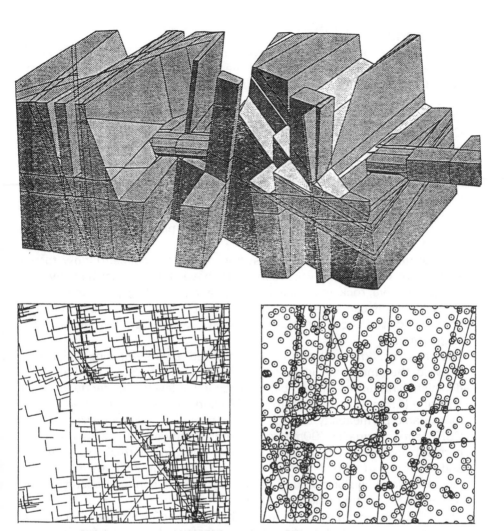

Fig. 12. 3DEC results for Validation Drift, showing cut-away 3D view (from
67°/165°), principal stresses (max. 57.9 MPa), and normal stress
increases on a steep joint plane 5m from the face. (Tinucci and
Israelsson, 1991)

Figure 12 shows a selection of Itasca's 3DEC results for the Validation
Drift (Tinucci and Israelsson, 1991). The right-hand lower figure shows
normal stresses on an oblique, steeply dipping joint about 5m from the face.
Note the increased diameter of the normal stress circles on the joint plane
closest to the face, which indicates at least a doubling of stress magnitude
compared to the far-field stress. This of course can explain some of the
observed flow reduction compared to the D-holes, but it is not a sufficient
explanation. Disturbance of flow channels by block shearing is perhaps also
required.

CONCLUSIONS

1. Two-dimensional modelling is obviously limited to the joints that strike parallel or sub-parallel to the axis of the modelled tunnel, in this case Sets A and C. Conclusions from the UDEC-BB modelling must therefore be used with caution in interpreting the three-dimensional behaviour of the drift.

2. Coupled H-M modelling cannot at present be performed with 3DEC but can be used in UDEC-BB. Consequently it is wise to perform both types of analysis for obtaining the best presently available understanding of how jointed rock masses behave. An essential ingredient in this understanding is the performance of coupled stress-closure-shear-dilation-flow tests on dominant joints, preferably at laboratory (CSFT) scale and on larger block tests *in situ*. Characterisation is essential in placing the results in the correct statistical perspective.

3. Several items of evidence for possibly explaining reduced inflow to the small, 3m × 2m × 50m validation drift have been unearthed by the comprehensive rock mechanics testing and modelling performed. The following items can stand alone, or can complement other explanations of possible disturbance by blasting, blast gasses, or two-phase flow (degassing).

4. The 1m × 1m × 2m block test on a mineralized, chlorite-bearing Stripa joint showed that moderate increases of normal stress (3 to 7 MPa) could diminish joint permeability to below measurable levels. Moderate levels of shearing (1.8mm) under a 10 to 11 MPa normal stress did not re-establish measurable permeability due to the joint planarity and block size.

5. CSFT tests on smaller 200mm joint samples showed significantly reduced permeabilities with stress and with time, following normal stress cycling and shearing of about 1mm. Creep of the joint infill material under increased load and/or shearing could have been a common occurrence in the newly excavated validation drift.

6. The UDEC-BB models, although limited to two-dimensions, do show reduced apertures and inefficient pore pressure dissipation for the near-field joints that are highly stressed by the excavation. Shear magnitudes in the 3m × 2m drift (0.3 to 0.9mm) were not sufficient to mobilise dilation. Potentially increased permeability was therefore only evident in an axial sense, due to channel formation at block corners. Block shearing in 2D could displace flow channels in the third dimension, i.e., in the H-zones.

7. Maximum stresses were twice as high in the jointed H-M model as compared to the continuum model (82.2 MPa compared to 43 MPa), so isotropic elastic models need cautious application.

8. Set B joints crossing the validation drift nearly at right angles (as for Zone H) were shown in 3DEC studies to bear at least a 100% increase in normal stress in the first 0.5 to 1m from the drift.

94

9. Closure-shear-flow (CSF) coupling along the mineralized joints at Stripa can therefore readily explain the principal SCV phenomenon of reduced flow to the drift. Geotechnical modelling should be used to actively supplement uncoupled, but three-dimensional geohydrological (network) modelling in future programmes.

REFERENCES

Bandis, S., A. Lumsden and N. Barton, 1981, "Experimental studies of scale effects on the shear behaviour of rock joints", Int. J. of Rock Mech. Min. Sci. and Geomech. Abstr. 18, pp. 1-21.

Bandis, S., A.C. Lumsden and N. Barton, 1983, "Fundamentals of Rock Joint Deformation", Int. J. Rock Mech. Min. Sci. and Geomech. Abstr. Vol 20, No. 6, pp. 249-268.

Barton, N. and V. Choubey, 1977, "The shear strength of rock joints in theory and practice", Rock Mechanics, Springer, Vienna, No. 1/2, pp. 1-54. Also NGI-Publ. 119, 1978.

Barton, N., 1982, "Modelling rock joint behaviour from *in situ* block tests: Implications for nuclear waste repository design", Office of Nuclear Waste Isolation, Columbus, OH, 96 p., ONWI-308, September 1982.

Barton, N. and S. Bandis, 1982, "Effects of Block Size on the Shear Behaviour of Jointed Rock", Keynote Lecture, 23rd U.S. Symposium on Rock Mechanics, Berkeley, California.

Barton, N., S. Bandis and K. Bakhtar, 1985, "Strength, Deformation and Conductivity Coupling of Rock Joints", Int. J. Rock Mech. & Min. Sci. & Geomech. Abstr. Vol. 22, No. 3, pp. 121-140.

Barton, N., A,. Makurat, K. Monsen, G. Vik and L. Tunbridge, 1992, "Rock Mechanics Characterization and Modelling of the Disturbed Zone Phenomena at Stripa", SKB Stripa Project TR 92-12.

Cundall, P.A., 1980, "A generalized distinct element program for modelling jointed rock. Report PCAR-1-80. Peter Cundall Associates; Contract DAJA37-79-C-0548, European Research Office, U.S. Army.

Gale, J. and A. Stråhle, 1988, "Site Characterisation and Validation - Drift and Borehole Fracture Data Stage I", SKB Stripa Project IR 88-10.

Gale, J., R. MacLeod and P. LeMessurier, 1990a, "Site Characterisation and Validation - Measurement of Flowrate, Solute Velocities and Aperture Variation in Natural Fractures as a Function of Normal and Shear Stress, Stage 3", SKB Stripa Project TR 90-11.

Gale, J. R. MacLeod, A. Stråhle and S. Carlsten, 1990b, "Site Characterisation and Validation - Drift and Borehole Fracture Data, Stage 3", SKB Stripa Project IR 90-02

Hakami, E., 1989, "Water Flow in Single Rock Joints", SKB Stripa Project TR 89-08.

Harding, W., and J. Black, 1992, "Site Characterization and Validation - Inflow to the Validation Drift", SKB Stripa Project TR 92-14.

Herbert, A., 1989. Private Communication.

Makurat, A., N. Barton, L. Tunbridge and G. Vik, 1990a, "The measurement of the mechanical and hydraulic properties of rock joints at different scales in the Stripa project", International Symposium on Rock Joints. Loen 1990. Proceedings, pp. 541-548, 1990.

Makurat, A., N. Barton, G. Vik, P. Chryssanthakis and K. Monsen, 1990b, "Jointed rock mass modelling", International Symposium on Rock Joints. Loen 1990. Proceedings, pp. 647-656, 1990.

McKinnon, S. and P. Carr, 1990, "Site Characterization and Validation - Stress Field in the SCV Block and Around the Validation Drift, Stage 3", SKB Stripa Project TR 90-09.

Monsen, K., N. Barton and A. Makurat, 1992, "Fully-Coupled Hydro-Mechanical Modelling of the D-Holes and Validation Drift Inflow, SKB Stripa Project TR 92-11.

Tinucci, J.P., and J. Israelsson, 1991, "Site Characterization and Validation - Excavation Stress Effects Around the Validation Drift. SKB Stripa Project, TR 91-20.

Vik, G. and P.M. Johansen, 1990, "Determination of shear strength of rock joints at two dam sites and at Stripa Research Mine", International Symposium on Rock Joints. Loen. Proceedings, 1990.

Borehole Seismics

Calin Cosma

Vibrometric Oy

Abstract

The paper reviews the gradual development and application of the borehole seismic techniques within the Stripa Project. The topics covered are: seismic equipment for boreholes, methods of enhancement and analysis of reflections from features of arbitrary orientations and seismic tomography. Regarding the equipment, the main concern has been to design probes for slim holes and relatively high frequency. The processing techniques were initially developed to cope with the specific problems posed by tests in crystalline rock. Variants of these techniques have been further developed and used in other surveys. As examples, VSP surveys performed at TVO's sites in Finland, tunnel surveys done at NAGRA's Grimsel Site in Switzerland and at the Äspö Hard Rock Laboratory in Sweden are presented.

1. Role of Remote Sensing Methods in Site Characterization

Remote sensing techniques (radar and seismics) were initially included in the group of site characterization methods as a means for extrapolating the information obtained from boreholes. It was assumed that the features detected by remote sensing could be linked with certain anomalies (e.g. an increase of the fracture frequency) observed in the boreholes, the main task of seismics and radar techniques being to give these anomalies a spatial dimension.

During the Stripa Project, the remote sensing methods became more flexible, allowing surveys to be carried out with various experimental setups and higher resolution. The processing routines acquired capabilities of detecting with increased reliability also features which did not intersect boreholes, tunnels or the ground surface. The geometrical model obtained by remote sensing gradually became the supporting frame for the conceptual model of the site.

Most of the development work concerning remote sensing techniques was directed towards improving the prediction of the geometry of the rock features. The significance of the physical parameters used to assess the geometry was given comparatively less attention. This approach was adopted due to the prevailing opinion that the main task for geophysics was to predict the position and orientation of the major fracture zones.

Most of the standard seismic methods and data collection equipment were designed for the large scale surveys required by the prospecting industry. On the contrary, radar is a detailed quality check method, the investigation range being limited to surveys of relatively small scale.

For these reasons, in the initial Stripa Project approach radar was meant for small scale studies (50 - 200 m) while seismic methods were intended for larger scale surveys (200 - 1000 m). This division of tasks was based on the rather optimistic assumption that each of the two techniques could produce by itself a non-ambiguous and relevant description of the structure of the rockmass at its respective investigation scale.

The large scale experiment with seismic tomography performed at the Gideå site indicated a rather complex structure of the rockmass. Meanwhile, radar reflection surveys at the small scale site in the Stripa Mine found a relatively simple structure consisting of extended planar features. Later it was recognized that the straight line reflections and the patchy trends in the tomograms were both reasonable approximations of the reality. This idea found a firm ground when seismic investigations were conducted at the "small scale" site in the Stripa Mine and an excellent match was obtained between radar and seismic tomograms /5/.

The lack of a seismic reflection technique similar to the one used with radar prevented the comparison of the two methods in reflection mode. This was done later when the results obtained within the programme of development of high resolution seismic techniques started to be used for site characterization surveys.

2. Development of Seismic Methods

The principal goal of the development project has been to build up a seismic methodology suitable for the characterization of the nuclear waste disposal sites. In this endeavor we have tried to create processing and interpretation tools usable for various site conditions.

The borehole seismic source developed during the Project is a resonant piezoelectric device able to work in continuous and coherent burst modes /1/. It consists of two axial stacks facing each other, the distance between the stacks being adjustable. The frequency is adjusted to bring to resonance the water column between the two transducers. Figure 1. shows schematically the principle of operation of the source. Short wave trains can be emitted with a high rate of energy transmission to the rock, due to the resonance. Different resonance frequencies are obtained by varying the distance between the transducers. If a wider frequency band is desired, several records with different center frequencies can be added. The device operates in a frequency range of 3 - 7 kHz allowing the reception of the signals at a distance of 200 - 300 m . The minimum borehole diameter in which the instrument fits is 56 mm.

In crystalline rocks the structural features representing the main seismic targets are concentrations of fractures which can not be accurately described as interfaces between geological units with different physical properties. The fracture zones can be wider or thinner and often have diffuse boundaries. Consequently, the reflected field is very weak.

The reflection methods developed within the Stripa Project were initially intended for VSP surveys, due to technical difficulties of placing both seismic sources and detectors in the same hole. The procedure has been extended later for other experimental layouts. The same processing scheme has been used for measurements along tunnels, on ground surface and offshore.

A VSP profile is normally displayed as a group of seismograms (traces) arranged in increasing order of depth of the stations forming the measuring array. As the other axis is the time history, a VSP section is a two-dimensional space with the axes representing depth and time. Therefore, the tangent of a direction in this space can be treated as a velocity. This apparent velocity is not identical to the physical velocity of a given wave front but, with various degrees of success depending on the application, it can help in the discrimination of different wave types. The emphasis in our work has been placed on the development of multichannel filtering techniques which permit the weak reflected signals to be enhanced.

We have started the study of the multichannel filtering techniques with the FK and tau-P transforms. In parallel, we worked on the development of the "Image Point" method, a new technique using the physical wave propagation velocity rather than the apparent velocity defined by the slope of an event in a given profile. This gives more flexibility in choosing the test configuration and allows the treatment of structures with any orientation in space. When the apparent velocity is used, like with the other multichannel filtering methods, there is a risk of rejecting parts of a real reflected event. This conclusion is illustrated in Figures 3.a-d.

Figure 3.a presents a synthetic VSP section containing typical elements for a survey in crystalline rock: a curving first arrival line due to the source offset and five reflection patters with different inclinations and curvatures. The representation in FK, tau-P and IP space are shown in Figures 3.b, 3.c and 3.d, respectively. In the IP representation from Figure 3.d the five reflection events collapse in as many well resolved spots. The tau-P transform in Figure 3.c produces a more distorted image. The distortion is larger for events displaying a stronger curvature in the original time-depth profile. The FK transform from Figure 3.b does not offer any clear means for the separate identification of any of the five reflectors.

So far, Image Point filtering seems to be the most efficient method for enhancing weak reflections in crystalline rock. An intuitive understanding of the procedure is that each reflection pattern can be constructed starting from an "image source", from which the signals propagate to each detector on a direct path. Image sources are virtual, like mirror images in optics. The mirror on which the image source is formed is a planar reflecting feature, e.g. a fractured zone.

A detailed descriptions of the Image Point method is given in reference /2/. Figure 2. shows schematically how the image source is obtained. One can notice that the image space has an axial symmetry, all image points disposed on a circle like the one presented in Figure 2. producing synchronous arrivals at all the detectors. This characteristic permits most of the computations to be done in two dimensions. During the Project we have also developed an interpretive processing method to locate the reflectors in three dimensions. The procedure is an extension of the IP method and consists of combining several two-dimensional transformed sections in a single three dimensional representation.

The seismic approach for the Site Characterization and Validation (SCV) Project consisted of applying in parallel two methods: two-dimensional crosshole tomography and the new three-dimensional reflection imaging technique developed during the project.

The reflection method detects changes in acoustic impedance, which is an accurate way of finding the boundaries of rock features. It permits to determine the position of the features in three dimensions. The tomographic method maps variations of the rock properties, like wave velocity and attenuation. The tomographic method gives a good estimate of the average values of the local rock parameters, while the reflection method gives a better image of the geometry of the zones. Figure 4. presents the geometrical model of the SCV site obtained by the tomography-reflection combined approach /3/.

The identification of a the three dimensional structure in a two dimensional tomographic section is sometimes difficult. Even well defined fracture zones may give a distorted image, if they are cut by the tomographic plane at an unfavorable angle. In order to discriminate between hardly recognizable but real features and possible processing artifacts, it is desirable to perform tomography in parallel on more than one parameter.

The first experiment with seismic tomography at Stripa was performed independently for P- and S-wave velocities. The assumption was that both the compressional and the shear strength of the rockmass decrease in fracture zones, which leads to a lower propagation velocity for both P- and S-waves /5/.

Radar tomography is normally done for velocity and attenuation. For a closer complementarity of the seismic and radar results, seismic velocity and attenuation tomograms were produced for the SCV Project /3/. An example of P-wave attenuation tomography is given in Figure 5. So far, the correspondence between high attenuation and fractured rock is qualitative. It seems that, in order to obtain a reliable image of the fracture zones by tomography, a more quantitative approach and a better understanding of how fractured rock responds to seismic waves are needed.

3. Applications of Techniques Developed in The Stripa Project

Seismic reflection surveys in boreholes were carried out for Teollisuuden Voima Oy at five sites in Finland /4/. The purpose was to detect large scale fractured zones, lithological contacts and other anomalies in the structure of the rockmass and to determine their position and orientation. All the sites were investigated by multi-offset VSP. The key processing method has been Image Point Filtering.

The position and orientation in three dimensions of the seismic reflectors have been interpreted with the aid of borehole information, single-hole geophysical logging, radar survey and hydraulic tests. The main difference between the surveys carried out within this project and the ones at Stripa is of scale. The task here has been to describe a rock volume of several cubic kilometers at each site. Most of the investigation methods can not cope with such a large experimental scale, so that seismics had an important role in building the geometrical models of the sites. Figure 6. shows the model obtained for the Konginkangas-Kivetty site /6/.

In 1988 Nagra - the Swiss National Cooperative for the Storage of Radioactive Waste - decided to start a new research and development program to use geophysical remote sensing techniques to look ahead of the tunnel face while the tunnel is under construction /1/.

The axial symmetry of the tunnel provides in principle the conditions for a VSP layout, with shotpoints and detector arrays placed along the tunnel. Due to the cylindrical symmetry of the layout (zero offset), it is not possible to determine the position of a reflector uniquely in three dimensions, except for reflectors nearly perpendicular to the tunnel. For features making a sharp angle with the tunnel additional information is needed for a non-ambiguous three-dimensional fix. The distinguishing feature of this experiment is that the IP method had to be closely connected with other data describing partially the rockmass structure. This was possible due to the special attention given in this project to the corroboration of the information obtained by different investigation methods.

An example of the results obtained in this project is given in Figure 7. Two profiles were processed with the same detector layout in the tunnel and source points separated by 70 meters. The interpretation was carried out independently for the two profiles. Generally, when data recorded from more shotpoints are available, it is normal to perform a combined interpretation in three dimensions. This requires that the experimental layout is also truly three-dimensional, i.e. that the detector array and the shotpoints are not co-planar. In a tunnel it is difficult to fulfill this condition.

Seismic surveys were carried out for SKB at the Äspö Hard Rock Laboratory in Sweden during 1992. The purpose has been to check the assumed position of zone NE1 which created severe problems when being intersected by the access tunnel.

The survey was carried out by recording in the tunnel the signals received from small explosions placed in shallow boreholes drilled from the ground surface. As seen in Figure 8., eight shot positions were set on Äspö Island, in two lines parallel to the direction of the tunnel.

The Image Point method was used for processing and interpretation. For this survey, it was not enough to determine the average orientation of Zone NE1 as a planar approximation. The consequences of re-intersecting the same zone during the excavation of the spiral ramp may be very serious, so a more detailed analysis was called for.

Variations in the orientation of the reflector may appear because different parts

- and possibly different branches - of the reflector are seen from each shotpoint. It is then important to calculate and represent graphically each actual reflecting element, as seen from the respective shotpoint. Figure 9. shows such a representation and confirms that the position of Zone NE1 has been determined within the region of interest for future excavations.

4. Final Comments

From the number of contemporary and later applications of the methods developed in the Stripa Project we can conclude that the scientific and technical achievements of the project have the flexibility required in the beginning of the project.

Of course, the equipment built in order to validate the methods has been designed specifically for the experimental conditions at the Stripa site. Even so, we tried to concentrate on new ways of producing and recording seismic signals, rather than scaling down standard seismic industry tools.

It seems that describing the fracture zones as a set of quasi planar features extending through the site volume is a valid starting approach, if the orientation and position of these planes can be accurately determined. However, a better control of the extrapolation process is needed in later phases. A truly deterministic approach should place the detected anomalies in the region of the site volume where they were actually detected. When all the data is taken in this proto-model, the extrapolation is to be done by simultaneous fitting of all parameters. This would also solve some difficulties in combining the reflection and tomographic results.

Another way to increase the relevance of the geophysical results is to focus on the physical meaning of the results and derive quantitative relations between these results and parameters used in rock mechanics and hydrogeology. Geometry, when deprived of its physical support, is not the safest ground for extrapolation.

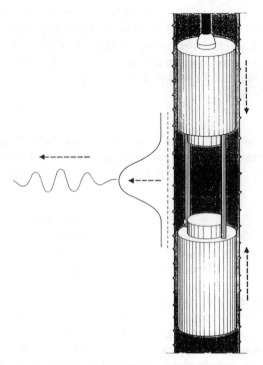

Figure 1. **Operation principle of the resonant seismic source /2/.**

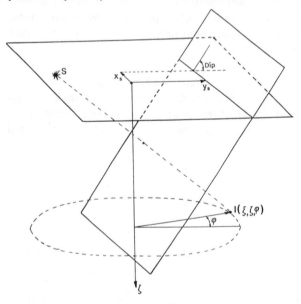

Figure 2. **The position of the image source (I) related to the real source (S) and a planar reflector.**

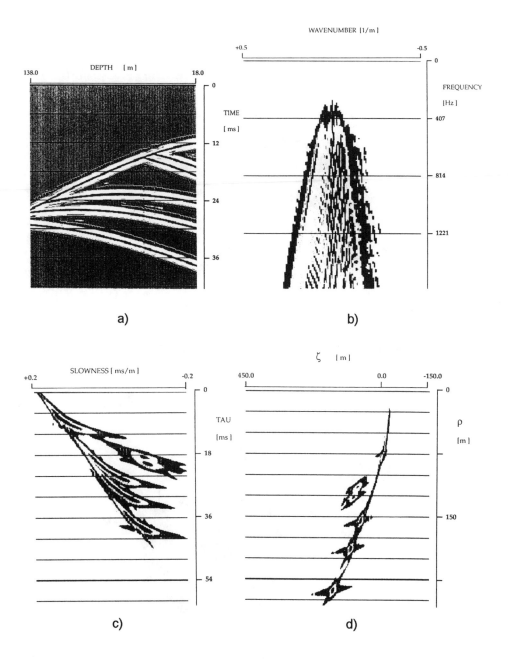

Figure 3. a) **Synthetic VSP section.** b) **FK transform.** c) **Tau-P transform** d) **IP transform.**

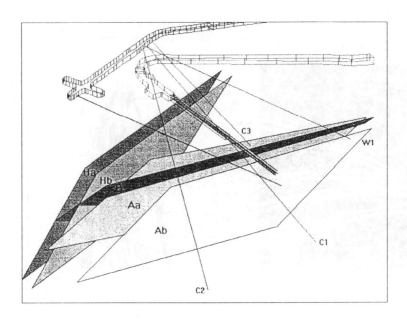

Figure 4. **Geometrical model of the SCV site obtained by seismics /3/.**

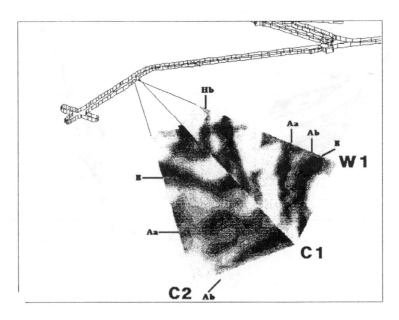

Figure 5. **Seismic amplitude tomograms from the SCV site /3/.**

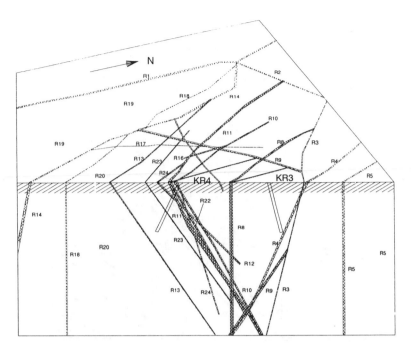

Figure 6. **Structural model of the Konginkangas-Kivetty site /6/.**

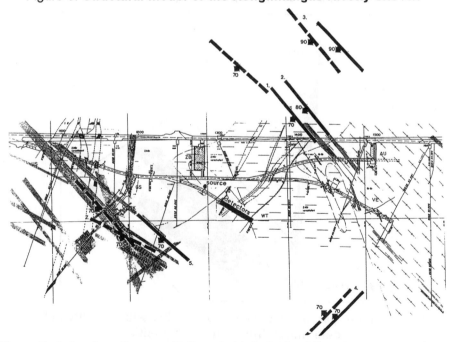

Figure 7. **Seismic reflectors (full and thick dashed lines) superposed on known geological features at Grimsel /1/.**

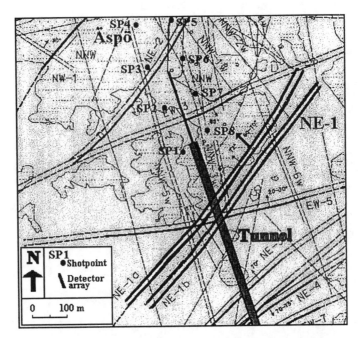

Figure 8. **Map of the Äspö site showing zone NE1.**

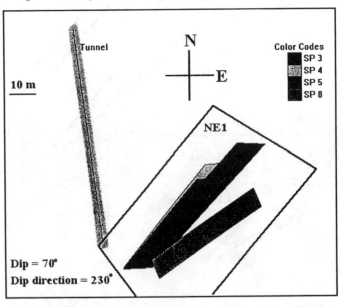

Figure 9. **In-depth perspective view of zone NE1 with reflective elements determined from four shot positions.**

References:

1. Bluemling P., Cosma C., Gelbke G., Cassell B., Korn M.: Geophysical Methods for the Prediction Ahead of the Tunnel Face, NAGRA-report NTB 90-07, 1992

2. Cosma C., Heikkinen P., Pekonen S.: Improvement of High Resolution Borehole Seismics, Part I: Processing Methods for VSP Surveys, Part II: Piezoelectric Signal Transmitter for Seismic Surveys. Stripa Project technical report 91-13, 1991.

3. Cosma C., Heikkinen P., Keskinen J, Korhonen R.: Site Characterization and Validation - Results from Seismic Crosshole and Reflection Measurements. Stripa Project technical report 91-07, 1991.

4. Keskinen J., Cosma C., Heikkinen P.: Seismic VSP and HSP Surveys on Preliminary Investigation Areas in Finland for Final Disposal of Spent Nuclear fuel. Report YJT-92-19, 1992.

5. Olsson O., Black J., Cosma C., Pihl J.: Crosshole Investigations - Final Report. Stripa Project technical report 87-16, 1987.

6. Teollisuuden Voima Oy: Final Disposal of Spent Fuel in Finnish Bedrock - Preliminary Site Investigations. Report YJT-92-32, 1992.

Borehole Radar and its Application to Characterization of Fracture Zones

Olle Olsson

Conterra AB, Uppsala, Sweden

Abstract

The RAMAC borehole radar system was developed within the framework of the International Stripa Project. The system can be used in three different measuring modes; single-hole reflection, cross-hole reflection, and cross-hole tomography. The radar operates in the frequency range 20-80 MHz which gives a resolution of 1-3 m in crystalline rock. The most recent development is a directional receiver antenna which makes it possible to determine the azimuth to a reflector with an accuracy of about 3°. Hence, the orientation of fracture zones can be determined from measurements in a single borehole. This can significantly reduce the number of boreholes required to adequately characterize a site.

Reflection measurements basically provide geometric information on geologic or manmade features located at some distance from the borehole. The magnitude of electrical properties of the rock is best obtained through tomographic imaging of cross-hole data. Repeated tomographic measurements (difference tomography) have been used to map the transport of saline tracers through the rock mass.

The radar data have provided a consistent description of the fracture zones at Stripa in agreement with geological, geophysical, and hydrogeological observations. The RAMAC system has also been successfully applied to a number of other sites.

Introduction

Experience from site investigations in crystalline rock related to radioactive waste disposal made in the 70'ies and early 80'ies revealed the need for better tools for finding and characterizing groundwater flow paths. In crystalline rock, fractures and fracture zones constitute the major flow paths and one of the main tasks of a site characterization program is to provide an adequate representation of the fracture system. A number of borehole tools exist which yield information about the occurrence, orientation, and properties of fractures in the immediate vicinity of the borehole. These methods provide little or no data on the presence and properties of fractures in the rock volume between boreholes and the large scale orientation of fracture zones has to inferred through extrapolation. In early 80'ies the need for characterization tools which could provide high resolution data on fractures and fracture zones between boreholes was recognized. Radar and seismic methods, which are both based on the propagation of waves through the rock, were considered most promising and comprehensive research and development was initiated within the framework of the International Stripa Project.

Basic principles

The radar method is based on propagation of radar waves through the ground. Radar waves are electromagnetic waves, just like radio waves and light. In a certain frequency band, radar waves can propagate appreciable distances through rock. Frequencies in the range 10-1000 MHz are commonly used in geological applications.

Radar-wave propagation through rock depends on dielectric permittivity (ε), magnetic permeability (μ), electrical conductivity (σ) of the material, and the transmitted frequency (ν). In most rock types variations in magnetic permeability can be neglected unless they contain significant amounts of ferromagnetic materials. The parameter $Q=2\pi\nu\varepsilon/\sigma$ determines whether a medium will support radar-wave propagation or not. In a medium where $Q>1$, electromagnetic energy will propagate as waves and radar can be used as an investigation tool. In this case, Q is a measure of the number of wavelengths a radar wave can propagate in a given medium before it is significantly attenuated. If $Q<1$, there will be no wave propagation. Sandberg *et al.* (1991) give data on the Q-values for some different rock types.

The relative dielectric permittivity, ε_r, of many minerals is in the range 5-7 and about 80 for water. In many geological environments, water will be the only material with significant contrast in dielectric permittivity and hence the bulk dielectric constant will be a measure of the water content. This implies that radar velocity ($v=c/\sqrt{\varepsilon_r}$) is a measure of water content or porosity. Radar attenuation depends both on conductivity and dielectric permittivity and cannot simply be related to a specific rock property.

Features with large contrast in electric properties will cause strong reflections of radar waves independent of their thickness. If the contrast in properties is low, as for fracture zones, a certain minimum thickness is required to give a reflection that stands out above the background. This is referred to as the detectable limit and is sometimes taken as about 1/30 the dominant wavelength. For borehole radar applications this corresponds to about 10 cm. Hence, the borehole radar will detect fracture zones, clusters of fractures, and tectonized zones rather than single joints.

The RAMAC system

In Phase 2 of the Stripa Project a new short-pulse borehole radar system (RAMAC) was developed which could be used both for single-hole and cross-hole measurements (Olsson *et al.*, 1987a). In phase 3 the development continued and a directional radar system was constructed which makes it possible to uniquely determine the orientation of a fracture zone from measurements in a single borehole (Falk, 1992). This had previously not been possible.

The main components of the new directional borehole radar system are shown in Figure 1. The computer unit is used for control of measurements, data storage, presentation, and analysis. Timing control and stacking is handled by the control unit. The electronic equipment necessary to create, receive, amplify, and register the radar pulses is placed in the borehole probes. The measured signals are transmitted to the control unit by optical fibers in digital form. Power to the transmitter and receiver is supplied through downhole batteries which last for 10 hours of continuous operation.

Omnidirectional transmitter and receiver antennas have been designed for center frequencies in the range 20-60 MHz. The antennas are broadband and allow transmission of a well-defined pulse. The directional receiver antenna operates at a center frequency of 60 MHz. These frequencies correspond to wavelengths in the range 1-5 m in granitic rock.

The transmitter in the directional radar system is equipped with a dipole antenna which has an omnidirectional radiation pattern relative to the borehole. The directional receiver is sensitive to the direction of the incoming radiation and this makes it possible to find the location of a reflector relative to the borehole. The directional antenna consists of an array of four loop antennas. The directional receiver also includes a direction indicator which senses the orientation of the antenna array relative to gravity or the total magnetic field vector (Falk, 1992).

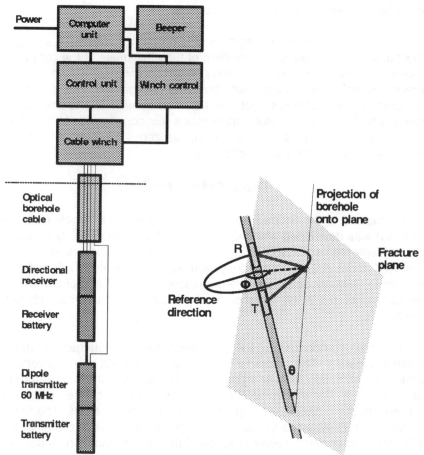

Figure 1   The components of the RAMAC directional borehole radar
             system and its principle of operation.

The directional antenna produces four composite signals, one from each
loop antenna. The signals are then decomposed to recover the three physically
meaningful entities; a dipole signal and two orthogonal directional components.
As four signals are measured there is a redundancy which is used to define a
fourth variable, called the checksum. The checksum is essential as it provides a
quality control on the directional antenna system. For a properly operating
system the checksum should be close to zero. The two directional signals can
be combined to produce a synthetic signal corresponding to an arbitrarily
oriented loop antenna. The radiation lobes of a loop antenna are broad, but the
minima are distinct and can be used to locate a reflector. There are two minima
of opposite directions, but the ambiguity can be resolved by comparing the
phase of the directional and the dipole signal.

Application modes

Single-hole reflection measurements

In a single-hole reflection measurement the receiver and transmitter are moved down a borehole at a fixed distance from each other. The signal is measured at regular intervals, usually in steps of 0.5 or 1 m. The signal reflected from a fracture zone or a point scatterer will produce a characteristic pattern that is used to identify the reflector in the radar map and to estimate its position and orientation.

Interpretation of directional radar data is made by an interactive program called RADINTER. The program calculates the radar reflection maps for the dipole, and the directional signal for every 10 degrees. Reflectors are marked in the radar maps and models corresponding to plane or point reflectors are adjusted to fit the observed reflector (Figure 2, left part). The depth of intersection and the angle between the borehole and the reflector is displayed automatically. Then the azimuth is changed until the minimum directional signal is found for a particular reflector (Figure 2, arrow right part). There are two minima and the correct azimuth is determined by comparing the phase of the directional signal at maximum signal strength with the phase of the dipole signal. Knowing the orientation of the borehole the dip and strike of the fracture plane can be calculated. The azimuth can normally be determined with an accuracy of $\pm 5°$ for data with good signal to noise ratio. For weak reflectors in a noisy environment it may not be possible to find the azimuth with sufficient accuracy.

An example of a typical single-hole reflection result from the Stripa Mine is shown in Figure 3. The linear reflections indicated by arrows correspond to planar fracture zones, most of which intersect the borehole. These zones consist of brecciated and mylonitized rock and have widths ranging from 1 to 8 m. The reflections which appear to intersect the borehole close to its start are caused by other boreholes, 76 mm in diameter and water filled. The hyperbolic reflection observed 40 m from the borehole at a borehole length of approximately 170 m is caused by a drift roughly perpendicular to the borehole. The radar probing range obtained at Stripa with the 60 MHz antennas was generally 50-60 m, as exemplified by Figure 2, while the 20 MHz antennas yielded probing ranges in excess of 100 m. With the frequencies used a resolution of 1-3 m is obtained.

The hydraulic testing results from borehole F4, also in Figure 3, show that fracture zones A and X are transmissive while there are no significant hydraulic anomalies associated with the other zones. In fact, zone C, which is not transmissive in this borehole, was found to be transmissive at its intersection with most other boreholes and hence one of the main hydraulic conduits at the

115

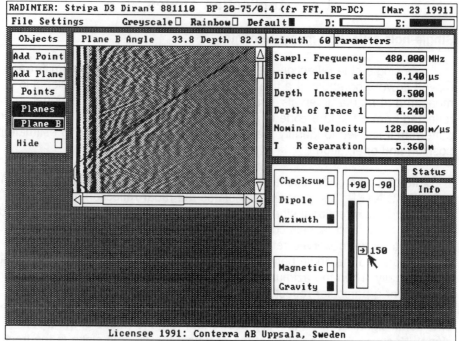

Figure 2   Screen display of the interactive interpretation program
            RADINTER.

Cross-hole site (Olsson et al., 1987b). These results demonstrate the local
variability in transmissivity of fracture zones but also that the radar detects
those fracture zones which are the potential paths for groundwater flow. Some
zones, such as zone K, may show little increased hydraulic conductivity,
possibly due to clay mineralization. However, it is important to note that the
radar identified practically all the hydraulic features and no hydraulically
significant features were missed (Olsson et al., 1992).

The reflectors observed with the directional antenna can be plotted at the
locations in space where the reflections actually have occurred (Figure 4).
During a reflection measurement in a borehole the reflection point traces a line
on the fracture plane (cf. Figure 1). It is along this line that we actually have
data verifying the existence and orientation of the fracture plane. In Figure 4
these lines are represented by circular disks to emphasize that the orientation
is actually known. This figure gives an honest image of what was actually
observed with the radar system and looking at the image with from different
views with a CAD-program the correlation of observations from different
boreholes can be evaluated.

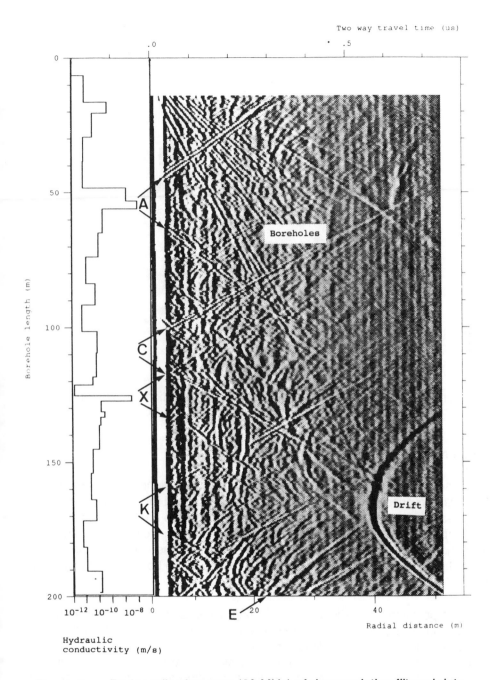

Figure 3   Radar reflection map (60 MHz) of deconvolution filtered data
            and hydraulic log from borehole F4 at the Crosshole Site.
            Reflections indicated by arrows are caused by fracture zones.

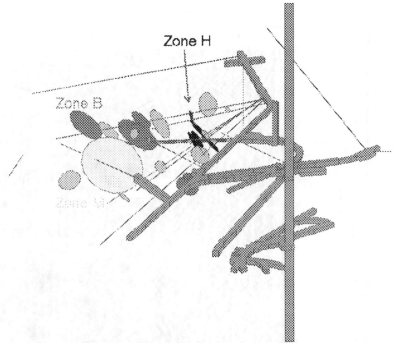

Figure 4   The distribution in space of radar reflections derived from single borehole directional radar data from the SCV site.

Cross-hole reflection

If antennas are placed in separate borehole, the orientation of a fracture zone can, in principle, be determined uniquely provided the boreholes are not in the same plane. Cross-hole reflection measurements are often an effective complement to the standard single-hole measurements. Velocity calibration at a new and unknown sites is normally made by placing one antenna on the ground near a borehole while the other is moved in the borehole, usually called Vertical Radar Profiling (VRP).

The possible orientations of a reflecting plane, as determined by a cross-hole reflection measurement, may be plotted as a curve in a Wulff diagram. Similar curves can also be plotted for reflections observed with the omnidirectional antenna in single-hole reflection measurements. The point of intersection for the curves determines the orientation of the fracture plane. This interpretation procedure (described by Olsson *et al.*, 1992) requires that several boreholes are located within the probing range of the radar. The use of cross-hole reflection data and the Wulff diagram interpretation technique has almost become obsolete since development of the directional antenna was completed.

Tomography

In tomography, small variations in time and amplitude observed in the directly propagated signal are used to map the properties of the rock between the boreholes. A large number of rays must be measured to produce a tomographic image, and the time required to complete measurement of a borehole section of typically about 1000 rays (source/receiver combinations) is about 6 hours. The complete processing of a data set takes about 2-4 hours. Most of the time is spent for data quality control, data corrections, etc. while the numerical tomographic inversion takes only a couple of minutes on a standard PC/AT computer and can be performed in the field. The tomographic inversion is made with the iterative Conjugate Gradient method (Ivansson, 1986). Tomograms have been produced using both arrival time and peak-to-peak amplitude of the radar pulses (Sandberg *et al.*, 1991, Olsson *et al.*, 1992).

Figure 5 shows slowness and velocity tomograms obtained from the borehole section W1-C2 at the SCV site. The dominating feature at the SCV-site, fracture zone H, is clearly seen in both tomograms. The tomograms show details of the shape of the fracture zone and its variability in properties, information not readily available from reflection measurements.

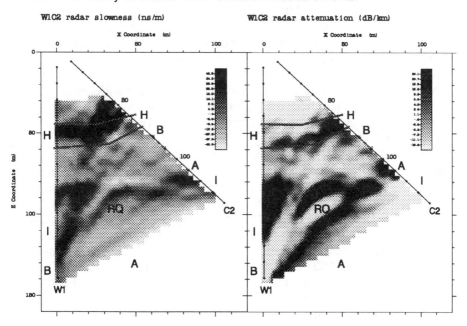

Figure 5   Tomograms for the section W1-C2 using 60 MHz radar; a) slowness, b) attenuation.

Difference tomography

Repeated tomography measurements can be used to show how a tracer moves through rock. The basic idea is to perform a tomographic reference measurement establishing the distribution of electric properties before tracers are injected. Then a conductive tracer (saline water), which increases radar attenuation, is injected at a suitable location and tomographic measurements are repeated at selected intervals. The increase in radar attenuation is proportional to the tracer density in the rock. Hence, the tracer mass distribution at a specific time can be obtained from tomographic inversion of the difference in radar amplitudes before and after tracer injection.

An experiment of this type was performed at the SCV-site with the objective to study changes in the flow regime due to drift excavation. A saline tracer with a concentration of 2 % was injected through a borehole where it intersected a fracture zone, denoted H. The injection rate was approximately 200 ml/min. In the first phase of the experiment a cylindrical array of boreholes acted as the hydraulic sink and in the second phase the borehole array had been replaced by a drift. Radar tomography was performed in three planes surrounding the injection point. The planes were nearly perpendicular to the fracture zone. After start of tracer injection tomographic measurements of the three planes were repeated 7 times during the first phase and 3 times during the second. In both phases measurements were made during a period of 700 hours after start of tracer injection. The time required for measurement of each tomographic plane was approximately 6 hours. The hydraulic conditions of the experiment were adjusted to assure that there was no significant tracer movement during the 6 hour measurement period as this would have resulted in "blurred" tomographic images (Olsson et al., 1991).

Figure 6 shows when and where tracer transported within fracture zone H arrived to the three tomographic planes at their lines of intersection with the fracture plane. Naturally, the largest and most rapid increase in attenuation were observed close to the injection point (borehole C2) along the W1-C5 and C1-C5 lines. After approximately 330 hours, significant amounts of tracer were observed essentially all along the W1-C5 line. The observation of tracer in the radar tomograms close to borehole W1 was verified by observed tracer arrival in borehole T1 after 180 hours and in borehole T2 after 400 hours. The radar difference tomograms also indicated transport of tracer outside of zone H. The radar results provided valuable data on tracer transport velocity and distribution in rock between the tracer sampling points (i.e. the boreholes).

Another important aspect of radar difference tomography is that it provides an independent technique for estimating flow porosity. Conventionally, flow porosity is estimated from mean travel times observed in tracer experiments. Such estimates are uncertain as the volume in which transport takes place is

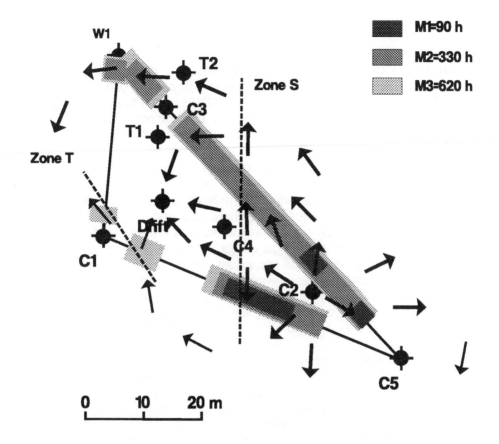

Figure 6   Conceptual model of saline tracer transport within zone H
            during the second Radar/Saline Tracer Experiment. The grey
            shades indicate when and where tracer was first observed in
            significant amounts.

normally not known. Radar attenuation is a function of pore volume and the
conductivity of the pore fluid and the increase in attenuation can be used to
estimate flow porosity. If we assume that the pore fluid is replaced by the more
conductive saline tracer only in the pores accessed by the moving water the
flow porosity (Φ_k) can be estimated from the following equation

$$\Phi_k = \frac{\Delta\alpha}{\alpha} \frac{\sigma_w \Phi}{\sigma_t - \sigma_w} \qquad (1)$$

where $\Delta\alpha/\alpha$ is the relative increase in attenuation, σ_w conductivity of groundwater, σ_t conductivity of tracer, and Φ total porosity (Olsson *et al.*, 1991). Based on data from this experiment at the SCV site, the flow porosity in fracture zone H was estimated to 0.7×10^{-4}. This compares favorably with the flow porosity estimate based on the mean travel time which was 1.1×10^{-4}.

Conclusions

The directional borehole radar system developed within the Stripa Project has truly given new capabilities in site characterization. The main contribution is that it provides the capability to determine uniquely the orientation of fracture zones from measurements in single boreholes. This significantly improves reliability in defining the location and extent of fracture zones at a site. The current radar system operates at wavelengths of a few meters. This gives the capability to identify small fracture zones with thicknesses larger than 10 cm. The system could be modified to operate at smaller wavelengths in order to detect smaller features, but this would lead to a reduction in range.

Borehole radar has been found to work extremely well in the Stripa granite. The high resistivity of the rock results in long probing ranges ($\approx$60 m in single hole reflection mode) which implies that large volumes of rock can be investigated from a few boreholes. In addition, the structures found at Stripa are fairly simple, in that we have a few essentially planar features with a contrast in electric properties relative to the background. If the host rock is more conductive, we would expect shorter ranges, and in highly conductive rocks ranges will be too short to be of practical interest. When probing ranges decrease, smaller volumes of rock will be characterized from each borehole. However, as the directional radar still yields the orientations of the fracture zones it is possible to extrapolate data and confirm occurrence and orientation in adjacent boreholes.

The borehole radar can also be used in cross-hole mode to produce tomographic images of the rock. These can display porosity variations in planar sections of the rock mass. Furthermore, if tomographic imaging is repeated, it is also possible to map the transport of saline tracer in the rock and fissures between the boreholes.

Acknowledgement

The development of the borehole radar system was performed by a research group originally at the Swedish Geological Company and later at ABEM AB. The creativity and dedication of this group was a prerequisite for the success of the radar development work. In addition to the author the group consisted of Lars Falk, Olof Forslund, Bo Hesselström, Lars Lundmark, and Eric Sandberg.

The basic research work was funded by the International Stripa Project. Additional refinements of the radar system have been funded by the Swedish Nuclear Fuel and Waste Management Co. (SKB).

References

Falk, L., 1992. Directional borehole antenna - theory. Stripa Project TR 92-16, SKB, Stockholm, Sweden.

Ivansson, S., 1986. Seismic borehole tomography - theory and computational methods. Proceedings of the IEEE 74, 328-338.

Olsson, O., Falk, L., Forslund, O., Lundmark, L., Sandberg, E., 1987a. Crosshole investigations - Results from borehole radar investigations. Stripa Project TR 87-11, SKB, Stockholm, Sweden.

Olsson, O., Black, J. H., Cosma, C., Pihl, J., 1987b. Crosshole investigations - Final report. Stripa Project TR 87-16, SKB, Stockholm, Sweden.

Olsson, O., Andersson, P., Gustafsson, E., 1991b. Site characterization and validation - Monitoring of saline tracer transport by borehole radar measurements, final report. Stripa Project TR 91-18, SKB, Stockholm, Sweden.

Olsson, O., Falk, L., Forslund, O., Lundmark, L., Sandberg, E., 1992. Borehole radar applied to the characterization of hydraulically conductive fracture zones in crystalline rock. Geophysical Prospecting, 40, 109-142.

Sandberg, E. V., Olsson, O. L., Falk, L. R., 1991. Combined interpretation of fracture zones in crystalline rock using single-hole, cross-hole tomography and directional borehole-radar data. The Log Analyst, 32(2), 108-119.

Innovations in the Characterisation of Fractured Rocks Developed within the Stripa Project

John Black and **Mark Brightman**,
Golder Associates, Nottingham

David Holmes
British Geological Survey,Nottingham
(United Kingdom)

Abstract

The hydraulic characterisation of repository host rocks has progressed during the 15 years of the Stripa Project. This has included developments in philosophy, organisation, equipment and interpretation.

The philosophy of hydrogeological characterisation has been focussed on a fractured rock approach. This has been consistently refined, though some porous medium concepts have been introduced and rejected. The philosophy of the SCV project was to conceive the system in two parts: major fracture zones and averagely fractured rock. The hydrogeological testing was organised to reflect this approach. Simple borehole testing concentrated on measuring the properties of the average fractured rock whilst geophysically-identified major fracture zones were targetted by specific crosshole tests.

The equipment developments concerned cost and data reliability. A computer-controlled set of test equipment was developed which reflected overall test strategy . Known as "focussed packer testing" the approach involves testing at an initial coarse level and only going to more detail [by inflating intermediate packers] if pre-ordained threshold values of transmissivity were exceeded. This necessitated close control over in-situ pressures and on-site test analysis. Other developments included a testing "manifold" to allow packer string movement without the usual severe loss of borehole pressure.

Interpretation of results included developments in the application of the "partial dimension" concept to both single and cross-borehole tests. Additionally some of the problems of multiple cross-borehole results in a highly heterogeneous fractured rock system were addressed. Validation of the conceptual model was broadly achieved.

Overall the work highlighted the impact of time and cost on testing a complex fractured rock system. Applying relevant concepts consistently through the various levels of design and interpretation proved difficult. A cost effective approach evolved.

Introduction

Hydrogeological testing has been a major component of work at Stripa since it became an underground research laboratory 15 years ago, in 1977. The testing has been organised and influenced by various groups with differing aims and constraints. This paper aims to answer two questions:
- has 15 years of work formed a coherently developed programme?
- are we now more capable of characterising the rocks at Stripa (and presumably elsewhere) than we were in 1977?

Naturally, in reviewing this work, there is a tendency to describe recent work in most detail since it has not yet been seen in the larger context. Hydrogeological characterisation is only a means to an end and it is easy to envisage that safety case modellers (the end-users) will have altered their requirements during 15 years. They will have influenced the direction of development. Similarly, technology has made some things possible.

In order to see the work in a larger context, we adopt the following format:

philosophy - was there an underlying philosophy?

testing organisation - did the philosophy affect the design of programmes of tests (mainly concerning Phase 3)?

testing development - did the programmes produce progress in relation to equipment? (mainly Phase 3)

interpretation - did interpretation evolve and was it consistent with the needs of end-users?

The answers to these questions should determine whether hydrogeological testing at Stripa has been a worthwhile enterprise.

Phases and philosophies

Phases. Fifteen years of activity has contained four phases of work programmes. The main hydrogeological projects (excluding hydrochemistry) are outlined below:

Swedish-American Cooperation (1977-80, 3 years)
Ventilation experiment, single borehole tests for boundary conditions coupled thermal-mechanical tests, scale effects

OECD/NEA Stripa Project Phase 1 (1980-82, 3 years)
Major fracture zone pump testing, small-scale test development [parallel work on single fracture tracer test].

OECD/NEA Stripa Project Phase 2 (1983-86, 3 years)
Crosshole Project, linked hydraulic and geophysical measurements [parallel work on 3D Migration test].

OECD/NEA Stripa Project Phase 3 (1986-91, 5 years)
Site Characterisation and Validation (SCV) Project, linked geophysical, hydraulic and tracer measurements, phased work with predictions, single borehole and cross borehole, large and small scale, Simulated Drift Experiment for excavation related effects.

Philosophies. The development of philosophy has accompanied the phases of work. For the SAC period the approach was clearly stated by Gale and Witherspoon, 1978 and involved quantifying the factors that control permeability in three dimensions - orientation, spacing, aperture and continuity - and from these derive the permeability tensor. The results were then to be checked against the results of larger scale tests to see if the approach "scales up" and produced sensible predictions. This approach also required knowledge of the hydraulic parameters describing individual fractures. The approach clearly focuses on conceiving the hydrogeological system as a homogeneous network of fractures. It was halted before completion for political reasons.

Stripa Phase 1 had a much less clear philosophy. It took the view that major fracture zones were the most significant features at Stripa and that to understand them was the solution to predicting nuclide transport. This used a test interpretation approach, derived from the oil industry, which included explicit geometry and porous medium ideas (Carlsson and Olsson, 1985 (a) and (b)). The approach was not linked to geophysics or tracer testing so that results were sparse and the derived values not really put to any test.

Some of the lessons of Phase 1 were included in the Phase 2 philosophy. In the Crosshole Project, hydraulic characterisation was intimately linked to much improved, geophysical remote sensing. Hence crosshole hydraulic tests were well targetted with respect to major fracture zones. A new test technique involving sinusoidal variation of pressure was introduced with an interpretation approach strongly founded in anisotropic porous media. As results became available the difficulties of interpreting a sparse fracture network came to the fore and the "fractional dimension" approach was conceived (Barker, 1988).

Stripa Phase 3 had a clear philosophy from the outset, linked to probabilistic fracture network models. However, it was effectively a development of the approach begun in Phase 2 as it also used geophysics to target fracture zones. Hence the underlying philosophy of the SCV Project was to conceive the system in two parts: major fracture zones and averagely fractured rock. The major fracture zones were to be identified geophysically and characterised deterministically. The averagely fractured rock was to be treated probabilistically. Applying the fractional dimension approach to the results completed the fracture based approach to characterisation.

Over the 15 years of activity, hydrogeological characterisation can be seen as having started off fairly well focussed on a fracture-based philosophy. This has been refined consistently, though some elements of porous media continua have been introduced and rejected.

The last five years (Phase 3) has seen a more mature philosophy evolve which explicitly identifies the major hydrogeological components of the Stripa site and how to treat them. The basics of the Phase 3 approach and how it became a practical testing scheme are described below.

Organisation of hydraulic testing

Hydrogeological characterisation of the SCV site was based on the assumption that a bi-modal distribution of hydraulic properties represented a reasonable model of the rock mass. This model implies that there are major fracture zones with good hydraulic connections over a scale of hundreds of metres separated by intervening blocks of averagely fractured rock. The hydraulic characterisation programme of the SCV experiment was therefore organised to measure the properties of such a rock mass. Three types of hydraulic tests were performed:
-   single borehole testing
-   small scale crosshole testing
-   large scale crosshole testing

The single borehole testing measured the hydraulic properties of the rock close to the borehole. It aimed to measure the transmissivity of a significant number of single fractures by using short straddle packers. The testing was biased towards the measurement of high transmissivity features which could be assigned to fracture distributions to yield a description of the "average rock" fracture network.

A key element of the hydraulic characterisation strategy was to maintain a close link with the geophysical characterisation of the SCV site. The small scale and large scale crosshole tests were designed to investigate the structural features identified by the preceding round of remote sensing. The small scale crosshole testing measured the variability of the hydraulic properties of major fracture zones between boreholes less than 10 metres apart. The large scale crosshole testing measured the hydraulic properties of the major fracture zones between boreholes located across the SCV block, thus yielding a description of the large scale fracture network.

Development of hydraulic testing methods and equipment

The hydraulic characterisation of the SCV site resulted in significant development of hydraulic test methods and equipment for single borehole testing. Ideally discrete fracture network models require that the hydraulic properties of many single fractures are measured in order to characterise the distribution of properties throughout the fracture network. This can only be achieved using straddle packers to isolate very small sections of borehole, containing single fractures. However testing several entire boreholes with short, straddle packer equipment is extremely time consuming and consequently very expensive. In order to collect a data set suitable for defining fracture network properties in a reasonable time period an approach known as "focussed packer testing" was developed.

The "focussed packer" testing concept involves making an initial measurement of transmissivity at a coarse scale with packer straddle lengths of several metres. If the transmissivity exceeds a pre-set threshold then the straddle is shortened and further measurements made. The concept can include a number of stages with progressively shorter straddles. The technique thus focuses on measuring the properties of short sections with higher transmissivities because these fractures are the most significant in determining the behaviour of the fracture network.

A profile of hydraulic conductivities similar to those previously found at Stripa was used to assess the potential impact of the technique. It was estimated that using a fixed 1 m straddle it would take three to four months to test a 200 m borehole. Using two packer spacings, an initial coarse 10 m fixed straddle followed by a detailed testing of appropriate intervals with a fixed 1 m straddle reduced the testing period to two months. This could be reduced further to one month by the use of multipacker string consisting of six packers. In this scenario intervening packers were inflated to reduce the straddle length and thus only one "pass" was required to test the entire borehole. Figure1 shows the application of this technique for a five packer string.

The favoured equipment design was obviously the multipacker approach but this raised a number of significant design problems:

- The test section would contain uninflated packers which are a potential source of substantial equipment compressibility thus limiting the resolution of the system.

- Inflating packers in low transmissivity sections which were already isolated could cause massive pressure increases in the section under test.

- A complicated series of downhole shut-in valves would be needed to allow the completion of pulse tests in multiple sections.

- The test operator would need to analyse the data in real time in order to evaluate whether the transmissivity exceeded the pre-set threshold.

The key to the success of the equipment which was built (Holmes and Sehlstedt, 1991) was that it was fully computer controlled and able to regulate water pressures quickly, reliably and accurately. For example, the problem of inflating packers in test sections was overcome by inflating the packers under computer control. If the observed pressure in the test section increased by a given threshold then inflation was suspended until the pressure had decayed. The other design challenges outlined above were overcome in the following ways:

i) The introduction of additional equipment compressibility was minimised by constructing a packer which consisted of a simple rubber sealing element bonded to a steel inner tube. Thus when deflated the thin rubber seal collapses against the steel minimising compressibility.

ii) The downhole valve and transducer assembly was difficult to design to meet the narrow (76 mm) borehole diameter. Ultimately this reduced the packer string to five packers compared to the initial design target of six.

iii) Real time test analysis was achieved by linking a second analysis computer to the test control computer. When the control computer was not busy it communicated with the analysis computer and downloaded the latest test data. Analysis software on the second computer allowed the test operator to perform analysis of slug and pulse tests within minutes of receiving additional data. This not only allowed the operator to take decisions regarding reducing straddle lengths but ensured that only good quality, analysable data was collected.

The computer controlled pressure regulation system allowed the equipment to easily perform slug, pulse, constant head injection or abstraction, constant rate injection or abstraction and sinusoidal tests. Tests could be programmed to take place at any time under the sole control of the computer, thus something approaching 24 hour testing was possible with a single operator.

During focussed packer testing the borehole was sealed by a steel manifold fitted with packer sealing elements through which the rods and tubing passed in order to ensure the borehole attained something near equilibrium pressure prior to testing. This had to be removed each time the packer string was moved which necessited a period of recovery prior to each test. This was developed further for the SDE inflow measurements to produce an adjustable seal which allowed the equipment to be moved in and out of the borehole without venting the borehole pressure. The test commenced immediately without waiting for re-equilibration.

The equipment used for the small scale crosshole and large scale testing used the flexible testing capabilities of the single borehole equipment to generate the hydraulic signals. Other packers and pressure monitoring systems were used to record the responses.

The single borehole testing equipment successfully performed focussed packer testing in eight boreholes during the SCV experiment. It proved to be a highly cost effective method of collecting good quality hydraulic data for the characterisation of fracture networks.

Interpretation

The Site Characterisation and Validation Project was a staged project which included three stages of characterisation and two of prediction (and comparison with measurement). The conceptual model of the site developed during stages 1 (see Olsson and others, 1988) and 3 (Black and others, 1989) saw the site comprising large scale major fracture zones with intervening averagely fractured rock. The aim of the hydrogeological testing was to characterise these two components.

The small scale fracture network

The aim in characterising the "averagely fractured rock" was to describe the rock in terms of the occurrence of hydraulically "active" fractures. This involved identifying all active fractures and assigning them to sets in terms of probability distributions of orientation and frequency. This was achieved by core logging (Gale and others, 1990). The extensiveness of fractures was derived from fracture mapping. Single borehole straddle packer testing was designed to yield the probability distribution of transmissivity. This meant that for each tested interval it was necessary to derive a transmissivity and then ascribe it to the intersecting active fractures. Active fractures were identified in the core as "coated" fractures. The intention of the focussed packer testing system was to test the most transmissive fractures on the shortest possible straddle interval. Hopefully a significant proportion of the highly transmissive test intervals would contain only one active fracture. This proved a vain hope since the average frequency of active fractures turned out to be about 3 per metre compared to the minimum 1 metre straddle interval. Only a few tests tested identified single fractures.

This left the problem of assigning transmissivity to the multiple active fractures identified in each test zone. Different groups took different approaches. The approach reported in Holmes, 1989 was equal assignment whilst that used by Herbert and Lanyon, 1992 was to sample from a log normal distribution. Dershowitz and others, 1991 adopted an approach based on assuming that both the occurrence and extensiveness of actively flowing fractures was likely to be less than "as mapped". They then forward modelled to match the distribution of measured transmissivities.

131

It should be borne in mind that all the single borehole tests in question were based on the slug or pulse technique. As such they are virtually unable to distinguish between different flow geometries. The transmissivities were derived assuming cylindrical flow and a porous medium. Some more detailed constant head tests were conducted to examine the possible influence of boundaries. These tended to show a well connected fracture network (Doe and Geier, 1991) with evidence of flow geometries more linear than cylindrical.

Major fracture zones and their flow dimensions

Qualitative assessment. The major fracture zones were identified largely on the basis of crosshole radar and seismic tomography. They were then hydraulically characterised by a series of crosshole tests involving setting flow sources in boreholes where the major zones intersected and where they were found to be transmissive. The results were assessed qualitatively in the first instance. In other words, did crosshole responses occur largely where expected (see Figure 2). On the whole, responses were more widespread than anticipated and fracture zones were seen to interconnect (Ball and others, 1991). The zones A and B behaved almost "as one". The quantitative assessment was based on a development of the "fractional dimension" idea of Barker, 1988.

Flow dimension. Analysing pumping tests in aquifers and reservoirs is a well developed skill based on assuming that the geometry of the flow system is well defined. For example, in the classic Theis example, it is assumed that flow converges evenly from all around the pumped well. The confining beds ensure that there are no flows except those converging on the pumping well. This specific flow configuration has a flow dimension of two. All places within the region under consideration have a well defined role. For instance, the bounding layers do not take part in flow and all places within the aquifer behave equally and contribute to the radially convergent two dimensional flow.

In fracture networks, it should not be expected to be as easy to assign a flow configuration as in the aquifer case. For instance, a three dimensional network of fractures in a crystalline rock cannot be expected to behave with two axes of significant flow and one of zero (ie. equivalent to the Theis "aquifer".) Also one can expect there to be large regions of virtually no flow.

The idea of flow dimension put forward by Barker, 1988 is that flow configurations can be generalised. Hence, flow dimension is the change in flow conduit area with distance from a source point (Figure 3). In one dimensional flow (linear flow) the area of the conduit is proportional to r^0. The area does not change with distance. For cylindrical and spherical flow geometries, the areas are proportional to the r^1 and r^2 powers of distance respectively. By extension of this logic, a conduit of "fractional dimension" is simply a conduit whose area is proportional to a non-integer power of distance from the source.

A second aspect of the flow dimension approach becomes obvious: space filling. How does one ensure that a fractional dimension flow system fills 3D space? The answer is one doesn't. In fact, it is logical to expect that flow in a fracture network will not affect large portions of the available rock. This is what is seen in many tracer tests; it should be expected (see Doe, 1991). Furthermore, is it logical to expect a fracture network to exhibit a single value of flow dimension? The Stripa Phase 3 results started to answer some of these questions.

SCV Crosshole results: interpretations of flow dimension. The interpretation of pumping tests, where flow dimension is unknown, is different from conventional analysis. In conventional analysis, the flow configuration is assumed known and the hydraulic diffusivity derived from the test is separated into hydraulic conductivity and specific storage. This is based on knowing how much flow occurs everywhere within the system. Hence in the flow dimension approach the first step is to derive flow dimension.

Flow dimension was derived in two ways. Firstly, the distance drawdown approach was adopted. For a given source point all drawdowns at a given time after start of pumping are plotted in terms of their normalised distance from the source (Figure 4). A separate plot for each value of flow dimension is required. As can be seen in Figure 4, a complete range of responses is to be expected. The flow dimension of the system (in the case of Figure 4, the system is major Fracture Zone H) is the type curve which best matches most of the responding intervals. Since the flow system being tested is Fracture Zone H, intervals identified as being in fracture Zone H carry most weight in the subjective interpretation. Other intervals will be classified as "over-connected" or "under-connected" depending on their relationship to the major trend of responses. Once flow dimension is determined then values of hydraulic conductivity should be determined by assuming that specific storage is known.

The second method of deriving flow dimension is to use the pressure time type curves to find a best fit to individual responses. In the case of the crosshole tests at Stripa a distribution of responses was derived (Figure 5 from Ball and others, 1991). Interestingly they have two peaks, one at about 1.7 the other closer to 2.3.

The distance drawdown plots indicate that the major fracture zones at Stripa have a flow dimension less than two. This means that, if they are assumed to behave as slabs of porous medium (in a porous medium model), the model will underestimate the speed of travel of a given solute. However, it is felt that this represents common experience in fractured rock systems.

Conclusions

Hydrogeological work at Stripa has shown consistent development. The need has been to find a concept which both adequately describes the fracture network - groundwater flow system and allows practical testing and interpretation. The elements of the hydrogeological approach to fractured crystalline rocks developed at Stripa are:

1. The hydrogeological system consists of "major fracture zones" and "averagely fractured rock".

2. The "major fracture zones" are treated deterministically. They are identified by geology and geophysics and confirmed qualitatively.

3. The "major fracture zones" are investigated by targetted crosshole tests which are analysed by the flow dimension approach.

4. The "averagely fractured rock" is treated probabilistically. Distributions of key parameters for network models are derived from single borehole tests.

This approach is consistent in general with the overall philosophy.

In addition to the development of the overall approach, there have been a number of developments in terms of testing approach and hardware design. These have concentrated on the use of computer control to improve test result reliability and reduce testing time.

In summary, hydrogeological work at Stripa mine has caused examination of the ideas concerning flow in networks of fractures. It has shown the progress possible by properly targetting hydrogeological testing through superb geophysical remote sensing. The impact of thorough integration of remote sensing, hydrogeology and computer code development should not be underestimated.

Acknowledgements

The authors would like to thank the Stripa Project (in particular Bengt Stillborg, Hans Carlsson and Per Eric Ahlström) for the privilege of having worked on the Project. We rubbed shoulders with really competent people at the mine like Gunnar Ramqvist and found some friends.

References

Ball, J.K., Black, J.H., Brightman, M.A. and Doe, T.W. 1991. Large scale crosshole testing. Stripa Project TR 91-17, SKB, Stockholm, Sweden.

Barker, J.A. 1988. A generalised radial flow model for hydraulic tests in fractured rock. Water Resources Research, Vol. 24:10 pp 1796-1804.

Black, J., Olsson, O., Gale, J., Homes, D., 1991. Site characterization and validation, stage IV - Preliminary assessment and detail predictions. Stripa Project TR 91-08, SKB, Stockholm, Sweden.

Carlsson, L and Olsson, T., 1985. Hydrogeological and hydrochemical investigations in boreholes - Final Report. Technical Report of the OECD/NEA Stripa Project No. TR 85-100.

Dershowitz, W., Wallman, P., Kindred, S., 1991b. Discrete fracture modelling for the Stripa site characterization and validation drift inflow predictions. Stripa Project TR 91-16, SKB, Stockholm, Sweden.

Doe, T.W., 1991. Fractional dimension analysis of constant pressure well tests. SPE Paper 22702 presented at 66th Conf. at Dallas, Tx, October 6-9, 1991.

Doe, T.W., Geier, J.E., 1991. Interpreting fracture system geometry using well test data. Stripa Project TR 91-03, SKB, Stockholm, Sweden.

Gale, J.E., Macleod, R., Stråhle, A. and Carlsten, S., 1990. Site characterization and validation - Drift and borehole fracture data, stage 3. Stripa Project TR 90-02, SKB, Stockholm, Sweden.

Gale, J.E. and P.A. Witherspoon, 1978. An approach to the fracture hydrogeology at Stripa - preliminary results. Sem. on In-situ heating experiments in Geological Formations. OECD/NEA, Ludvika, Sweden. Also Lawrence Berkeley Laboratory report LBL-7079, SAC-15. Berkeley, California.

Herbert, A.W., Lanyon, G.W., 1992. Modelling tracer transport in fractured rock at Stripa. Stripa Project TR 92-01, SKB, Stockholm, Sweden.

Holmes D. 1989 Site characterization and validation - single borehole hydraulic testing. Stripa Project TR 89-04, SKB, Stockholm, Sweden.

Holmes D., and Sehlstedt, M., 1991 Site characterization and validation - Equipment design and techniques used in single borehole hydraulic testing, simulated drift experiment and crosshole testing. Stripa Project TR 91-25, SKB, Stockholm, Sweden.

Olsson, O., Black, J., Gale, J., Holmes, D., 1989. Site characterization and validation stage II - Preliminary predictions. Stripa Project TR 89-03, SKB, Stockholm, Sweden.

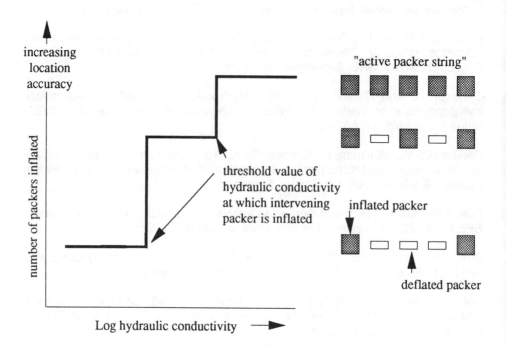

Figure 1: **The way in which an active packer string allows location accuracy to be dependent on the value of hydraulic conductivity**

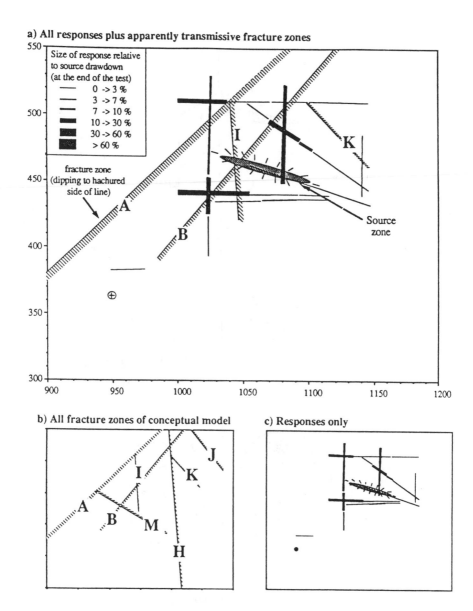

a) All responses plus apparently transmissive fracture zones

Size of response relative to source drawdown (at the end of the test)

- 0 -> 3 %
- 3 -> 7 %
- 7 -> 10 %
- 10 -> 30 %
- 30 -> 60 %
- \> 60 %

fracture zone (dipping to hachured side of line)

Source zone

b) All fracture zones of conceptual model

c) Responses only

Figure 2: **Crosshole testing (based on a 24 hour period sinusoidal test) using the borehole D6 (28-100) as a source zone - essentially a test of zone B**

137

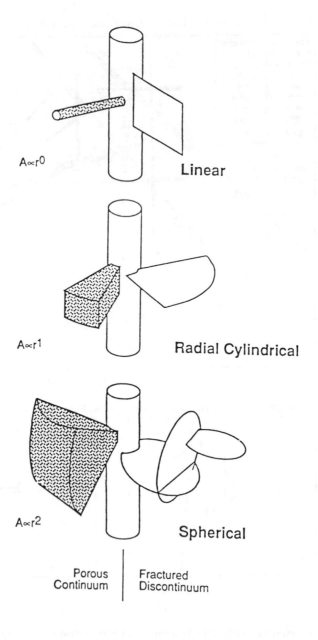

$A \propto r^0$ Linear

$A \propto r^1$ Radial Cylindrical

$A \propto r^2$ Spherical

Porous
Continuum

Fractured
Discontinuum

Figure 3: **The integer cases of flow dimension**

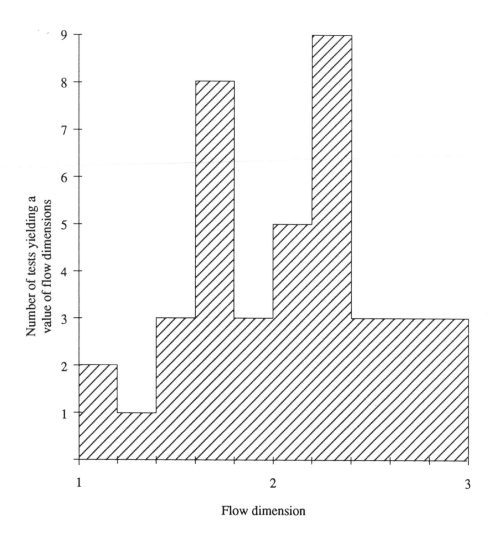

Figure 4:   **Distribution of flow dimension derived from each responding interval during crosshole tests at Stripa**

Responding intervals marked in terms of the fracture zone with which they are associated (* = type curve match point, O = no associated fracture zone)

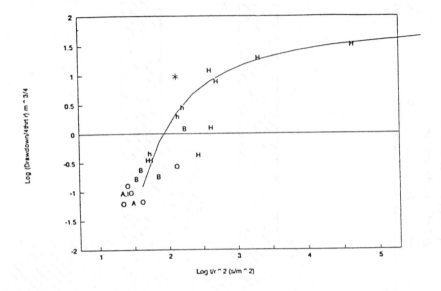

Figure 5:   **Distance-drawdown plot of the test D3-H, major fracture zone H, dimension = 1.75 = 3 days**

Studies of Groundwater Chemistry at Stripa: an Evaluation

Stanley N. Davis
University of Arizona

Abstract

Concentrations and distributions of 24 elements together with 8 stable nuclides, 15 radionuclides, and 12 organic compounds were studied in samples from the Stripa mine. Most stamples were of groundwater, but careful attention was also paid to the minerals which may have been in contact with both modern and ancient groundwater. Primary objectives of the work were, first, to develop sampling and analytical techniques which might be useful in evaluating repositories for radioactive waste and, second, to try to reconstruct the history of the groundwater using chemical and isotopic analyses. The development of sampling and analytical techniques was an outstanding success. The reconstruction of groundwater history was generally successful but hampered by uncertainties about the effects of antecedent mining. It is now clear that geochemical sampling must be the first activity scheduled in the exploration process.

Introduction

Chemical studies of water and rocks from the Stripa mine in southcentral Sweden were completed by a large number of research teams during a ten-year period from 1978 through 1988. Detailed results of these studies are discussed in numerous publications including those of Nordstrom et al. (1985), Davis and Nordstrom (1992), and a series of papers in Volume 53, Number 8 of Geochimica et Cosmochimica Acta published in 1989. A short summary of the last phase of the chemical studies was presented by Davis and Andrews (1990). The present paper will touch on a number of highlights from these published reports. The overriding emphasis, however, will be to evaluate the utility of the chemical work in assessing sites which are considered for repositories for radioactive waste.

Although most researchers involved with the Stripa project will undoubtedly agree with the general geochemical interpretations which will be presented, the relative merits of various analyses are open to divergent judgments. In this regard, then, it must be emphasized that the present paper is not a statement of a consensus of the various researchers. The present author takes full responsibility for all statements made, particularly those in the sections dealing with the evaluation of methods.

Objectives of chemical studies

The primary objectives of the chemical studies reported in this paper were, first, to develop sampling and analytical techniques which could be useful in evaluating repositories for radioactive waste and, second, to demonstrate how to reconstruct the history of groundwater using chemical and isotopic analyses. In terms of predicting the future movement of radioactivity once it has escaped the local confinement of a repository, a thorough understanding of the subsurface hydrology of a region is absolutely essential. Outside of human intrusion, groundwater transport is the most likely way by which radionuclides from a repository can move to the land surface in significant amounts. If the time span of interest is less than about 200 years, then transport modeling based on historical measurements of hydrogeologic variables may be the most reliable basis for necessary predictions. However, slow changes in topography, vegetation cover, and climate will introduce important changes in the hydrogeologic system which will be difficult to model. After perhaps as short a time as 500 years, water history as deduced from chemical and isotopic measurements becomes more reliable than modeling as a method of predicting the possible future transport of radionuclides. In short, the past is the ony reliable key to the future, and the reliability is a direct function of the time span considered. Data of hydrodynamic variables gathered over a few decades cannot be used alone to predict long-term transport of potential contaminants. Besides the transient nature of hydraulic heads which are affected by changing surface topography and water quantities available for recharge, changes in evaporation and surface soil conditions will drastically alter chemical factors which control the transport of radionuclides. Most important, changes in the redox potential can alter the transport velocity of some transuranics by several orders of magnitude.

One of the most important contributions of geochemistry to the evaluation of repositories is to answer the question of possible long-term migration of radionuclides from the repository under natural conditions. If groundwater has been essentially static in the host rock of the repository for thousands of years under fluctuating climatic conditions, then the danger of offsite migration of radionuclides is probably minimal. It is my opinion that geochemical data are far more reliable than any other types of information which can be used for predictions of natural groundwater migration over periods of hundreds as well as thousands of years.

Besides the desire to reconstruct water history, numerous reasons exist for making careful hydrochemical studies during the exploration phase of repository evaluation. These reasons will not be discussed at length but should be noted carefully, because the interpretation of groundwater history is not entirely a separate add-on cost. Chemical studies are required even if this critical use is somehow bypassed. First, in many regions the water collected in deep subsurface excavations is brackish or saline and cannot be discharged into fresh water at the surface. The chemical nature of this deep waste water must be determined at the earliest possible stage in exploration. Second, by contrast in some arid regions, water recovered from repository drainage may actually be essential for construction or for sanitary purposes. Chemical analyses of major and certain trace constituents would be needed in order to certify these uses. Third, chemical analyses must be available to predict the chemical stability of all construction and confinement materials in contact with subsurface water. Fourth, if containers of waste are breached, initial mobility and excursion of radionuclides will be in large part a function of water chemistry. The hydrochemistry of the native water must be known in order to design chemical buffering systems to reduce radionuclide transport. Fifth, heating with subsequent cooling over hundreds of years will shift the mineral-water solubilities which can either increase or decrease the preexisting permeability of the rock through dissolution or precipitation of minerals in pores and along fractures. This effect relates particularly to deposits of calcite and opaline silica and to the dissolution of all carbonate and silicate minerals. Sixth, careful monitoring is needed for legal purposes. Although a well-designed repository will not cause regional changes in the quality of groundwater, careful sampling and accurate analyses of a host of constituents are needed to answer questions of regulating agencies and to guard against misdirected legal actions. Seventh, the overall chemical quality of pore water must be known for the proper interpretation of some types of geophysical measurements. Eigth and last, the concentrations of a number of gases must be known in order to design a proper ventilation system if full-scale excavation is undertaken. Natural gases which commonly are of concern are H_2S, CO_2, CH_4, and Rn.

Some of the salient results of the study

The development of field sampling and laboratory analyses was very successful. A number of researchers in various countries undertook difficult and often unique measurements of various nuclides and compounds. A large number of measurements of a geochemical nature taxed existing methods because of the low concentrations encountered, because of only slight differences among samples from various locations, or because the constituent had never, or rarely, been measured before. In addition, special field sampling methods had to be devised, particularly for the separation of dissolved gases and for the capture of trace organic fractions.

Gases in groundwater are rarely studies in detail. The work at Stripa has been an outstanding exception. Isotopes of dissolved helium, argon, krypton, and radon were given extra attention. Natural levels of Ar-37 in deep groundwater were measured for the first time. Furthermore, dissolved constituents which are sometimes in equilibrium with a gas phase, such as H_2S and CO_2, were measured.

One of the earliest scientific questions associated with the chemical studies at Stripa was the origin of the natural salinity in the deep groundwater. All research groups concluded that the salinity did not originate from geologically recent seawater or even Pleistocene seawater. Researchers, however, did not reach a clear consensus on the actual origin of the salinity. Three possiblities considered seriously were (1) slow diffusion of ions from fluid inclusions and other micropores, (2) dissolution of minerals, and (3) slow migration, either recent or ancient, of waters from evaporite beds which are no longer present in the region. Some combinations of these sources, of course, could be compatible with most of the geochemical data. Most researchers postulated that some, or maybe most, of the salinity originated by one or more of the three mechanisms, but the leptite surrounding the Stripa granite was the host rock for the salinity and that the movement into the granite was geologically recent, perhaps even induced by mining operations.

Another point of consensus is the fact that most (by volume) of the deep groundwater had a surface origin during a time when local climatic conditions were different. This is an observation quite distinct from the question of the origin of the bulk of the dissolved solids in the groundwater. The water originated as rainfall or snow melt and had very low total dissolved solids at the surface. The bulk of the dissolved solids, in constrast, came from subsurface sources including the soil horizon as well as small amounts of more saline water mentioned already which mixed with the water of surface origin. Because the δ0-18 is about 3 o/oo lighter and the δH-2 is about 15 o/oo lighter in the deep groundwater than in the modern surface water, most researchers have concluded that the deep groundwater originated from surface recharge during one of the colder periods of the Pleistocene.

One of the most spectacular scientific results of the work at Stripa came from the clear demonstration for the first time that a number of so-called "cosmogenic radionuclides" are actually generated in the subsurfarce in quantities which overhelm the real cosmogenic components. Most of the production comes from thermal neutron activation with some coming directly from the natural fission of U-238. The neutrons in turn are generated by natural (α, n) reactions on light elements with a minor component of neutrons from spontaneous fission of U-238. Subsurface production is a very important source of Cl-36, Ar-37, Ar-39, Kr-85, and I-129. Additionally, owing to the high concentration of uranium and thorium in the Stripa granite, H-3 and C-14 could be produced in measurable amounts. Proof of a subsurface source of H-3 and C-14, however, is as yet lacking.

Distinct differences exist in the general water chemistry as well as in the concentration of isotopes of various elements in samples from distinct horizons from different holes. This was generally recognized by all researchers and is a condition which is expected in groundwater flowing through fractured rocks. Despite these local differences, a general increase with depth of total dissolved solids and chloride exists within the Stripa granite. This is interpreted to mean that circulation of groundwater generally decreases with depth because if vigorous circulation were to have existed, conservative ions would have been flushed out of the system within a relatively short time.

A major objective of the geochemical work at Stripa was to date the groundwater. In the context of hydrogeology, dating of water is much different from normal geochemical dating of solid matter. Dating the water is an attempt to determine how long the water has been isolated from the surface water and atmospheric portions of the hydrologic cycle. Actually, all dating, except for H-3 dating, determines the average age of the dissolved matter in the water. The age of the water itself is only determined by trying to relate the history of the dissolved matter to the complex history of the water. For the first time, a very large number of techniques were used in a single study. These techniques involved isotopic disequilibrium, buildup of radiogenic gases, decay of cosmogenic radionuclides, buildup of in situ-produced radionuclides and matching changes in concentrations of stable nuclides with known climatic fluctuations.

Results were judged only partially successful, owing to the natural complexity of the fracture-flow system and to the disruption of the system by past mining operations. Nevertheless, the conclusion of most of the researchers who addressed this problem was that the bulk of the deepest groundwater had been isolated from the surface for thousands of years.

More than 50 mg/L of dissolved organic matter had been measured previously in one of the groundwater samples from Stripa. Because water in granite was expected to have less than 2 mg/L of dissolved organic matter, the large concentration gave rise to considerable speculation. Phase II work, however, showed that the large concentrations were from human sources and that only less than 1 or 2 mg/L of natural organic matter was dissolved in the water. This reconfirmed previously held ideas concerning expected levels of dissolved organic matter in water deep within plutonic rocks.

A critical question not answered by the research at Stripa was that of the significance of trace amounts of tritium from deep within the Stripa granite. Concentrations in these deep waters at various levels in the V2 borehole average from about 0.3 to 0.7 tritium units. Four explanations appear to be almost equally reasonable. First, the traces of tritium may come from subsurface production through natural neutron activation of Li-6. Second, systematic errors may be present in laboratory work even though similar results are obtained by more than one laboratory. Third, the tritium may come from a slight contamination within the mine during its operation. Fourth, groundwater velocities may have been greatly increased due to sampling at depth. The small amount of tritium may be from the early arrival of filaments of high-tritium water advancing rapidly along selected fractures and then mixing with the preexisting groundwater. The importance of a correct interpretation of tritium data is essential. If similar levels of tritium were found at an actual potential repository site and if the first theory is correct, then tritium analyses would not indicate that the repository site should be automatically rejected. In constrast, the last theory, if correct, would present an insurmountable obstacle to the acceptance of the host rock as an effective geologic barrier. As mentioned later, analyses of water in order to detect anthropogenic compounds such as Freons might have helped resolve the question at Stripa.

Recommended geochemical program for the evaluation of proposed sites for the disposal of radioactive waste

The following recommendations are not only based on the work at Stripa but also on a number of other studies, most notably that of Pearson et al. (1991). The recommendations are directed towards the objective of reconstructing the history of the groundwater with particular emphasis on water dating. Some consideration of economics is implied, although this is not an overriding factor in the selection process. For the purpose of this discussion, we will assume that the spatial distribution of samples is statistically and hydrogeologically sound, that the techniques of sampling are correct, and that the laboratory analyses are accurate. All of these items are of utmost importance, but limitations of space dictate that they should be covered elsewhere. In the tables which follow (Tables 1, 2, and 3), a number of constituents are listed for analyses. The importance of most of the consistuents is common knowledge, so extended comments on the reasons for measuring pH, for example, are not given. Comments, however, are given on a number of constituents which are not as well understood. A very general guide to the usefulness of most of the measurements, however, is given in Table 4.

Bromide. If proper analytical methods are used, the negative ion of the element bromine can be one of the most useful dissolved constituents to measure. Unfortunately, published analyses, until a few years ago, were generally unreliable at concentrations below about 1.0 mg/L. Bromide is almost as conservative as chloride, so groundwater from a given source maintains a nearly constant Cl/Br ration unless mixed with water from other sources.

Silica. Dissolved silica is commonly omitted from water analyses. Silica concentrations must be measured if the hydrochemical system is to be understood.

Rock characteristics. Normal rock-forming minerals will react with groundwater, so the chemistry and mineralogy of the solid phase must be determined. Also, in order to calculate the subsurface production of "cosmogenic" radionuclides, concentrations of elements with high cross sections for the capture of thermal neutrons must determined.

Table 1

**Constituents which must be measured
in order to make minimal geochemical evaluations
of repositories for radioactive waste.**

Gases in water: Oxygen, nitrogen, helium and radon

Dissolved inorganic material: Sodium, potassium, calcium, magnesium, iron, chloride, bromide, carbonate, bicarbonate, nitrate, sulfate, and silica.

Stable nuclides in water: S-34/S-32, H-1/H-2, O-18/O-16, C-13/C-12.

Radionuclides in water: H-3, C-14, Cl-36, U-234, U-238.

Organic chemicals: Total dissolved organic carbon. Specific analyses for any organic material used with drilling fluids.

Other measurements related to groundwater: Specific electrical conductivity, pH, Eh, total dissolved soilds, and water temperature.

Mineralogy of primary and secondary minerals.

Bulk chemistry of major rock units and of secondary fracture fillings: Chemical analyses must include all common elements plus U and Th and elements with high neutron capture cross sections, for example, B, Gd, Sm, and Cl. Also, will need Sr, Li, and Rb.

Table 2

**Constituents which should be measured
in order to make thorough geochemical
evaluations of repository sites**

(Items are to be added to those in Table 1.)

Gases in water: Neon, argon, krypton, xenon, methane, hydrogen, hydrogen sulfide, and ammonia.

Dissolved inorganic material: Boron, strontium, iodine, fluoride, aluminium, manganese, and lithium.

Stable nuclides: N-15/N-14, Sr-87/Sr-86, Ar-36/Ar-40, He-3/He-4, and B-10/B-11.

Radionuclides: Ar-39, Kr-85, I-129, C-14 of selected organic fractions.

Suspended solids in groundwater: Colloids and viable organisms.

Constituents leached from crushed minerals and rocks: Analyses for major dissolved substances, isotopic analyses for O-18/O-16, Cl-36, and S-34/S-32 in leachate.

Isotopes useful for dating solid mineral matter in contact with groundwater: List would depend on site-specific circumstances but may include K-40/Ar-40, C-14, Rb-87/Sr-87 and others.

Table 3

Constituents that may prove important in the future

(Development work is still in progress on some of the items in the list)

Dissolved gases in water: Freon-11/Freon-12.

Stable nuclides: Li-6/Li-7, Cl-37/Cl-35, Ne-21/Ne-20

Radionuclides: Si-32, Ar-37, Ca-41, Kr-81.

Organic compounds: Amino acids.

Table 4

Some potential uses of chemical data

Indication of surface environmental conditions at the time of recharge: H-2/H-1, O-18/O-16, Ar, Kr, Xe, and amino acids.

"Fingerprinting" groundwater in order to trace movement of distinctive types of water: Chloride/bromide, S-34/S-32, Cl-36, U-234/U-238, Sr-87/Sr-86, B-10/B-11, Li-6/Li-7, and Cl-37/Cl-35.

Key to origin of salinity: Chloride/bromide, S-34/S-32, Cl-36, Sr-87/Sr-86, He-3/HE-4, and Cl-37/Cl-35.

Indication of subsurface diagenetic changes and biological activity: Methane, hydrogen sulfide, N-15/N-14, C-13/C-12, pH, Eh, nitrate, sulfate, S-34/S-32, bicarbonate, ammonia, and amino acids. Also, the presence of viable microorganisms.

The presence of water which has infiltrated rapidly from the surface: Freon-11/Freon-12, Kr-85, H-3, and anthropogenic organic compounds. If most of the carbon is picked up in the modern soil horizon, then C-14 values above 100% modern would also be diagnostic.

Extent of rock-water interaction: Silica, pH, calcium, magnesium, Sr-87/Sr-86, Ar-40, O-18/O-16, Li-6/Li-7, and iron.

Dating of old groundwater: C-14, Cl-36, Ar-39, Kr-81, U-234/U-238, He (accumulation), Ne-21 (accumulation), and Ca-41.

Noble gas concentrations. The use of noble gas concentrations to reconstruct surface temperatures at the time of groundwater recharge has been attempted in many projects with varying success. In theory, the method should be an excellent way to reconstruct past environmental conditions and to help date the groundwater by matching paleotemperatures with the known Pleistocene record. The method should be used wherever water samples can be obtained which are not unduly affected by modern gases introduced through drilling or by uncontrolled degassing of the water prior to sampling. In general, analyses of noble gases from water with high methane concentrations have been difficult to interpret.

Chlorine-36. The successful use of chlorine-36 analyses in hydrogeology began in 1979 and continues to increase in importance today. Because chloride migrates easily with groundwater and because its geochemistry is relatively simple, it has a very broad application. Subsurface production, however, complicates the interpretation of the analyses. At Stripa, for example, the extent of buildup of chlorine-36 above atmospheric (cosmogenic) levels can be used to infer groundwater ages whereas in other groundwater systems subsurface production is small and radioactive decay will reduce chlorine-36 concentrations below atmospheric concentrations. One major advantage of chlorine-36 work is that samples can be taken and stored without problems of contamination.

Boron isotopes. Recent advances in mass spectroscopy have generated an interest in the isotopic composition of boron in groundwater. Because of the large natural fractionation of boron and because many dissolved chemical species are nearly conservative in water, analyses of boron isotopes should be useful in helping to decipher flow directions and geologic origins of different masses of groundwater. Also, where diagenetic alteration of marine sediments is active, this should be reflected in changes of boron concentrations as well as isotopes.

Freon-11/Freon-12. Busenberg and Plummer (1992) have recently reviewed the use of Freons (chlorofluorocarbons) in hydrogeology. They also presented the results of a thoroughly document field study. Because both Freon-12 (CCl_2F_2) and Freon-11 (CCl_3F) are anthropogenic and increasing in concentrations in the atmosphere, they serve both as a method to trace recent water and to date the water. Freon measurements would be particularly important where the possibilities of subsurface production of both Kr-85 and H-3 exist. Zero Freon concentrations coupled with measurable Kr-85 would suggest subsurface production.

Lithium isotopes. Owing to potential fractionation in the lithosphere, isotopes of lithium might prove to be useful naturally induced tracers. This suggestion is speculative.

Stable chlorine isotopes. Chloride is one of the most concertative ions in natural waters within the normal pH range of groundwater. Slight variations in chlorine isotopic ratios, if measured accurately, could prove to be the best naturally introduced tracers. Development work on the use of chlorine isotopes is continuing.

Neon-21. Theoretically, the reaction 0-18 (α, n)Ne-21 should produce measurable changes in the natural Ne-21/Ne-20 ratio in groundwater within fractured granite after a period of a few hundred thousand years (Davis and Murphy, 1987). Thus, the buildup of Ne-21 would be a function of time. Uncertainties in diffusion rates of neon from the solid minerals into the water, however, would prevent precise dating. Qualitative information on ages of very old water, nevertheless, should be obtainable.

Argon-37. One of the very interesting discoveries at Stripa was the presence of significant amounts of argon-37 in deep groundwater (Loosli et al., 1989). Because its half-life is only 34.8 days, a subsurface source, Ca-40(n, α)Ar-37 is certain. Two items are of interest; one is that Ar-37 is an indirect measure of the neutron flux and the other is that the higher the Ar-37 concentration is, the more likely that Ca-41 from Ca-40(n, γ)Ca-41 would be present.

Calcium-41. Currently developed technology is probably not able to isolate useful amounts of Ca-41 of near-surface origin which might be present in groundwater. However, should the techniques be developed, Ca-41 measurements could be useful in understanding the Ca-bicarbonate system and perhaps in identifying secondary calcite which is less than a half-million years old.

Krypton-81. Of all the radionuclides potentially available for the dating of very old water, Kr-81 may eventually be the most useful (Lehmann et al., 1991). This is because subsurface production is probably very small (Lehmann et al., 1992) and krypton will not react chemically with solids. Limitations might be a slight sorption of krypton on natural materials, difficulty of collection and analysis of krypton-81, and scarcity of krypton in samples which have been outgassed through subsurface methane production or through other processes.

Amino acids. Owing to the great variety of amino acids which originate from distinct settings and react differently to environmental conditions and maturation effects of time, reliable analyses of trace amino acids in groundwater would yield valuable information on the origin of the water as well as its age and geochemical history (Davis and Murphy, 1987). Also, selective C-14 analyses on individual acids would be possible provided enough amino acid could be separated from groundwater. Development of reliable collection and separation method is, nevertheless, required.

Conclusions and discussion

Most objectives of the geochemical and hydrochemical investigations at Stripa were achieved. Techniques related to the analysis of most of the specific constituents listed in Tables 1 and 2 were employed in the research. In order to complete some of the work, new techniques were developed. Virtually all the analyses were useful in reconstructing the history of the groundwater migration and chemical interaction with the minerals in the subsurface rocks. The bulk of the water at depths greater than 500 m appeared to have been isolated from the atmosphere for several thousand years or more. Antecedent mining operations at Stripa introduced uncertainties in the dating of the groundwater, however.

Future exploration activities related to the evaluation of possible sites for repositories for radioactive waste should make full use of geochemical techniques. Becaues contamination from drilling, excavation, and other exploration activities can compromise the geochemical interpretations, advice from geochemists is essential during the planning stages for the exploration. In general, all drilling fluids including compressed air should have one or more chemical tracers added in order to facilitate their identification in samples of water which are collected for analyses. Furthermore, most geochemical sampling should be scheduled as the first activity in the exploration process in order to avoid the time-dependent intrusion of gases and liquids which are not native to the horizons being sampled. With some constituents such as tritium, Freons, noble gases, and amino acids sampling should be completed the instant drilling fluids have been effectively purged from drill holes. In many settings, delays of a few hours could seriously compromise the usefulness of a sample. Valuable information is irreparably lost if geochemical sampling is left as an ancillary afterthought.

Acknowledgments

Without the support of the Division of Research and Development of the Swedish Nuclear Fuel and Waste Management Company, the geochemical research at Stripa would have been impossible. Special thanks are due to Dr. Bengt Stillborg of this division for his patient encouragement throughout the last part of this project. In addition to the host of excellent researchers involved with the project who have actually completed the scientific investigations, I have personally gained very useful ideas and information from extended conversations with Drs. J. Andrews, J. Fabryka-Martin, P. Fritz, B. Lehmann, E. Murphy, and D.K. Nordstrom. Any misconceptions which may be presented, however, are mine. In constrast, the outstanding success of the geochemical research is to be credited to all of those mentioned above.

References cited

Busenburg, E., and Plummer, L. N., 1992, Use of chlorofluorocarbons (CCl_3F and CCl_2F_2) as hydrologic tracers and age-dating tools: The alluvium and terrace system of central Oklahoma: Water Resources Research, v. 28, no. 9, p. 2257-2283.

Davis, S. N., and Andrews, J. N., 1990, Results of Phase II geochemical studies at Stripa, International Stripa Project: Proceedings of the 3rd NEA/SKB Symposium on In-Situ Experiments Associated with the Disposal of Radioactive Waste, p. 147-160.

Davis, S. N., and Murphy, E., 1987, Dating ground water and the evaluation of repositories for radioactive waste: U.S. Nuclear Regulatory Commission, NUREG/CR-4912, 181 p.

Davis, S. N., and Nordstrom, D. K., editors, 1992, Hydrogeochemical investigations in boreholes at the Stripa mine: SKB, Stripa Project, Technical Report 92-19, 178 p.

Lehmann, B. E., Davis, S. N., and Fabryka-Martin, J., 1992, Atmospheric and subsurface sources of stable and radioactive nuclides used for groundwater dating: Unpublished manuscript submitted to Water Resources Research.

Lehmann, B. E., Loosli, H. H., Rauber, D., Thonnard, N., and Willis, R. D., 1991, Kr-81 and Kr-85 in groundwater, Milk River aquifer, Alberta, Canada: Applied Geochemistry, v. 6, p. 419-423.

Loosli, H. H., Lehmann, B. E., and Balderer, W., 1989, Argon-39, argon-37, and krypton-85 isotopes in Stripa groundwaters: Geochim. et Cosmochim. Acta, v. 53, p. 1825-1829.

Nordstrom, D. K., Andrews, J. N., Carlsson, L., Fontes, J.-C., Fritz, P., Moser, H., and Osson, T., 1985, Hydrogeological and hydrogeochemical investigations in boreholes-final report of the Phase I geochemical investigations of the Stripa groundwaters: SKB, Stripa Project, Technical Report 85-06, 250 p.

Pearson, F. J., Jr., Balderer, W., Loosli, H. H., Lehmann, B. E., Matter, A., Petters, Tj., Schmassmann, H., and Gautschi, A., 1991, Applied isotope hydrogeology, a case study in northern Switzerland: Amsterdam, Elsevier Science Pub., Studies in Environmental Science 43, 459 p.

Conceptual Model Development within the SCV Project

Olle Olsson
Conterra AB, Uppsala, Sweden

John Black
Golder Associates, Nottingham, United Kingdom

John Gale
Fracflow Consultants, St.Johns, Nfld, Canada

Abstract

A major aim in site characterization for nuclear waste repositories is to be able to predict groundwater flow and radionuclide transport from a future repository to the biosphere. A flow model of a site is based on a conceptual model of the site which identifies the main hydraulic conduits. The basic assumption in the conceptual model of the SCV site was that flow was concentrated to fracture zones which could be described deterministically. A minor portion of the flow would occur in single fractures which have to be described stochastically.

Data on rock properties from the boreholes showed that a binary representation of the rock mass was relevant. A Fracture Zone Index (FZI) was introduced and used for identification of fracture zones. A structured approach for constructing the conceptual model was devised based on the use of the FZI, remote sensing geophysics, and cross-hole hydraulic testing. Emphasis was put on an iterative procedure with successive testing of the model with independent data. The selected approach provides a reasonably well documented and objective procedure for constructing a conceptual model of a site.

Introduction

The aim of site characterization activities is to be able to predict the performance of a site in response to different stimuli. In this case we are interested in the hydrogeological and solute transport performance. To achieve this we must consider all possible ways in which flow might occur within the site together with the possible processes and associated geometries. For instance, it is possible to view the site as a homogeneous porous medium or as a number of discrete fractures. Each concept would require its own geometrical specification.

Ultimately, it is necessary to decide on a basic conceptualization and then gather data suitable for that concept. The decision is commonly based on the perceived geology. From that point on the investigation focusses on representing the chosen concept in a numerical form suitable for predictive modelling. An outline of the successive levels of conceptualization in relation to the work within SCV Project is given below.

A flow model of a site is usually based on what is termed a conceptual model of the site as a whole. This is commonly a generalized description of the main geological features in terms of lithological units and fracture zones, their geometry and properties. Of particular significance for groundwater modelling is, of course, identification of the main hydraulic conduits. Since flow modelling necessarily averages various processes at different scales (flow in channels within single fractures, flow in fracture zones, etc.), there is some diversity between different research groups and individuals as to what the conceptual model of the site should contain.

To circumvent the problem of nomenclature we have, in the SCV Project, termed the model of the site based on the characterization data a "structural conceptual model". This model identifies the major structural features at the SCV-site and assigns average properties to these features. It provides a structured approach to determine what features are important (for the hydrology) and where they are located. The model incorporates and describes the large scale features while smaller scale components of the flow models, like flow distribution within fractures, are not included in the site model.

Successive steps of conceptualization

The most basic assumption, made in this work, is that flow occurs in fractures surrounded by a poorly permeable rock matrix. The fractures are assumed to be irregularly distributed throughout the rock mass and concentrated within what generally are referred to as "fracture zones". Hence, it assumed that it is useful to divide the description of the rock mass into two parts, "fracture zones" and "averagely fractured rock". Another implicit assumption is that these "fracture zones" are the pathways for a significant portion of the groundwater flow through the rock. If they were not, there would be no purpose in specifically including "fracture zones" in a hydraulic model. The work within the SCV Project has shown that through the use of remote sensing geophysical methods (radar and seismics) these features can be identified and their three-dimensional extent described. Hence, it is feasible to describe these features deterministically with respect to geometry. However, information density on how properties (e.g. hydraulic transmissivity) vary along the two-dimensional extent of the features is normally too sparse for a deterministic description. Here there are two alternatives, either the properties have to be assumed to be spatially homogeneous based on some average of values obtained at sampling points, or they must be described stochastically.

Fractures in "averagely fractured rock" are observed in drifts, in core, and by single borehole testing methods. However, from these observations it is not possible to know the lateral extent of the fractures or their properties away from the boreholes. Neither, do we obtain any information about the presence of smaller fractures some distance from the boreholes. The presence of fractures, together with their size and properties in "averagely fractured rock" between boreholes thus has to be inferred from observations in boreholes and drifts. Such inferences are made by stochastic methods to provide a stochastic description of fracture geometry and properties.

Hence, one of the basic building blocks of a groundwater flow and transport model is the description of the **geometric structure and properties of the fracture system**. This description consists of two parts:

- the "fracture zones" where the geometry is known and described deterministically while their properties are described stochastically or by some averaging procedure.
- the "averagely fractured rock" where both geometry and properties of the fractures have to be described stochastically.

The next step in the conceptualization is to describe **how flow and transport occurs in the fracture system.** It is normally assumed that flow through fractures is laminar and that Darcy's law is valid. However, additional assumptions have to be made in order to assign transmissivities to fractures or fracture zones. This is an intricate problem as the flow and transport parameters used in the numerical models (e.g. single fracture transmissivities) are generally not those measured in the field (e.g. transmissivity of a borehole interval containing several fractures). A conceptual model has to be used to transform measured flows and pressures into values of hydraulic conductivity which can be used as model input. Also, there are the assumptions made concerning the small scale distribution of flow within a fracture and at fracture intersections. The flow and transport model can also include physical processes assumed to be significant but not measured directly in the field. For example, there is the relationship between stress and permeability where data normally is obtained from laboratory experiments.

In order to model the groundwater flow within a limited volume the **boundary conditions** need to be known at the limits of the modelled volume. In this context the boundary conditions consist of head and/or flow data on the bounding surfaces of the model. Such data can be obtained through measurements of head, in practice available only for a very limited number of points, or synthesized by means of a regional flow model.

In order to model the groundwater inflow and transport to the Validation Drift, a conceptual model of **drift excavation effects** was required. There was a need to identify and model the processes which make the flow into a drift different from what it would have been into a set of equivalent boreholes.

Another step in the conceptualization was to make the **numerical approximations** required to run the hydraulic models on reasonable sized computers and within a reasonable time frame. In this context we can include decisions about how many and what fractures to include in the model, the method for representation of fractures, the numerical method used to solve the problem (e.g. finite element), element sizes, interpolation schemes, and other numerical parameters.

Figure 1 represents an attempt to display graphically the components of the flow and transport modelling and to outline briefly the data required for each basic component in the model. The following discussion concentrates on the structural conceptual model of the SCV-site.

The approach to constructing the conceptual model

Basis for a binary structural model

The appropriateness of a binary division of the rock mass into "fracture zones" and "averagely fractured rock" needs to be justified. First of all, there is a need to show that the rock properties are distributed in such a manner that a binary division is meaningful. Secondly, some quantitative procedure for

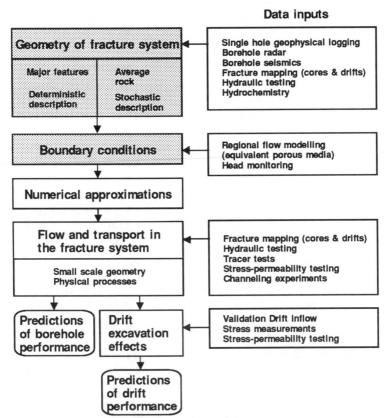

Figure 1   Basic components of flow and transport models and the data
            required to construct it. The components belonging to the
            structural model are shaded.

discriminating between the two parts using measured data is required. In
addition, one of the parts should have a dominating influence on the
groundwater flow system for the binary model to be hydrogeologically relevant.

    A structural model of a site has to be based on measured data. The data
collected within the SCV Project which can be used for this purpose are
basically of two types;

-    measurements of rock properties in the vicinity of the boreholes (with a
     range less than 1 m). These include core logging (detailed fracture
     statistics), tests on core samples, geophysical single hole logging, and
     hydraulic single hole testing.
-    remote sensing data provided by radar and seismic investigations.
     Reflection measurements yield data on the geometric shape of geologic

structures but can also be used to estimate physical properties. Tomographic measurements give data on geometry as well as electric and elastic properties of the rock between boreholes.

Hydraulic connections across the site are derived from cross-hole hydraulic testing and head monitoring even though such tests cannot be used independently to construct a geometric model. However, the hydraulic data should be consistent with the geometric model based on radar and seismic data. In addition, data on groundwater chemistry and geology have to be compatible and consistent with the geometric model based on the two data types mentioned above.

It is natural to base a binary representation of the rock mass on physical properties measured in the vicinity of the boreholes. Hence, the location and width of "fracture zones" is defined where they intersect the boreholes. The extent and geometry of the features at larger distances from the boreholes can then be described using remote sensing methods.

During the course of the SCV Project a comprehensive set of single hole data have been collected. A subset was selected for identification of the fracture zones on the following grounds:

-   the data should represent properties at the measurement location.
-   the measured properties should be known to be fracture dependent in a general sense.
-   only one data set should be included for each type of physical property to avoid biasing in favor of a specific property.
-   data should exist for all boreholes.

Based on these considerations the following data sets were chosen; normal resistivity, sonic velocity, hydraulic conductivity, coated (open) fractures, and single hole radar reflections.

A basic problem associated with data sets provided by different single hole techniques is that data represent different scales of investigation from a few millimeters up to a few meters. In order to make use of different methods the data have to be transformed to a common scale. For the purposes of the SCV Project it was considered relevant to use a 1 m resolution along the boreholes.

The procedure developed within the SCV Project for classification of the rock into "fracture zones" and "averagely fractured rock" is based on principal component analysis of the single hole methods listed above (Olsson, 1992). First, the logarithm is taken of the normal resistivity, sonic velocity, and hydraulic conductivity data. Then the data are normalized by subtracting the mean value and dividing by the standard deviation for each parameter. A matrix of correlation coefficients is formed and the eigenvectors found for that matrix. Each eigenvector represents a weighting of the data, and new parameters (principal components) can be produced by multiplying an eigenvector by the normalized data values. The parameter obtained by multiplying with the eigenvector corresponding to the largest eigenvalue should represent the most important characteristic of the rock.

For the SCV-site we expected the new parameter generated by the eigenvector belonging to the largest eigenvalue to correspond to the fracturing of the rock in some way. This is because there is essentially only one rock type at the site and all anomalies observed are essentially caused by increases in fracturing or tectonized rock, i.e. mylonites and breccias related to faulting. Hence, we will refer to this parameter as "fracture zone index" (FZI).

The information content in the principal components is related to the relative magnitude of the corresponding eigenvalue (Table 1). The information content of the parameters corresponding to the three smallest eigenvectors has been examined and found to be essentially noise. The second eigenvector is dominated by information on radar reflectors. For the "fracture zone index" (eigenvector 1) we find that almost equal weight is given to normal resistivity, hydraulic conductivity, and fracture frequency. Lesser weight is given to sonic velocity and radar reflections.

Table 1   The eigenvectors obtained from the matrix of correlation coefficients based on data from all boreholes.

Eigenvector	1 = FZI	2	3	4	5
Eigenvalue	1.9268	0.9835	0.8866	0.7099	0.4932
Normal resistivity	0.5594	0.0708	0.3579	-0.1570	-0.7275
Sonic velocity	0.3685	-0.3001	-0.7868	0.3360	-0.2055
Hydraulic conductivity	0.5229	-0.1109	-0.1301	-0.6866	0.4754
Coated fractures	0.4982	-0.0670	0.4078	0.6200	0.4434
Radar reflection	0.1722	0.9424	-0.2637	0.0820	0.0767

The frequency distribution of the "fracture zone index" values for all holes shows a skewed distribution (Figure 2). It is not bi-modal but it can be considered to consist of two parts. One part consists of a more or less normal distribution centered around a mean value which is slightly less than zero. Then there is a tail of higher values. The dividing line between the tail and the normal distribution occurs approximately at a value of 2 for the "fracture zone index".

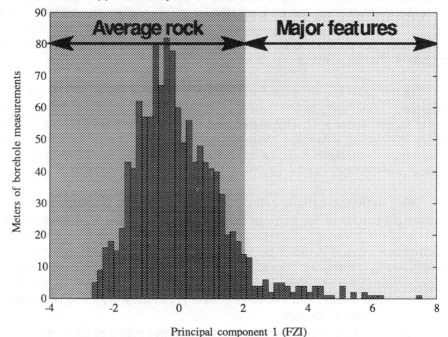

Figure 2   Frequency distribution of "fracture zone index" (principal component 1). Values from the tail of the distribution (FZI>2) are designated as "fracture zones" while values less than 2 are designated as "average rock".

Based on the frequency distribution of the "fracture zone index" we considered it justifiable to use a binary description of the rock mass whereby it is divided into "average rock" (FZI < 2) and "fracture zones" (FZI > 2). Thus, using the FZI we can define the points in the boreholes which are considered to represent the occurrence of "fracture zones".

The "fracture zone index" compresses the information gathered in the single hole investigations into a single parameter which describes the most significant properties of the rock. It simplifies interpretation as it allows a single parameter to be used for identification of the anomalous sections in the boreholes. As the "fracture zone index" has been obtained through a quantitative and well defined procedure it provides an objective means of classifying the rock into the two classes; "averagely fractured rock" and "fracture zones".

The "fracture zone index" is also considered to be better for definition of the hydraulically significant features than the single hole hydraulic conductivity data alone. The basic reason is that single hole hydraulic tests yield parameters which are applicable within a very small volume surrounding the borehole. In the fractured rock at Stripa hydraulic properties vary by more than an order of magnitude over small distances (Holmes *et al.*, 1990). Hence, a weighted parameter including several types of data should be preferable for defining the hydraulically important features. In the definition of the "fracture zone index" the hydraulic conductivity is included as just one of several measurements and the weighting is determined by the data set itself. It is anticipated that the fracture zone index will be less sensitive to small scale variations in the rock mass. Consequently, it should give a better representation of groundwater flow through the rock mass than the single hole hydraulic conductivity data alone.

The anomalous portions of the boreholes, defined as locations where the FZI is greater than 2, occupy 7 % of the length of the boreholes and that part accounts for 76 % of the measured transmissivity. Hence, the "fracture zones" defined this way account for the bulk of the flow at the SCV-site.

Procedure for constructing the structural model

Having provided and tested the conceptual basis of the structural model, the next step was to use this to produce a quantitative model of the structure of the SCV site. The integration of all relevant data to produce a geometric model of a site is a relatively complex task. The procedure used may appear obscure at a first glance. For the sake of credibility and traceability the interpretation procedure should be structured, understandable, and as quantitative as possible. The procedure developed is outlined in Figure 3 which shows how data from different investigations were integrated and the iterative steps required before the model could be finalized.

The basis for the construction of the conceptual model was the identification of anomalies within the single hole data. The quantitative procedure based on the "fracture zone index", as defined above, was used for this purpose. The "fracture zone index" identifies a number of anomalous sections in the boreholes distributed in space and thought to be associated with specific major features. The geometric extent of these features (i.e. connections between anomalous sections if any) is unknown from the single hole data but can be outlined using radar and seismic remote sensing data.

The next step was to prepare a table of all single hole anomalies (based on the FZI) together with radar and seismic reflection data. For each borehole, the intersections of reflectors were listed together with the geometrical parameters obtained from the radar and seismic data. In boreholes where directional radar data were available, the orientation of the corresponding

163

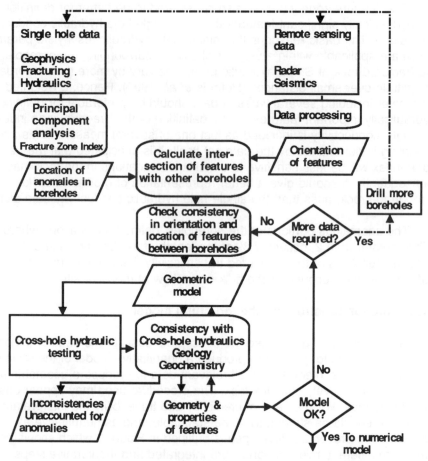

Figure 3   Outline of procedure used for construction of the structural
conceptual model of the SCV-site.

reflectors could be determined uniquely. In such cases the intersections with
neighboring boreholes were calculated. A check was then made as to whether
there were anomalous sections in the neighboring boreholes close to the
predicted locations (allowing for uncertainties in orientation data). The
neighboring boreholes were also checked for the presence of radar and seismic
reflectors and whether their orientations agreed. If agreement was found with
respect to the fracture zone index and the presence and orientation of
reflectors, the anomalies were considered to belong to the same feature and
were subsequently marked in the list. Tomographic data were also used in this
process as they yield connections of anomalous sections between boreholes.

The procedure was then repeated for each reflector and its intersections
with neighboring boreholes sought. Eventually, the geometry of as many

"fracture zone index" anomalies as possible were explained and the list of "unexplained" anomalies became small. Naturally, a few anomalies remained for which data were not available to describe their orientation. Also, some were not sufficiently extensive to intersect another borehole. There were also a number of radar and seismic anomalies which intersected the boreholes which were not associated with anomalies in the "fracture zone index". These were considered to represent minor features in the rock mass which did not qualify as fracture zones in the binary conceptual model.

Hence, the radar and seismic data yielded the orientation and extent of major features and made it possible to identify intersections between features and boreholes. At these intersections, data on the physical properties of the features were obtained from single hole measurements, which could be used to quantify variability in properties and width. Radar and seismic data also provided information on variability in width and orientation between the boreholes.

The success of this approach will depend on the extent to which all single hole anomalies can be explained. If the majority of anomalies have been accounted for we have been successful and the conceptual model provides a consistent and comprehensive description. If many anomalies remain the description will be incomplete. It should also be expected that the "fracture zone index" will identify, to some degree, features of such limited extent that they are intersected by only one borehole.

In addition to the "fracture zone index" and remote sensing data there are sets of data which cannot be used to define the geometry but which provide checks on the geometric model. A basic condition is that the geometric model should be compatible with the crosshole hydraulic responses and the head monitoring data. The crosshole hydraulic tests will also provide data on the large scale hydraulic properties of the extensive features required for hydraulic modelling. Groundwater chemistry data will also provide information on the overall flow pattern. Furthermore, when the geometry is known, it is possible to correlate geological characteristics of the features at the borehole intersections such as fracture orientations, fracture minerals, alteration, and degree of tectonization. Such data can serve as a verification of the geophysical interpretation and facilitate creation of a tectonic model of the site.

It should be emphasized that the construction of the structural model of a site is an iterative process as outlined in Figure 3. If the quality of data is good and sufficient amounts and appropriate types of data have been collected the iterative process should eventually converge. When model changes become minor, after checking consistency of fracture zone locations and properties between boreholes and different data types, the iterative process should stop. The unaccounted-for anomalies and/or uncertainties should be assessed. If the description, for some reason, is considered inadequate drilling of additional boreholes and input of new data to the process may be required.

Fracture zones at the SCV-site

The structural conceptual model of the SCV-site contains three major features or fracture zones named A, B, and H and three minor features I, M, and K. A perspective view of the features is shown in Figure 4. Data on geometry and properties of these features are summarized in Table 2.

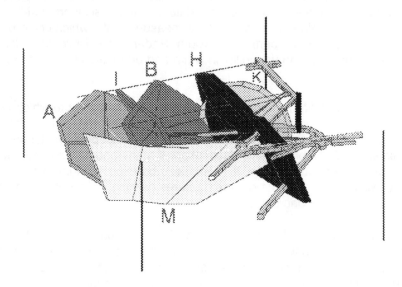

Figure 4   Perspective view of the deterministic features contained in the conceptual model of the SCV site.

The properties and width of the three major features or fracture zones are highly variable where they are observed intersecting the boreholes. Thicknesses of these features vary from 2 m to 12 m with an average of approximately 6 m. At the borehole intersections the features generally exhibit anomalous properties compared to "averagely fractured rock". This is demonstrated by the fracture zone index which has an average of approximately 2.5 for these zones. Fracture zone index anomalies are generally smaller in the southern part of the site and larger towards the north. The borehole with the largest FZI anomalies is W2 which is probably caused by the proximity of W2 to the intersections of A, B, and H. These major features are

166

Table 2   Summary of geometry and properties of fracture zones contained in the structural model of the SCV site.

Zone	Dip	Strike	Real width[1]	FZI[1]	Transmissivity (m^2/s)			Extent
	(deg)	(deg)	(m)		Fraction[3] %	Single hole[2]	Cross-hole	
A	48	47	6.2	2.72	18.3	9.9 10^{-10}	5.7 10^{-8}	≈3 km
B	43	40	5.0	2.37	22.9	1.1 10^{-8}	2.5 10^{-8}	≈3 km
H	76	355	5.8	2.39	33.5	3.8 10^{-8}	3.3 10^{-8}	≈1 km
I	63	356	3.7	1.99	1.0	9.8 10^{-10}	-	≈100 m
M	87	300	1.5	1.69	0	4.6 10^{-10}	-	≈100 m
K	65	305	3	2.27	2.6	9.1 10^{-8}	-	≈30 m

[1] Arithmetic mean   [2] Geometric mean
[3] Fraction of total transmissivity measured in single hole tests contributed by zone.

important for the groundwater flow system across the SCV-site in that they account for 75 % of the hydraulic transmissivity as measured by single hole hydraulic tests. For zones B and H there is approximate agreement between the geometric mean of the single hole transmissivity and the transmissivity determined from cross-hole testing.

Consistency with independent information

Following the procedure outlined above the consistency of the structural model with geological, geochemical, and hydrogeological data must be demonstrated.

Scrutiny of geological data showed general agreement between the interpreted extension of the fracture zones at the SCV site and the regional fracture system as well as fracturing in the small scale. Comprehensive cross-hole hydraulic testing resulted in hydraulic responses at expected locations consistent with the various source positions within the conceptual model. The conceptual model implies a specific flow system under the prevailing boundary conditions. This flow system was shown to be consistent with the chemical characteristics of the groundwater. Finally, the unaccounted for anomalies were few and appeared to be insignificant relative to the features contained in the model. Hence, the structural conceptual model was considered to give a realistic representation of the flow system at the SCV site.

Conclusions

The rationale behind the structural conceptual model of a site is an important issue with respect to performance assessment. A site characterization program includes several scientific disciplines, which results in data of disparate types which have to be combined into a common description. Even though the data collected are normally well documented and presented it is often less obvious how the data were combined into the resulting conceptual model. To address this problem a structured approach has been devised in order to produce the conceptual model. To meet the needs of performance assessment, it is essential that the procedure is objective (i.e. the result should essentially be independent of the person doing the work), traceable, and to the greatest extent possible, quantitative.

The procedure devised within the SCV Project is conceptually simple. It is based on the principle that any quantitative model needs information on properties (parameter values) and the geometric distribution of these properties. It recognizes that single hole investigations give data on physical properties only along the lines defined by the boreholes, and that remote sensing techniques can give reliable information about the geometry of geological features. Finally, the hydrogeological significance of the features is verified through cross-hole hydraulic testing. The conceptual model should also be corroborated by geological and geochemical data.

A binary representation of the rock mass was shown to be appropriate for the SCV site. A "Fracture Zone Index", based on principal component analysis of single hole data, was used to identify locations where fracture zones intersected the boreholes. At the SCV site, 80-90% of the flow was through the fracture zones and a binary representation of the rock mass was considered relevant. It is considered that the approach adopted within the SCV Project can, with minor modifications, be applied at other sites and at different scales.

References

Holmes, D., Abbot, M., Brightman, M., 1990. Site characterization and validation - Single borehole hydraulic testing of C boreholes, simulated drift and small scale hydraulic testing, stage 3. Stripa Project TR 90-10, SKB, Stockholm, Sweden.

Olsson, O. (ed), 1992. Site Characterization and Validation - Final Report. Stripa Project TR 92-22, SKB, Stockholm, Sweden.

SESSION I (Cont'd)

SÉANCE I (Suite)

Chairmen - Présidents
V. Ryhänen (Finland)
R. Jackson (United States)

In-situ Tracer Experiments in Single Fractures, Fracture Zones and Averagely Fractured Rock

Lars Birgersson, Hans Widén, Thomas Agren
Kemakta AB, Stockholm, Sweden

Ivars Neretnieks
Royal Institute of Technology, Stockholm, Sweden

ABSTRACT

This paper is based on the results and interpretations from five tracer migration experiments performed in the Stripa mine. The experiments are:

- In-situ diffusion experiment
- Migration in a single fracture
- 3-D migration experiment
- Channeling experiments
- Tracer Migration Experiment in the Validation drift

These tracer experiments have given valuable information for the general understanding of flow and transport in crystalline rock. The experiments have had different purposes and have studied migration in as small scale as a few tens of centimeters and up to about 50 m.

The experiments have been performed at the 360 and 385 m levels in the Stripa mine. Some of the basic phenomena like diffusion into the rock matrix, channeling within fracture planes, flow and tracer distributions within the average fractured rock as well as in a fracture zone are phenomena that are most likely to be valid for other sites situated in similar geological formations.

BACKGROUND AND INTRODUCTION

In recent years interest has increased considerably in the area of flow and transport in low permeability fractured rock. The reason for this is that many countries are considering to site final repositories for nuclear waste in such environments, at depths ranging from a few tens of meters for low and intermediate level waste, up to 500 m or more for high level waste.

There is considerably less information and experience on depths below a few hundred meters than at shallower depths. To assess if a repository is sufficiently isolated, information in several fields is needed. The flow rate and flow distribution at repository depth will strongly influence the rate of dissolution of many radionuclides. The flow paths and velocities will influence their travel time. This will in turn determine the decay of the radionuclides. Axial dispersion will dilute the species in time but also allow a fraction of the nuclides to travel faster. Channeling has the same effect. Transverse dispersion will cause dilution but also exchange species between fast and slow flow paths. For those nuclides which sorb on fissure surfaces and/or diffuse into the rock matrix, the frequency of water conducting channels and their exposed surface directly influences the contact area between flowing water and rock.

Most of the nuclides are expected to be cat-ionic or neutral in the waters present in the crystalline rocks investigated. They will adsorb and/or ion-exchange on the negatively charged surfaces of the rock minerals. These processes may considerably retard the radionuclides which in some instances are expected to move many orders of magnitude slower than the water. For a given flow rate the retardation will be larger for a nuclide if there are more exposed surfaces with which the nuclides interact. One of the crucial questions thus is how much fracture surface the flowing water encounters. In addition to interaction with the fracture surface, the radionuclides may diffuse into the rock matrix and sorb onto the inner surfaces of the rock matrix. The inner surfaces are much larger (many orders of magnitude) than the fracture surfaces in contact with the flowing water. If the inner surfaces are accessed, the retardation may increase considerably.

In this paper we first give very short descriptions of the experiments. Then the observations of matrix diffusion and channeling in the experiments are discussed because they have so large impact on the transport of solutes and radionuclides especially. Finally in a section the impact of the findings in these experiments on radionuclide migration is discussed. Many other observations were also made and data collected. These are discussed in a separate section.

172

OVERVIEW OF TRACER MIGRATION EXPERIMENTS PERFORMED IN STRIPA

The tracer migration experiments were aimed at studying different transport properties in the averagely fractured rock as well as in a fracture zone. The experiments have studied migration in as small scale as a few tens of centimeters and up to about 50 m. The performed experiments and the major findings are summarized in Birgersson et al., 1992. The main aims for the experiments are given below:

In-situ diffusion experiment. A detailed description of the experiment is given in Birgersson and Neretnieks, 1982, 1983, 1988, 1990, Birgersson 1988. Laboratory experiments have shown that it is possible for sorbing as well as non-sorbing tracers to migrate into the rock matrix by diffusion, but the experiments were not carried out in "undisturbed" rock. It could not be ruled out that the release of the rock stresses, which occur when samples are taken, have irreversibly induced microfissures. For this reason, in-situ experiments in rock in the natural stress environment before a first stress release are necessary because a re-compression will not necessarily close the induced microfissures.

Three long 150 mm diameter holes were drilled to get far enough away from the drift. To avoid disturbances from the drilled hole, a thin hole was drilled from the bottom of the wide hole, see Figure 1. Tracers were injected for up to 3.5 years in the little hole. Mixtures of Uranin, Cr-EDTA and I⁻ were used in all three experiments. The tracers diffused and seeped into the surrounding rock.

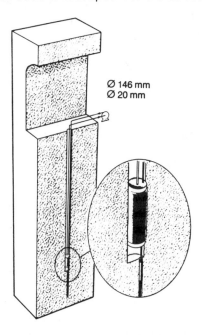

Figure 1. **In-situ diffusion experiment. Experimental layout.**

After termination of the injections, the packers were retrieved and the little hole was over-cored. The core was cut into 50 mm long cylinders from which a number of sampling cores with a diameter of 10 mm were drilled at different distances from the injection hole. Additional boreholes made it possible to study the tracer migration as far as about 400 mm outward from the injection hole. Diffusivities and hydraulic conductivities were evaluated from the concentration profiles.

Migration in a single fracture, the 2-D experiment. A detailed description of the experiment is given in Abelin et al., 1982, 1985. The main purpose of this experiment was to investigate if it is possible to apply results on sorption and retardation of radionuclides in granitic rock, obtained from laboratory experiments, to a real environment with migration distances up to 10 m. Further it was attempted to determine the extent of channeling within a single fracture.

Figure 2 shows the layout of the test site with two fractures and five injection holes intersecting the fracture planes between 5 and 10 m above the drift.

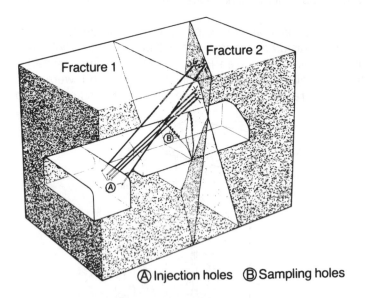

Figure 2. **2-D experiment. Schematic view of the test site.**

Sorbing as well as non-sorbing tracers were injected into the fracture which had a "natural" water flow toward the drift, where the water coming out of the fracture was collected. The sorbing tracers were continuously injected at one of the injection points for several months. As most of the sorbing tracers were expected to be sorbed in the vicinity of the injection point, part of the fracture close to the injection point was excavated. Elevated concentrations of the sorbing tracers were found on the fracture surface in the vicinity of the injection hole indicating sorption. It was also found that the sorbing tracers had migrated some millimeters into the rock matrix indicating diffusion from the fracture plane into the rock matrix.

The observations of water and tracer distributions clearly show that the investigated fracture plane has channels. There seems to be fairly limited mixing between the channels on the scale considered, 5-10 m, since sampling areas where large amounts of water are found do not necessarily have to account for the largest mass flow rates of tracers.

3-D migration experiment. A detailed description of the experiment is given in Abelin et al., 1987, 1991a, b. This experiment aimed at obtaining information on migration over long distances, up to about 50 m, in averagely fractured rock. Of special interest was to gather information on flow porosity, longitudinal and transverse dispersion and channeling. Some common models for tracer transport were tested against the experimental results.

The upper part of the test site, approximately 700 m^2, was divided into about 350 sampling areas each with an area of 2 m^2. Each sampling area was covered with a plastic sheet. The results from these measurements are presented in Figure 3, where it can be seen that the water inflow is limited to a few areas with large dry areas in between.

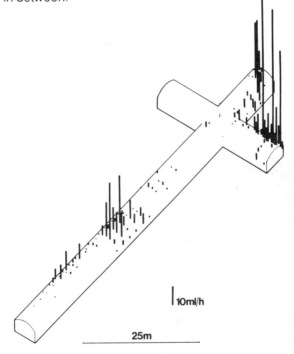

10ml/h

25m

Figure 3. **3-D migration experiment. Water inflow rates.**

175

Tracers were continuously injected for more than 20 months from nine different sections located between 10 and 55 m above the test site. Water samples were taken in the drift during the tracer injections plus another 6 months. The results give strong support to the notion that a non negligible portion of the flow takes place in pathways or channels which have little contact with each other.

Channeling experiments. A detailed description of the experiment is given in Abelin et al., 1990. The channeling experiments aimed at a detailed study of channeling properties within single fracture planes. This experiment gave a significantly more detailed description of channeling compared with the previous experiments which also, at least partly, aimed at studying channel effects in fracture planes. A number of fracture planes were investigated.

Large diameter holes, Φ 200 mm, were drilled along selected fractures to a depth of about 2.5 m from the face of the drift. A multipede packer was used to inject water with a constant pressure all along the intersected fracture plane. Flowrates were monitored with a resolution of 1/100 ml per hour in 50 mm sections along the fracture plane. These measurements gave information on the variation in the local conductivity along the fracture planes.

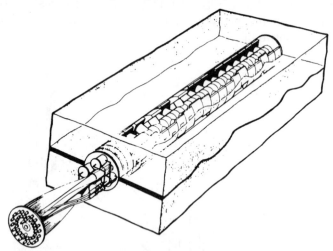

Figure 4. **Channeling experiments. Investigation borehole and multipede packer.**

In one of the fractures a second hole, parallel to the first, was drilled in the same fracture plane at a distance of almost 2 m. Hydraulic tests and tracer tests were made between the two holes to obtain information on connections in the fracture plane and residence time distributions in different paths (channels).

The results from the single hole experiments as well as the double hole experiment showed that, on the average, 25 % or less of the fracture planes investigated were open to flow.

<u>Tracer migration experiment in the validation drift.</u> A detailed description of the experiment is given in Birgersson et al., 1992a. These tests aimed at monitoring the water inflow rates and distributions into a drift partly intersected by a fracture zone, the H zone, and to obtain information about migration of tracers within the fracture zone. The output from this experiment was compared with modeling predictions within the SCV-project (Olsson 1992) to determine the applicability of these flow and transport models.

Solutions of a dye and a metal complex were injected from six sealed off borehole sections located 10-25 m from the Validation drift. Five out of the six injection sections were located within the H zone. The water inflow and tracer breakthrough were monitored in the Validation drift.

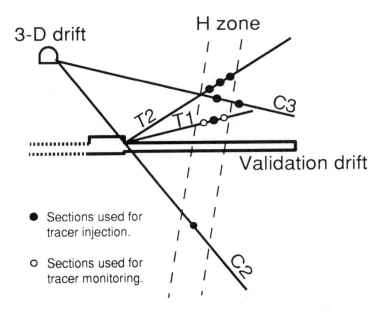

Figure 5. **Tracer test in the Validation drift. Schematic section view.**

The water inflow monitoring as well as the monitoring of tracer breakthrough indicated that water and tracers migrate in preferential flowpaths, channels, in the fracture zone as earlier found for the "averagely" fractured rock.

OBSERVATIONS OF CHANNELING

Channeling may have a very strong impact on the transport of solutes especicially of decaying radionuclides. The notion of channeling emerged during the 2D experiment and the awareness of the importance of this phenomenon grew during the 3D experiment. The Channeling experiment was specifically designed to study the phenomenon in more detail.

Uneven distributions of flow and transport properties, which may be described as channeling effects, has been observed in the tracer migration experiments when investigating single fractures, a large block of averagely fractured rock and a fracture zone. The channeling may have several causes. In a fracture with uneven surfaces, where the aperture varies and where the fracture may be filled with precipitated minerals and clays, the water will seek out the easiest paths for the prevailing gradient. The paths will change as the direction of the gradient changes (Tsang et al., 1988; Moreno et al., 1988). The paths may be clogged or opened by chemical or erosion processes in the fracture plane, e.g. dissolution channels as found in fracture zones in Switzerland (Nagra, 1985). In a three dimensional system, intersections between fractures are potential causes for channeling and have also been observed to make up high flowrate conduits (Neretnieks, 1987a, b).

A compilation of water inflow rates from the two fractures measured during the 2-D experiment gave the following results; Almost 80 % of the total amount of sampling points had low water inflows corresponding to less than 10 % of the total water inflow. Less than 5 % of the sampling points had between 70 and 80 % of the total water inflow. Injected conservative tracers arrived in only three of the sampling points, see Figure 6. After excavating the fracture, the injected sorbing tracers showed elevated concentrations in one direction which was not the direction of the shortest way towards the drift.

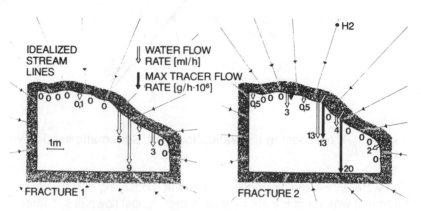

Figure 6. **2-D migration experiment. Water and tracer flow in fractures No 1 and 2.**

178

The water inflow to the 3-D drift was unevenly distributed among the 350 plastic sheets covering the entire upper section of the drift, as seen in Figure 3. The water inflow was found to be located to a few sampling areas with large dry areas in between. Measurable quantities of water emerged into 1/3 of the 350 sampling areas. 50 % of the total inflow came from approximately 3 % of the covered area and the sampling area with the largest water inflow gave about 10 % of the total inflow into the entire drift. A correlation between sampling areas with a high number of fracture intersections among the 100 largest fractures observed, and increased water inflow rates could be observed. No such correlations could be observed between water inflow and other fracture characteristics such as fracture length, fracture filling or fracture orientation. This indicates that fracture intersections may form conducting paths. Furthermore, one of the tracers arrived in a location where it must have passed through the pathways of the other without much mixing. Considerable quantities of one of the tracers injected 15 m above the drift was found in a location about 150 m from the injection section.

These observations indicate that there are more or less isolated flow paths in the rock volume investigated in the 3-D migration experiment which do not mix their water over considerable distances.

The Channeling experiments investigated hydraulic conductivity variations in fracture planes over the scale of 2 m. The flow rates were measured over 40 sections at each side of the fracture plane, see Figure 7.

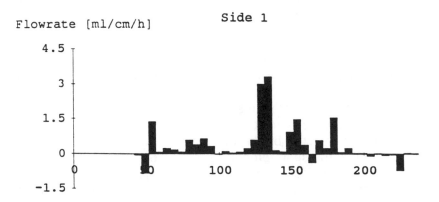

Figure 7. **Channeling experiments. Injected flowrate in one sides of a fracture in 50 mm sections along the hole.**

Photographs taken inside the boreholes along the 2 m intersections between fracture planes and boreholes gave information on visual fracture apertures, fracture intersections and fracture filling. No correlation could be

179

observed between injected flow rates and visual apertures. Nor was there any correlation between the injected flow rates and the number of fracture intersections or fracture filling. It was found during the single hole experiments, when injecting water into 2 m sections of fracture planes, that large sections of the fracture planes consists of dead-end pathways or closed parts. When it was possible to inject water into the fracture plane, a number of 50 mm sections with elevated water flow rates could be grouped into clusters, sometimes up to 0.5 to 1 m wide, separated by low flow sections. Only two of the five tracers injected during the double hole experiment arrived in the receiving hole. The two tracers seemed to have travelled mostly in separated pathways, although injected only 100 mm apart and emerging in the same section of the receiving hole. Smaller amounts of four of the five tracers emerged on the drift wall as colored spots with different tracer compositions. This together with the observed pressure dissipations indicates that a three dimensional network of pathways exists already over the 1.7 m separating the two boreholes in the same fracture plane.

The rock volume in the vicinity of the Validation drift, where the Tracer Migration Experiment in was performed, is hydraulically dominated by the presence of a fracture zone, the H zone. The total length of the Validation drift is 50 m and more than 99 % of the total water inflow into the drift emerged from the 6 m long intersection with the H zone. The water inflow into the boreholes intersecting the H zone also show an uneven distribution between high conductivity and low conductivity sections. The high conductivity sections were all located within the H zone, but was found to vary by several orders of magnitude between the different holes. Although the H zone had a high density of fractures, only a very limited number of fractures carried the bulk mass of water into the Validation drift. Two fractures within the fracture zone carried more than 90 % of the water flow to the Validation drift. The mass flows of the injected tracers were also concentrated to a few sampling areas. The three sampling areas with the highest mass flow rates for each tracer accounted for approximately 75 % of the mass flow. Four of the tracers were injected from injection sections within the fracture zone located in a rather limited region above the drift. The largest parts of the recovered tracer mass from these injections were collected in sampling areas intersected by the two main fractures, see Figure 8. Large differences in the transport properties exists between different injection sections, in spite of the fact that the tracers were injected near each other and also emerged close to each other in the drift. The transport properties seem to vary less between breakthrough curves from one injection section compared to breakthrough curves from different injection sections. This indicates the existence of separated pathways from the different injection sections to the sampling areas in the drift.

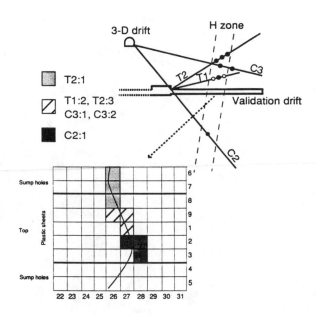

Figure 8. **Tracer migration experiments in the validation drift.
Sampling areas where the major mass flow rates
of the tracers were found.**

To sum up the observations of channeling it was found that, on the average 25 percent or less of the fracture plane is open to flow, with individual channel widths ranging from millimeters to a decimeter. These channels normally occur in clusters with cluster widths of decimeters. Individual fractures may have properties that strongly deviate from the average. It should be noted that the about ten investigated fractures were selected among more than 100 candidate fractures. Many of these seemed to be tighter. The investigated fractures have a clear bias, beeing more open than other fractures. Water flow and tracer monitoring in the Tracer migration experiment indicate that water flows in preferential flowpaths in a fracture zone as well.

OBSERVATIONS OF MATRIX DIFFUSION

Matrix diffusion is another mechanism which has a profound impact on radionuclide migration. The In situ diffusion experiment was specifically designed to study this mechanism under field conditions. It was also directly observed in the 2D experiment and indirectly observed in the 3D and the Tracer migration experiment.

181

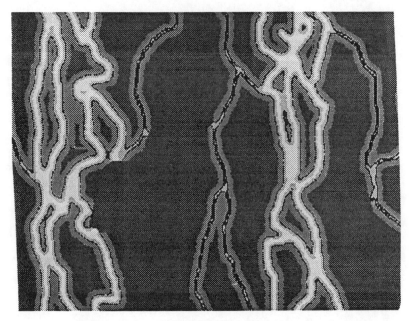

Figure 9. **Artist's view of single channels and clusters within a fracture plane.**

It has been observed that even seemingly dense crystalline rocks are porous. This porosity has a potential of acting as a sink for dissolved radionuclides and thereby withdrawing them from the flowing water (Neretnieks, 1980). Especially many positively charged species that sorb on the micropore surfaces in the interior of the rock matrix will be retarded and can therefore, for many of the nuclides, be expected to decay to insignificant amounts without reaching the biosphere.

Laboratory experiments have been performed for the purpose of determining diffusion coefficients for various tracers in granite and other crystalline rocks (Skagius et al., 1982; Skagius and Neretnieks, 1986a, b, 1988; Skagius, 1986; Bradbury and Stephen, 1985; Bradbury et al., 1982; Bradbury and Green, 1985, 1986). These laboratory experiments show that it is possible for sorbing as well as non-sorbing tracers to migrate in the rock matrix by diffusion, but the experiments were not carried out in "undisturbed" rock. It can not be ruled out that the reduction in the rock stresses, which occur when samples are taken, have irreversibly induced microfissures. For this reason, in-situ experiments in rock in the natural stress environment before a first stress release are necessary because a re-compression will not necessarily close the induced microfissures.

Three in-situ diffusion experiments in "undisturbed" rock were performed in Stripa with duration up to 3.5 years. It was noted that the difference in diffusivity could be as large as an order of magnitude for samples separated by just a few tens of centimeters. However, all these experiments clearly showed that the injected tracers had migrated through the disturbed rock in the vicinity of the injection hole and a considerable distance into the "undisturbed" rock. This implies that also radionuclides escaping from a repository at large depths in granitic rock will have access to the large surfaces within the rock matrix. Diffusivities were also determined in laboratory experiments that involved rock samples taken from the In-situ migration experiment. Similar diffusivities were found in the laboratory experiments and in the In-situ experiment which is an important observation since it indicates that diffusivities found in the laboratory also is valid in rock under natural stress conditions.

A part of the fracture plane used in the 2-D migration experiment was excavated to determine the migration of the sorbing tracers. Elevated concentrations of the sorbing tracers were found down to depths of 1.4 mm, see Figure 10. This is another indication that tracers can migrate from a fracture plane and into the pore system in the rock matrix.

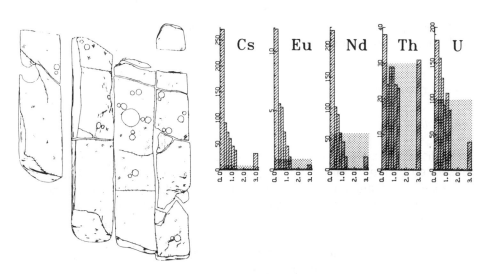

Figure 10. **2-D migration experiment. Concentration profiles into the rock.**

183

In the 3-D migration experiment the loss of tracers indicates that migration into the rock matrix could be a factor influencing the tracer migration. The indication was even stronger in the Tracer Migration Experiment in the Validation drift were the difference between the two tracer types injected simultaneously, one dye and one metal-complex tracer, was successfully modeled as caused by a very slight sorption within the rock matrix for the dyes. These observations are however indirect and other mechanisms such as diffusion into stagnant waters in the fractures may also have contributed.

OTHER OBSERVATIONS AND FINDINGS

Residence times and residence time distributions were measured in the experiments. From these observations it was possible to assess the porosity of the rock and of the fracture zone as well as estimate the fracture apertures of the single fracture experiments. The porosity values are not necessarily meaningful for these types of very heterogeneous rocks because they may vary considerably in different locations. Also measures of dispersivity in the sense of the advection-dispersion model were evaluated from the residence time distribution data. In most cases the dispersion was found to large or very large with Peclet numbers smaller than 4 or even much below 1 in many pathways. The frequency and average length of channels, in the sense that the water pathways make up a channel network was evaluated in the 3D- and in the Tracer migration experiment. This is a new model concept which has evolved based on the observations in primarily the 3D experiment. A channel network has flow and transport properties which differs from the-porous medium concept in several important respects. The network concept can better explain several observations in Stripa than the porous medium concept which usually is described by the advection dispersion model. A general observation is that a small fraction of the visible fractures carry most of the water.

IMPACT ON RADIONUCLIDE MIGRATION

Introduction

Radionuclides which escape from a repository in crystalline rock will be transported by the mobile water. The radionuclides decay and the longer their residence time is the more of the nuclide will decay. Most nuclides interact with the rock by sorption and will be retarded in relation to the water velocity. The retardation effects can be very large and the nuclides may move many orders of magnitude slower than the water. Even nonsorbing nuclides will be retarded because they can diffuse in the porous rock matrix and thus access a large volume of stagnant water. The added residence time while the nuclide resides in the matrix water may be several orders of magnitude larger than the residence time of the water flowing in the fractures. For a given flowrate of water the velocity and thus the residence time for sorbing species is not noticeably influenced by the water velocity. It is determined by the amount of sorption sites the nuclide has access to. The sorbing nuclides which diffuse into the matrix will have very large inner surfaces on which to sorb in addition to the fracture surfaces. The rate of uptake into the matrix will be influenced by the diffusivity, the sorption coefficient and in addition by the flow wetted surface. The latter is the fracture surface which the flowing water contacts as it flows through the rock. The larger this surface is the more of the nuclide can be "soaked" up by the rock matrix.

Water which flows in the fracture network in the rock has a residence time distribution. A part of the water moves faster than the mean velocity. The decaying nuclides which are in the faster water will have less time to decay. The residence time distribution is determined by the mixing processes along the flowpaths. When mixing is frequent of the sub–pathways of the flowing water the residence time distribution is narrow and centered around the mean residence time. If mixing is infrequent and if there are large velocity differences between the sub-pathways then a noticeable fraction of the water can move much faster than the mean. Infrequent mixing can occur when isolated flow paths, channels, develop in the network of fractures. Distinct channeling is found both in individual fractures and in networks of fractures. Below we illustrate by some very simple examples the impact of matrix diffusion and channeling. The illustrative examples are based on data and observations from Stripa.

Matrix diffusion

Water which flows in the channel is in contact with the surfaces of the channel and nuclides dissolved in the water will sorb on this surface by surface sorption and also diffuse into the rock matrix. The residence time of the water is the volume of the channel (capacity of the system to hold water) divided by the flow rate. The residence time of a nuclide is in addition determined by the capacity of the system to hold nuclide in the stagnant water in the matrix and sorbed on the inner surfaces in the rock matrix. Some nuclides will also sorb on the surfaces of the channel. The rate of uptake into the rock matrix is directly proportional to the flow wetted surface at early times when the penetration depth is small and the diffusion is directed perpendicularly to the surface. At longer times when the penetration distance is larger than the channel width the diffusion will more resemble radial diffusion from a circular hole. The channel geometry will have some impact on the uptake of the nuclides but in relation to other entities this is not one of the most important issues. Strongly sorbing nuclides will not penetrate more than a fraction of the expected distance between channels. Typical penetration depths are on the order of 10:s of centimeters for contact times of hundreds of thousands of years.

The volume of rock accessed by diffusion depends on the time the flow wetted surface is exposed to the species. The retardation will also be time dependent and the notion of a constant retardation factor is thus misleading.

For a single channel with no dispersion the effluent concentration from a channel with a flow rate Q, width W and length L for a species that is supplied to the inlet with a constant concentration C_0 which can be written as follows

$$\frac{C}{C_0} = \mathrm{Erfc}\left[\frac{LW}{Q}\sqrt{\frac{D_e K_d \rho}{t - t_w}}\right] \quad (1)$$

Equation (1) illustrates that the flow wetted surface, L*W, per flow rate, Q, in the channel plays a dominating role. The time since the species started to flow into the system t, the effective diffusion coefficient D_e and the sorption coefficient $K_d\rho$ also influence the results but less so because their values are raised to the power 1/2. Because the water residence time $t_w \ll t$ this entity has negligible influence for sorbing species and long exposure times.

The example has served to illustrate that matrix diffusion can retard nuclides very much, that the effect is time dependent and that the flow wetted surface plays a important role.

Channeling

The above discussion has highlighted the entities which are important for matrix diffusion. Dispersion effects were consciously not discussed and equation (1) does not account for it. Dispersion is often conceived as being caused by frequent mixing of the different fluid parcels with somewhat different velocities. This results in a Gaussian residence time distribution and is often described by the Advection–Dispersion equation where it is inherently assumed that a dispersion length α can be found which is constant. Compilation of field and laboratory experiments (Neretnieks, 1985) together with more recent results (Abelin et al., 1991a,b; Birgersson et al., 1992; Shapiro, 1988) show that the dispersion length increases proportionally to the distance between injection and monitoring points. Figure 11 shows this relation.

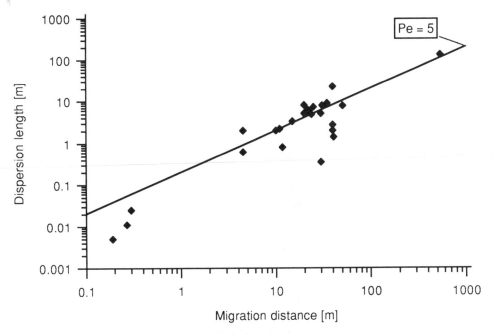

Figure 11. **Dispersion lengths versus distance between injection and observation points in the experiments.**

The observed effect is expected if there is a number of independent flow paths, channels, in the rock (Neretnieks, 1983). It has the consequence that in the residence time distribution of the water there is always the same fraction of water which moves with a constant percentage of the mean water velocity. This is in contrast to the hydrodynamics dispersion theory which implies that the residence time distribution must become narrower relative to the mean residence time. It will broaden in absolute values though. This may have a very large impact on the decay of radionuclides. To give an example, if a nuclide has a half life of 30 years, it will have decayed to 10^{-9} in 900 years. If 1 % of the nuclide moves with 3 times the average velocity this fraction has only 10 half lives in which to decay. It will decay to 10^{-3} only and the total outlet concentration will be 10^{-5} which is 10 000 times higher than what nuclide transported with the mean residence time.

In the chosen example the fast fraction of water is realistic for the Stripa rock judging from the obtained breakthrough curves. Because of the strongly nonlinear effects sorbing nuclides will broaden the residence time distribution very much in a channeling environment. Larger fractions of the nuclides will move faster than the mean residence time of the nuclides. A larger fraction may then have less time to decay. These effects have been demonstrated earlier

(Neretnieks 1983). Recently Moreno and Neretnieks (1991) have shown that if the channels are connected in a channel network with intersections and mixing then the residence time broadening effect may decrease considerably.

DISCUSSION AND CONCLUSIONS

Matrix diffusion is one of the dominating mechanisms which can retard radionuclides which move with the flowing water. The experiments at Stripa have shown that matrix diffusion takes place in granitic rock at repository depth. The magnitude of the flow wetted surface from which the nuclides are taken up into the rock is another important entity. The 3-D and Tracer Migration Experiments in the Validation drift indicate that the flow wetted surface may vary in a wide range depending on what assumption are made in the interpretations. The Channeling experiment together with other observations indicate that the lower values i.e. considerably less than 1 m^2/m^3 rock are more probable. Even quite low values can give a substantial retardation of sorbing species.

Fast long pathways have been directly observed at Stripa. This indicates that more or less isolated extensive channels are present in the rock which may transport a fraction of the nuclides faster than the average mass. The structure and the properties of the channel network is not known. Theoretical considerations indicate that some conceivable network structures may have very different transport properties. This is still an area which must be further investigated.

It might be discussed whether or not the obtained results are site specific. Effects of channeling and matrix diffusion have been observed in a number of investigations and have to be accounted for in all crystalline fractured rocks and not only in Stripa granite. The actual values of flow and transport properties are more likely to be site specific although estimations of the properties can be given depending on e.g. type of rock, rock stresses and fracture density.

In performance assessments for nuclear waste repositories there is a need to model regions and distances which are considerably larger than what time and effort has allowed us to study. It must be recognized that the question of scaling and extrapolating to larger distances has not been a question addressed in the performed investigations. It is still very much an open question.

It is known that for sorbing substances mechanisms are active which cannot be investigated by hydraulic testing and the use of non-sorbing tracers. For example, the magnitude of the specific flow wetted surface which dominates the migration of sorbing species has only been indirectly inferred and must be further studied with other experiments and observations.

REFERENCES

Abelin H., J. Gidlund and I. Neretnieks, Migration in a single fracture, Scientific Basis for Nuclear Waste Management V, p 529, North-Holland 1982.

Abelin H., I. Neretnieks, S. Tunbrant and L. Moreno, Final report of the migration in a single fracture - Experimental results and evaluation, Stripa Project Report 85-03, OECD/NEA, SKB, Stockholm 1985.

Abelin H., L. Birgersson, J. Gidlund, L. Moreno, I. Neretnieks, H. Widén and T. Ågren, 3-D migration experiment, Report 3, Part I, Performed experiments, Results and Evaluation, Stripa Project Report 87-21, Stockholm 1987.

Abelin H., L. Birgersson, H. Widén, T. Ågren, L. Moreno and I. Neretnieks, Channeling experiments, Stripa Project Report 90-13, Stockholm 1990.

Abelin H., L. Birgersson, J. Giglund, L. Moreno and I. Neretnieks, A Large-Scale Flow and Tracer Experiment in Granite 1. Experimental Design and Flow Distribution, Water Resources Res., 27, p 3107-3117, 1991a.

Abelin H., L. Birgersson, L. Moreno, H. Widén, T. Ågren and I. Neretnieks, A Large-Scale Flow and Tracer Experiment in Granite 2. Results and Interpretation, Water Resources Res., 27, p 3119-3135, 1991b.

Birgersson L. and I. Neretnieks, Diffusion in the matrix of granitic rock. Field test in the Stripa mine. Part 1, SKBF/KBS Technical Report 82-08, Stockholm 1982.

Birgersson L. and I. Neretnieks, Diffusion in the matrix of granitic rock. Field test in the Stripa mine. Part 2, SKBF/KBS Technical Report 83-39, Stockholm 1983.

Birgersson L. and I. Neretnieks, Diffusion in the matrix of granitic rock: Field test in the Stripa mine, Scientific Basis for Nuclear Waste Management VII, Mater. Res. Soc. Symp. Proc., 26, 247, 1984.

Birgersson L. and I. Neretnieks, Diffusion in the matrix of granitic rock: Field test in the Stripa mine, Scientific Basis for Nuclear Waste Management XI, Mater. Res. Soc. Symp. Proc., 112, 189, 1988.

Birgersson L. and I. Neretnieks, Diffusion in the matrix of granitic rock: Field test in the Stripa mine, Water Resources Res., 26, p 2833-2842, 1990.

Birgersson L., Diffusion in the matrix of granitic rock. Field test in the Stripa mine, M.S. Thesis Dep. Chemical Engineering, Royal Institute of Technology, Stockholm, Sweden, May 1988.

Birgersson L., H. Widén, T. Ågren, I. Neretnieks and L. Moreno, Site Characterization and Validation - Tracer Migration Experiment in the Validation Drift, Report 2, Part 1: Performed Experiments, Results and Evaluation, Stripa Project Report 92-03, Stockholm 1992a.

Birgersson L., H. Widén, T. Ågren and I. Neretnieks, Tracer Migration Experiments in the Stripa Mine 1980-1991, Stripa Project Report 92-25, Stockholm 1992b.

Bradbury M.H., D. Lever and D. Kinsey, Aqueous phase diffusion in crystalline rock, Scientific Basis for Nuclear Waste Management V, Ed. W. Lutze, Elsevier, 1982, p 569-578.

Bradbury M.H. and A. Green, Measurements of important parameters determining aqueous phase diffusion rates through crystalline rock matrices, J. Hydrology 82, 1985, p 39-55.

Bradbury M.H. and A. Green, Investigations into factors influencing long range matrix diffusion rates and pore space accessibility at depth in granite, J. Hydrology 89, 1986, p 123-139.

Moreno L., Y.W. Tsang, C.F. Tsang, F.V. Hale and I. Neretnieks, Flow and tracer transport in a single fracture. A stochastic model and its relation to some field observations, Water Resources Research, 24, p 2033-2048, 1988.

Moreno L. and I. Neretnieks, Fluid Flow and Solute Transport in a Network of Channels, submitted to Water Resources Res. 1991.

NAGRA projekt Gewähr 1985, Endlager für Hochaktive Abfälle: Das System der Sicherheitsbarrieren, Jan 1985.

Neretnieks I., Diffusion in the Rock Matrix: An important Factor in Radionuclide Retardation? J. Geophys. Res., 1980, 85, p 4379.

Neretnieks I., A note on fracture flow mechanisms in the ground, Water Resources Res., 19, p 364-370, 1983.

Neretnieks I., Transport in Fractured Rocks. Paper at IAH 17th congress, Tucson Arizona, Jan 1985, Memoires vol XVII, part 2, Proceedings, p 301-318, 1985.

Neretnieks I., Channeling in crystalline rocks. Its possible impact on transport of radionuclides from a repository, Colloque international Impact de la physico-chimie sur l'étude, la conception et l'optimisation des procédes en milieu poreux naturel, Nancy; 10-12 Juin 1987a.

Neretnieks I., Channeling effects in flow and transport in fractured rocks - Some recent observations and models, Paper presented at GEOVAL symposium, Stockholm, Proceedings, p 315-335, 1987b.

Olsson O. (editor), Site Characterization and Validation - Final Report, Stripa Project Report 92-22, Stockholm 1992.

Shapiro A.M., Interpretation of Tracer Tests Conducted in an Areally Extensive Fracture in Northeastern Illinois, Symposium Proceedings of International Conference on Fluid Flow in Fractured Rocks, Georgia State University, Atlanta, Georgia, May 15-18, p 12-22, 1988.

Skagius K., G. Svedberg and I. Neretnieks, A study of strontium and cesium sorption on granite, Nuclear Technology, 59, 1982, p 302-313.

Skagius K. and I. Neretnieks, Porosities and diffusivities of some nonsorbing species in crystalline rocks, Water Resources Res., 22, p 389, 1986a.

Skagius K. and I. Neretnieks, Diffusivity measurements and electrical resistivity measurements in rock samples under mechanical stress, IBID 22, p 570, 1986b.

Skagius K. and I. Neretnieks, Measurements of cesium and strontium diffusion in biotite gneiss, Water Resources Res., 24, p 75-84, 1988.

Tsang Y.W., C.F. Tsang, I. Neretnieks and L. Moreno, Flow and tracer transport in fractured media - A variable aperture channel model and its properties, Water Resources Res., 24, p 2049-2060, 1988.

SESSION II

NATURAL BARRIERS - MODELLING

SÉANCE II

BARRIÈRES NATURELLES - MODÉLISATION

Chairmen - Présidents
K. Dormuth (Canada)
B. Levich (United States)

Stripa Project
Modelling in the Site Characterization
and Validation Program

Gunnar Gustafson
Chalmers University of Technology
(Sweden)

Abstract

In the Site Characterization and Validation (SCV) project a selected volume of the Stripa granite was characterized geohydrologically by methods developed in the Stripa project. This formed a basis for a conceptual geohydrological model, that was refined stepwise. In this context Discrete Fracture Flow Models were used to predict ground-water inflow and transport of solutes to an excavated drift in the SCV-block. Model calculations were compared to experimental results in the drift and through a formalized process, involving judgements from a peer group,the validity of the numerical models was evaluated.

Introduction

This paper gives a short introduction to the numerical modelling and the evaluation of the validity of these models in the Site Characterization and Validation (SCV) project. All modelling work and the evaluation of the models are presented on their own at this conference, and thus here only a short decription of the different activities and the philosophy behind is given.

The main objective for the SCV-project is "To integrate different tools and methods in order to predict and validate the ground- water flow and transport in a specific volume of the Stripa granite." An important part of the project is thus to develop ground-water flow and solute transport models for fractured rock, and to evaluate the validity of such models at the selected site in the Stripa Mine. To support this modelling an advanced characterization methodology was developed, that resulted in a conceptual model of the site that was used as a basis for the different modelling approaches. The specific purposes of the modelling effort were thus:

- to establish a formalized process for evaluating the validity of ground-water flow and transport models.

- to develop fracture-flow models and to evaluate their validity, along with the validity of the equivalent-porous-media approach, for simulating ground-water flow and transport at the SCV Site in the Stripa Mine.

Principal investigators heading the fracture-flow modelling groups were:

- William Dershowitz (USA: Golder Associates Inc)

- Alan Herbert (UK: AEA/Harwell Laboratory)

- Jane Long (USA: Lawrence Berkeley Laboratory)

These groups all had different approaches to fracture flow modelling, with somewhat different concepts, different approximations etc. In my opinion this was a significant strength of the project that made it to a broad test of the state of the art.

The reference equivalent porous-media modelling was made by John Gale (Canada, Fracflow Consultants Inc) who also was a principal investigator for characterization together with John Black (UK: Golder Associates Ltd) and Olle Ohlsson (Sweden, Conterra AB)

The Task Force on Fracture flow modelling

Three modelling groups were thus to interact with each other and with the characterization group, and upon this the validity of the models was to be evaluated. Several therefore felt the need for some peer group to review and guide the modelling work and in June 1988 the jJoint Technical Commitee (JTC) of the Stripa Project endorsed a task force on fracture flow modelling with the following charter:

-        to guide and review the development of numerical modelling of fracture
         flow within the Stripa Project.

-        to recommend criteria for the verification and validation of the
         numerical models.

-        to report on annual basis to the TSG and the JTC concerning the
         status and future conduct of such activities.

An initial meeting was held before the endorsement in February 1988 in Rutherford Ca. wich can be seen as the startpoint of the work.

Chairman of the Task Force was Paul Gnirk (USA: RE/SPEC) and delegates from the different countries were:

-        Canada: Tin Chan (AECL Research)

-        Finland: Veikko Taivassalo (Technical Research Center)

-        Japan: Yuozo Ohnishi (Kyoto University)

-        Sweden: Gunnar Gustafson (Chalmers University of Tech.)

-        Switzerland: Stratis Vomvoris (NAGRA)

-        United Kingdom:David Hodgkinson (Intera Sciences)

-        USA: George Barr (Sandia National Laboratories)

When the Task Force started to work we also decided, that except for the fulfilment of the demands of the charter we would try to be very open with the progress of the work, or lack of it. The reason was simply that we found out that very few similar projects had been carried through, and that the way to a possible failure or success would be as important to make known as the result. This is the reason why the minutes of all Task Force meetings are

197

published in the Stripa series[1]. David Hodgkinson acted as secretary and editor of them and it is to him we owe the thanks for the excellent documentation.

Model Validation and Validation Criteria

The approach was to work very much in parallell with the validation process and the modelling. This gave a very useful interaction between thoughts and ideas about validation and the realities of three major modelling projects. Furthermore it was not all that clear at the start what validation meant, and different views and opinions on validation existed in the task force. In the end, on the basis of IAEA[2] the following definition of validation was adopted:

-   A model is considered to be validated for use in a <u>given application</u> when the model has been determined by <u>appropriate measures</u> to provide a a representation of the process or system that is acceptable to an assembled <u>group of knowledgeable experts</u> for purposes of the application.

The given application is in this case ground-water flow and transport in the SCV Site. By appropriate measures we mean comparison of model calculations against field observations and/or data from experimental measurements. The group of experts is in this case the Task Force.

This definition lead to a set of validation criteria very much in the spirit of confidence building and applicability assessment of the used methods rather than a set of quantitative performance measures. After a thorough discussion between the Task Force and the Principal Investigators and at least one turn to the JTC the following formal criteria were adopted:

<u>MODEL VALIDATION</u>
<u>CRITERIA FOR EVALUATION OF MODEL VALIDITY</u>

<u>Task:</u> Evaluate (1) the validity of an approach for modelling ground-water flow and solute transport in the saturated, fractured rock mass of the SCV Site, and (2) the validity of the components of the modelling approach.

<u>First set of criteria</u> (as referenced to the Performance Measures)

-   <u>Quantitative:</u> Are the quantitative predictions of the correct order of magnitude as compared to the measurements?

-   <u>Qualitative:</u> Are the quantitative predictions of the "distribution patterns" reasonable when compared qualitatively to the observations?

Second set of criteria (from viewpoint of general applicability)

- Usefulness: Is the modelling approach useful for representing ground-water flow and solute transport in a geohydrologic environment similar to that of the SCV Site?

- Feasability: Can the characterization data required to fully support the modelling approach be collected in a feasible and timely manner?

Modelling Steps

When the modelling work started the different codes were in various states of development. At the first Task Force meeting in 1988 a strategy for the work was laid out. In short it said that we initially should concentrate on the flow problem and when this was achieved go on with transport modelling. This also complied with the experiment program. The idea was also to start with one or more training excercises before the validation excercise of each problem type. The comparisons, on which the validation judgements were based, should be made between predictive model calculations and experiment results not known to the modellers. This emphasized the comparability between experiment and calculations and one thing I have learnt from this project is that an agreed format for predictions and measurements simplifies a work of this kind a lot, but it may be hard to achieve despite a very good will from all involved.

The development of the conceptual model fo the SCV Site involved 5 stages from the preliminary site characterization to the detaled evaluation and validation. In concert with this the following modelling and validation exercises were carried through:

EVALUATION OF GROUND-WATER FLOW MODELS

- STEP 1 (TRAINING EXERCISE)
  Predict: (1) Rate of ground-water inflow to the D-boreholes at the SCV Site
  (2) Distribution of ground-water flow along the D-boreholes

- STEP 2 (VALIDATION EXERCISE)
  Predict: (1) Response of the ground-water head distribution in the SCV site to construction of the Validation Drift
  (2) Total rate and distribution of ground-water inflow into the Validation Drift
  (3) Rates of inflow to the Validation Drift from significant fracture zones

EVALUATION OF SOLUTE TRANSPORT MODELS

- ## STEP 1 (TRAINING EXERCISE)

  Predict: (1) Breakthrough curve for Saline Tracer-Transport Test No. 2 to Validation Drift
  (2) Saline breakthrough curves for boreholes T1 and T2 in the H-zone
  (3) Histograms of breakthrough concentrations

- ## STEP 2 (VALIDATION EXERCISE)

  Predict (1) Tracer concentrations and arrival times for the Tracer Test in the Validation Drift
  (2) Pattern of distribution of tracer arrivals in the Validation Drift
  (3) Tracer concentrations and arrival times for injection in borehole T2 and collection in borehole T1

For each of these steps model calculations by the three modelling groups were performed as well as the reference calculation with the EPM-model. Calculations and experimental results were then compared. These comparisons are reported in two separate reports[3,4]. Based on the comparisons a judgement on the validity of the models was given by the Task Force.

Some Conclusions

During the progress of the project we have seen an impressive development of discrete fracture flow modelling. Today it is quite clear that these models are versatile, useful tools for describing ground-water flow and transport of solutes in a fractured crystalline rock. We have also found that the rock in Stripa can be effectively characterized with existing methods in order to build a conceptual model as a basis for the numerical modelling.

The validation process applied in the project is of course somewhat artificial. In the long run the normal scientific process will show if we were right or not. However, we found it possible to follow a formalized process for evaluating the validity of the models. This process requires a close coordination between the modellers , the experimentalists and the peer group, that validation criteria are established in the context of the application of the model and if practicable that experiment results are unknown prior to the predictions.

References

[1] D Hodgkinson, 1992:"A Compilation of Minutes for the Stripa Task Force on Fracture Flow Modelling", Stripa Project 92-09, SKB, Stockholm

[2] IAEA , 1982:"IAEA - TECDOC-264"

[3] D Hodgkinson, N Cooper (1992):"A Comparison of Measurements and Calculations for the Stripa Validation Drift Inflow Experiment", Stripa Project 92-07, SKB, Stockholm

[4] D Hodgkinson, N Cooper (1992):"A Comparison of Measurements and Calculations for the Stripa Tracer Experiments", Stripa Project 92-20, SKB, Stockholm

Equivalent Porous Media Modelling of Flow and Transport in the Granitic Rocks at Stripa

J. Gale, R. MacLeod, G. Bursey
Flacflow Consultants Inc., St.John's, Nfld, Canada

A. Herbert
AEA Decommissioning and Radwaste, Harwell, United Kingdom

Abstract

The 3-D groundwater flow and transport finite element code, CFEST, was used in four levels of nested models to simulate flow and transport to both boreholes and drifts in the Stripa granite. The 3-D discrete fracture code, NAPSAC, was used to integrate the fracture and hydraulic data for the large scale equivalent-porous-media finite element model input parameters. Model results agreed well with the simulated drift experiment inflows and the tracer test field data. The computed flowpaths are consistent with the geochemical-isotopic interpretations of the local groundwater flow system. The coupled discrete fracture - porous media approach used in this study provides a powerful tool for modelling flow and transport in large fractured rock masses.

Background

The main objectives of the 3-D porous media finite element modelling using the CFEST code (Gupta et al., 1986) were to determine how well a large scale, porous media, model could predict (1) the observed pattern of hydraulic heads in the Site Characterization and Validation (SCV) site, (2) the flux measured during the Simulated Drift Experiment (SDE), and (3) the flux into the Validation Drift (Olsson et al., 1992). In addition, the large scale porous media model simulations were designed to provide the hydraulic head boundary conditions for the 3-D discrete fracture network modelling of flux to the D boreholes in the SDE and to the Validation drift. The porous media model also provided a means of incorporating the equivalent porous media properties, that were generated using the 3-D discrete fracture code NAPSAC (Herbert et al., 1991), into a larger scale model that included a reasonable approximation of the far-field hydraulic head boundary conditions.

Boundary conditions for the detailed borehole and drift models were developed from the regional flow models (Figure 1) resulting in four levels of nested models. The structural framework for the site was developed from the regional geological data (Lundstrom, 1983) and data collected during detailed core drilling. Based on these data, the rock mass was separated into major fracture zones and averagely fractured rock (Olsson, et al., 1992). For the Phase III study, three finite element meshes, MINE, MINE2, and SCV, were used for the equivalent porous media modelling of flow and transport to the D boreholes (SDE) and the Validation Drift (Herbert et al., 1991). The MINE2 mesh (Figure 2A) was similar to and developed from the MINE mesh to accommodate changes in the location and orientation of the major fracture zones as the SCV Project progressed. The SCV mesh (Figure 2B) was developed (1) to allow a better approximation of the inclined nature and horizontal and vertical interconnection of the major fracture zones in the SCV block for the tracer modelling experiments and (2) to provide finite elements that were on a scale consistent with the size of the simulation volume that could be used by NAPSAC to generate equivalent porous media properties. In addition, the detail provided by the small finite elements in the SCV model were required for flowpath and tracer simulations (MacLeod et al., 1991). The relevant details of the boundary conditions and the input parameters for the MINE2 and SCV models are summarized below and the details are presented in Herbert et al., 1991, and MacLeod et al., 1991.

Description of Porous Media Models

The MINE2 model, approximately 1 km^2 in surface area, is centred over the eastern half of the Stripa mine. This model consists of fourteen layers (Figure 2A) of varying thickness and extends to a depth of 600 m below ground surface. Four fracture zones have been included in the MINE2 model (Figure 2A). These include an east-west fracture zone located in the southwest corner of the model, the H zone, I zone, and the combined A-B zone. All of the H, I, and part of the A-B fracture zones are represented by narrow vertical elements, with a width of 5 m. The hydraulic conductivity of the fracture zones was adjusted to reflect the measured transmissibility for the assumed thickness of the actual fracture zone.

The held head conditions for the nodes on the vertical sides as well as the bottom of the MINE2 model at 600 m of depth were extracted from the hydraulic heads computed for the Sub-Region model, outlined in Figure 1, (Gale et al., 1987), that simulated the large scale hydraulic sink effects of the Stripa mine. Internal nodes, located on mine openings, were assigned hydraulic head values equal to the elevation head at each node. Hydraulic conductivities were assigned to each layer using the depth versus permeability relationship given in Gale et al., 1987.

The mesh for the SCV model, 0.36 km^2 in surface area, is given in Figure 2B. This model focuses on the vertical section that lies between the 310 m mine level and 490 m of depth. This section has been divided into twelve layers within and ten layers outside the immediate Validation Drift area. The four middle layers in the immediate Validation Drift area have been assigned a thickness of 2.5 m. The hydraulic head boundary conditions for the SCV model were extracted from the MINE2 model. The SCV model simulations were conducted to show the effects of the smaller element size, in the SCV model, on the flux into nodes representing the D boreholes and the Validation Drift.

The initial porosity and permeability values assigned to the MINE model were those developed from the pre-1986 data set for the entire mine area and used in the earlier regional flow modelling work (Gale et al, 1987). The MINE models were used to estimate or predict the flux for the SDE using very little site specific data. Actual predictions for the SDE and the Validation Drift were made using the MINE2 model. Porosity and permeability values for both the average rock and the fracture zones in the MINE2 and SCV models reflect the additional measurements made during the SCV Project, including those made during the detailed borehole hydraulic tests, both steady state, transient and interference tests. The MINE2 model mesh incorporated the most recent geometric data for the fracture zones in the SCV site. This included changing the permeability values of selected elements within the MINE2 model in order to examine the different ideas on distribution and interconnection of high permeability zones within and adjacent to the SCV block.

Comparison of Measured and Computed Heads

The initial head distribution before the opening of the D boreholes was computed using the MINE2 model with input parameters based first on the raw field data and second on the permeability tensor data from NAPSAC (Herbert et al., 1991). The heads computed using the MINE2 model with the NAPSAC tensor data are higher than the initial set of measured heads (Herbert et al., 1991), with the difference having a mean of 20.67 m with a standard deviation of 51.02 m, assuming a normal distribution. For this model the I and H zones are vertical and are continuous only between the 600 m and the 280 m levels. However, the A-B zone extends in a stair-stepping fashion to the surface. For each of the different MINE2 models, the highest computed hydraulic heads occur when the fracture zones are extended to the surface.

It should be noted that the initial set of measured heads (Herbert et al., 1991) are not steady state values while the computed heads are steady state values. Also, the computed heads have not been corrected for both drift excavation effects or for the difference produced by comparing point values computed using the model to the average heads for the packed-off interval lengths in the boreholes. Thus, one would expect the computed heads to be higher than the measured heads as is the case in this model. It should be noted that using the raw field values for the hydraulic properties, but with the same distribution of large fracture zones, gave a much better fit between the measured and computed heads (Herbert et al., 1991).

The predicted drawdowns for the SDE are those computed using the MINE2 model and the same input properties used to determine the initial heads. Good agreement was obtained between the measured and computed drawdowns (mean of 1.0 m and a standard deviation of 24.45 m) using the same model with the heads on the nodes, representing the D boreholes, equal to elevation heads. The differences between the measured and computed drawdowns (CASE18P) have a mean of 3.63 m and a standard deviation of 23.76 m (Herbert et al., 1991) when the elevation head boundary conditions on the SDE nodes equal 17 m, as imposed during the actual outflow tests. The drawdowns predicted for the Validation Drift model (CASE18S) agree very well with the heads measured during the SDE. The set of head data, measured after the Validation drift was excavated, are in reasonably good agreement with the model results, except near the drift itself.

When the initial set of measured heads were extrapolated to steady state conditions (Olsson et al., 1992), the CASE18P SDE model (Figure 3) gave a mean of 13.7 m and standard deviation of 22.2 m for the differences between the measured and computed drawdowns. In Figure 3, the drawdowns in metres are shown as small vertical columns in the approximate locations of the mid-point of the packer intervals in which the actual pressures were measured.

In summary, the MINE2 and SCV models have produced close agreement between the measured and computed initial pressure heads and close agreement between the measured and computed SDE drawdowns. However, the best agreement between the measured and computed pressure heads and drawdowns were found for the models in which the I and H fracture zones were limited in vertical extent. Since this conflicts with the groundwater geochemistry, that suggests relatively rapid movement of surface waters to the SCV site, we must assume that the hydraulic properties of the fracture zones are variable in the vertical direction as observed during hydraulic testing in the SCV site (Holmes, 1989). Given the scale of the porous media models and the uncertainty in the value of the measured steady state hydraulic heads, the models are considered to provide a good representation of both the total hydraulic heads and the drawdowns over the entire SCV block, except near the drift walls.

Predicted Flowrates for the SDE and Validation Drift

The predicted flowrates (Olsson et al., 1992) for the SDE and the Validation drift are tabulated in Table 1. A detailed discussion of each of the models referenced in Table 1 is given in Herbert et al., 1991. Figure 4 shows a schematic of the fracture zones within the SCV block and the computed flowrates for selected permeability distributions and fracture zone connections, relative to the measured SDE inflow for the first 50 m of the D boreholes. In all three cases shown, the I and H fracture zones extend from the 600 m level to the 280 m level. The A-B fracture zone extends from the base of the model to the surface. The CASE16 model includes several high permeability zones that were identified during the initial borehole injection testing program (Holmes 1989). The flux to each node is given in Figure 5 for the full 100 m of the SDE. The ellipse formed by the D boreholes and the Validation Drift has an approximate well radius of 1.33 m. The nodes simulating the D boreholes and the Validation Drift in the MINE and MINE2 model have an effective well radius of 2.5 m (Trescott et al., 1974). Similarly the D borehole and drift nodes in the SCV model have an effective well radius of 1.04 m. Hence, the computed fluxes have been corrected for the differences in effective well radius (Table 1).

The initial simulations using the MINE mesh, CASE1, and CASE2 models gave reasonable estimates (Table 1) of the measured flux to the D boreholes. The MINE model, which was based on the Stage 1 hydraulic data (Olsson et al., 1992), predicted a flux of 3.169 l/min for the SDE. It should be noted that when this flux is corrected for the difference between the drift size and the effective node (well) size, the flux predicted for the SDE is 2.619 l/min, less than a factor of 2 more than the measured flux.

The 3-D discrete fracture code NAPSAC (Herbert et al., 1991) was used to integrate the measured fracture geometry and borehole hydraulic data and generate the permeability tensor values for the elements in the large scale equivalent-porous-media finite element model. The permeability tensor values used in these in the porous media models were generated in the early part of the modelling study and are slightly lower than the final tensor values obtained from the network analysis (Herbert et al., 1991). Using these permeability data as input to the 3-D equivalent-porous-media CFEST code produced excellent agreement between the computed and measured flux results for the large scale simulated drift experiment.

A flux of 2.44 l/min (CASE18P) was predicted (Table 1) for the full 100 m of the D boreholes that were open during the SDE. The flux predicted for the first 50 m of the D boreholes in the SDE is 0.562 l/min. This represents close agreement between the measured and computed flux for the first 50 m of the SDE boreholes but a poorer agreement between the predicted and measured results for the full 100 m of the D boreholes. The lack of agreement between the measured and computed fluxes for flow to the full 100 m of the D boreholes indicates that the I and A-B fracture zones were not properly characterized and represented in the MINE2 and SCV models.

The flux predicted (Table 1) for the same 50 m of excavated drift, the Validation Drift, is 0.779 l/min (CASE18S) when the permeability of the elements bounding the drift nodes were corrected for the depth-stress effects (Herbert et al., 1991) using results from the stress modelling and the field measurements reported in Gale et al., 1987. Similar values (Table 1) for the flux to the D boreholes were calculated using the SCV model. The computed inflows to the drift were approximately a factor of eight greater than those measured (Olsson et al., 1992). The corrections for stress concentration effects around the Validation Drift produced an increase in flux to the drift rather than the expected decrease in flux.

The calculated fluxes for the additional MINE2 simulations used to examine the effects of connectedness of fracture zones are reported in Herbert et al., 1991. These flux calculations have not been corrected for the much larger effective well size that resulted from the high permeability elements that were located along the D-holes for both the SDE and the Validation drift simulations. The computed fluxes for the full 100 m and the first 50 m section of the D-holes (Herbert et al., 1991), when corrected for drift or borehole area of influence versus the effective volume influenced by each node along the D-holes, give flux values that are less than a factor of 5 greater than the measured flux values even when the high permeability zones

Table 1. Summary of predicted fluxes into the D-Holes (SDE) and the Validation drift for the MINE, MINE2 and SCV models. See Herbert et al., 1991, for description of each CASE.

MODEL	CASE	FLUX (L/Min)	
		COMPUTED	CORRECTED
MINE	CASE1	0.145	0.120
	CASE2	3.169	2.619
MINE2	CASE16		
	SDE 50 m	2.604	2.152
	SDE 100 m	5.034	4.160
	CASE18, kij		
	SDE 50 m	0.752	0.621
	SDE 100 m	3.152	2.605
	CASE18P		
	SDE 50 m	0.680	0.562
	SDE 100 m	2.952	2.440
	CASE18S		
	SDE 50 m	0.943	0.779
	SDE 100 m	3.333	2.755
SCV	CASE2		
	SDE 50 m	0.781	0.849
	SDE 100 m	1.998	2.171
	CASE2P		
	SDE 50 m	0.720	0.783
	SDE 100 m	1.871	2.034

are included.

In addition to computing the hydraulic heads and the flux to the boreholes and Validation drift, the hydraulic heads from the sub-region, MINE2 and SCV flow models were used to define specific pathways of groundwater flow across the Stripa mine site. Figure 6 shows a plan view of five pathlines computed using these models. The open circles show the recharge point or starting point for each flowline and the solid circles show the termination or discharge point for the pathline. Some of these pathlines could not be completed to their recharge origins at the surface or their discharge point in the mine because of numerical problems in defining the hydraulic gradients sub-parallel to the element boundaries. Pathways 1 and 2 illustrate the short-circuiting effect of the regional fracture systems on deep groundwater flow to the mine. Pathways 3, 4 and 5, however, bypass a regional fracture zone and reflect recharge near the tailings pond and discharge into the mine. The flow directions and pathways computed using the 3-D equivalent porous media models were consistent with the isotopic and geochemical data obtained from the boreholes drilled in the immediate site area.

Tracer Simulations

The SCV model was calibrated (MacLeod et al., 1991) to give the correct flux to the SDE and the D-holes plus the Validation drift (Figure 5) and to provide a reasonable match between the distribution of flux between the "average rock" and the individual fracture zones. The SCV-SDE model was then calibrated against the tracer data for the Saline I experiment to the D-holes and the Validation drift (excavation) model was calibrated against Saline II tracer experiment (Olsson et al., 1992). The initial porosity values used for calibrating the numerical model of the SCV site were those computed from field data (Gale et al., 1987). The porosity values were subsequently adjusted by multiplying the original estimate by factors that ranged from 1 to 4.75 in order to calibrate the model using the first and second saline tracer experiments. The initial values of longitudinal (D_L) and transverse (D_T) dispersion that were assigned in the transport simulations were obtained from the literature (Domenico and Schwartz, 1990; and Herbert et al., 1991).

The initial calibration of the SCV-SDE transport model consisted of obtaining as close a fit as possible between the D boreholes arrival times and the measured breakthrough curve for the first saline tracer test. Tracer breakthrough to the D boreholes was determined from concentrations at the nodes bordering on the H zone element which were used to simulate the collection interval for all of the D boreholes.

The first arrival of tracer in the D boreholes predicted by the model, using the initial porosity and dispersivities, was earlier than measured. However, the peak concentration C/C_o agreed very well with the measured values. The best match between the model results and the measured data for the first saline tracer test was provided by dispersivities of $D_L = 5$ and $D_T = 0.5$ m and porosities ranging from 0.00021 to 0.00038 for the "average rock" and from 0.00059 to 0.00074 for the fracture zones.

The Saline II tracer test was modelled by the SCV-Drift model using the final porosity and dispersivity values from the SCV-SDE Saline I model. It was found that these input properties predicted that tracer would arrive in the drift much earlier than was observed. This observation plus steeper slope on the breakthrough curve suggested that the dispersivities and/or porosities assigned were too small. It was determined that the simulation with the best overall match to the measured data was the simulation which used a dispersivity of $D_L = 1$ and $D_T = 0.1$ m and porosity of 4.75 times the originally assigned porosity ranging from 0.00048 to 0.00087 for the "average rock" and from 0.00130 to 0.00160 for the fracture zones. The small dispersivity length versus element size did not appear to produce any numerical problems in the simulation.

The validation phase of the porous media modelling exercise was to predict tracer movement to the drift from tracer injected into a series of packed off sections of individual boreholes. Constant concentrations of tracer were injected at low injection heads and low flow rates in each interval. The model node which most closely matched the coordinates of the injection interval was designated to represent the tracer injection point. Due to the scale of the SCV model and the close proximity of many of the actual injection intervals (< 5 m), some injection points (eg. T2-1 and T2-3, See Olsson et al., 1992) had to be represented by the same model node. The concentration of tracer and the injection flow rate were fixed at the node for a particular tracer test. However, using this approach a much larger mass of tracer was introduced into the rock mass due to the volume represented by the node compared to the actual injection point. Therefore, the measured concentration at the nodes had to be corrected by a dilution factor to account for the volume of flux through the node compared to the injection flux (MacLeod et al., 1991).

Tracer breakthrough curves for the Validation Drift experiments and steady state concentration values for each node along the simulated Validation Drift are given in MacLeod et al., 1991. The earliest arrivals and largest peak recovery were from tracers injected in the T1 borehole. This is probably due to the fact that borehole T1 is the closest borehole to the drift. Tracers injected in boreholes T1 and T2 and interval C3-1 all show early arrival of tracer and approach a peak "steady state" concentration within 400 to 500 hours. Tracer injected in interval C2-1 is slower to arrive at the drift, within the first 100 hours, and are present in low concentrations with the concentrations slowly approaching steady state concentrations after 4500 hours. Very little tracer from the C3-2 interval which is in "average rock" reaches the drift until 300 to 400 hours after injection and steady state concentrations are not predicted until after 4000 hours. Predictions of t_5 and t_{50}, time to reach 5% and 50% of the "steady state" breakthrough, and C_{ss}/C_0 the normalized steady state breakthrough concentrations for each test are given in Table 2.

Most of the tracer from the tests T1-1, T2-1, T2-3, and C3-1 was recovered from the H-zone. However, for the test with injection in C2-1, the tracer is only recovered in the three nodes representing the fist 20 metres of the drift and very little tracer is recorded in the H-Zone. The same is true for injection in interval C3-2, which is in the "average rock", where the bulk of the tracer is recorded in the last 20 m of the drift with very little intersecting the H-Zone. The actual measured tracer results are given in Olsson et al., 1992.

Summary

This study has shown that fracture geometry data can be collected from boreholes and drifts, analyzed and the resulting fracture statistics used to compute the average flow properties that are needed to build large scale flow and transport models. The computed heads, flowrates, except for the predicted drift inflows, and tracer travel times are in reasonable agreement with the field measurements. The computed flowpaths are consistent with the geochemical and isotopic interpretations of the groundwater flow system in the immediate Stripa mine area. This approach, in which the 3-D discrete fracture model is used to develop the properties for the larger scale model, provides a powerful tool for modelling flow in large fractured rock masses. Once the appropriate fracture porosity/velocity relationship has been determined, this approach will allow one to incorporate the spatial variability, anisotropy and unique flow and transport properties of discrete fractures into large scale 3-D flow and transport models of fractured rock masses.

References

Domenico, P. A. and Schwartz, F. W., 1990. Physical and chemical hydrogeology. John Wiley & Sons, New York, USA, 824 p.

Gale, J.E., MacLeod, R, Welhan, J., Cole, C. and Vail, L, 1987. Hydrogeological characterization of the Stripa site. Stripa Project TR 87-17, SKB, Stockholm, Sweden.

Gupta, S.K, Cole, C.R., Kincaid, CT. and Monti, A.M, 1986. Coupled: "Fluid, Energy, and Solute Transport (CFEST) Model: Formulation, Computer Source Listings, and User's Manual", Battelle Memorial Institute Technical Report BMI/OWNI XXXX, July 1986.

Herbert, A. W., Gale, J.E, Lanyon, G. W. and MacLeod, B., 1991. Modelling for the Stripa site characterization and Validation Drift inflow prediction of flow through fractured rock. Stripa Project TR 91-35, SKB, Stockholm, Sweden.

Holmes, D., 1989. Site characterization and validation - Single borehole hydraulic testing. Stripa Project TR 89 04, SKB, Stockholm, Sweden.

Lundstrom, I., 1983. Beskrivning till berggrundskartan Lindesberg SV. SGU Af 126, Geological Survey of Sweden, Uppsala, Sweden.

MacLeod, R, Gale, J. and Bursey, G., 1992. Site Characterization and Validation Porous media modelling of Validation Tracer Experiments. Stripa Project TR 92-10, SKB, Stockholm, Sweden.

Olsson, O. et al., 1992. Site Characterization and Validation Project -Final Report. Stripa Project TR 92-22, SKB, Stockholm, Sweden.

Trescott, P.E., Pinder, G.F. and Larson, S.P., 1976. Finite-difference model for aquifer simulation in two dimensions with results of numerical experiments. In Techniques of Water-Resources Investigations, US Geological Survey.

Table 2. **Results of individual tracer tests (after MacLeod et al., 1991)**

Injection Interval	C_{ss}/C_o *1000 corrected	T_5 hrs	T_{50} hrs
T1-2	0.14	50	150
T2-1	0.09	60	200
T2-3	0.16	60	200
C2-1	0.13	125	1100
C3-1	0.14	50	150
C3-2	0.17	250	1000

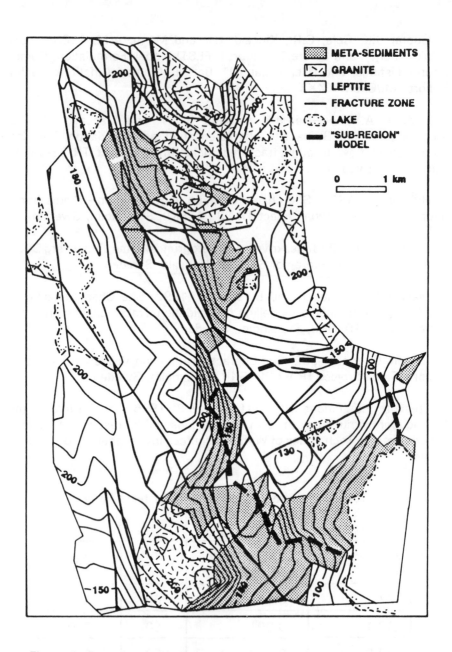

Figure 1. Boundary, simplified geology, and water table contours for the Stripa regional flow model. The boundary of the sub-region model is outlined in the lower right corner.

(A) **MINE 2 MODEL**

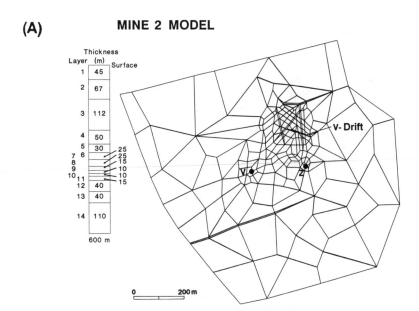

(B) **SCV MODEL**

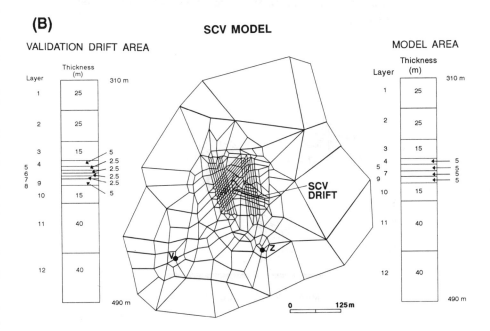

Figure 2. **A) MINE 2 model mesh showing layer assignments. B) SCV model mesh showing layer assignments for model area (10 layers) and Validation Drift area (12 layers).**

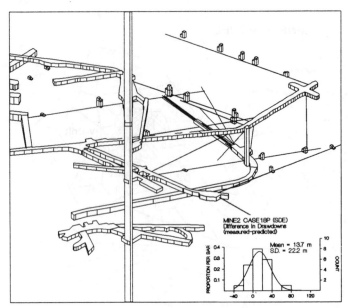

Figure 3. **Comparison of measured (left column) and computed (right column) drawdowns for the SDE. Pressure columns have 50 m sections.**

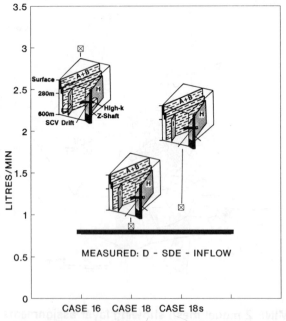

Figure 4. **Total flowrates predicted for SDE for different models using permeability tensor from NAPSAC.**

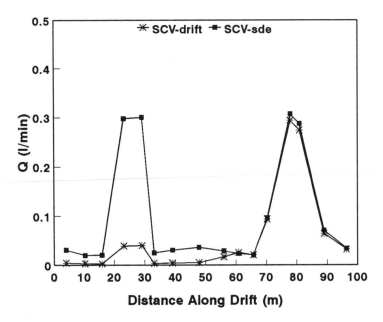

Figure 5. Distribution of flux by nodal point along the D boreholes and Validation Drift for the SCV Model simulation.

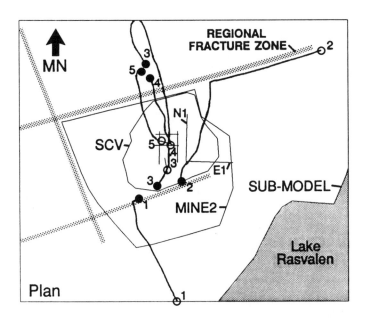

Figure 6. Plan of 5 pathlines, started from different locations, showing the potential sources (open circles) of groundwater and discharge points (filled circles) in the mine.

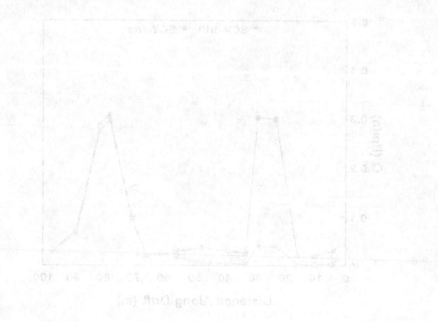

Discrete Fracture Network Modelling
for Phase 3 of the Stripa Project using NAPSAC

A.W. Herbert
AEA Decommissioning and Radwaste, United Kingdom

G.W. Lanyon
Geoscience Ltd.,

J.E. Gale and R. MacLeod
Flacflow Consultants Ltd., Canada

Abstract

During phase three of the Stripa project we have developed the discrete fracture network modelling approach for understanding flow and transport through fractured rocks. We describe the NAPSAC software package developed to incorporate this approach and how NAPSAC can be used in conjunction with convensional models as part of an integrated programme of site investigation and assessment. The usefulness and feasibility of the approach has been demonstrated by applying it to make successful predictions of a series of field experiments.

Introduction

The aim of our fracture network modelling is to improve our understanding of flow and transport through fractured rock systems. In fractured hard rocks such as Stripa granite the water flows through a discrete network of individual fractures. Conventional approaches to modelling such flow systems are based on averaging the flow and transport properties of the system over representative elementary volumes (REVs). The scale of these volumes must be large enough to be representative of the fracture network, and this can mean that the REV is difficult to characterise directly. Such approaches have been successful for flow predictions, but it is not clear that they always will be, nor is it clear at what scale such an approximation is valid for a given fracture network. Finally, effective-porous-media models have not succeeded in explaining nuclide transport in fractured rocks, where scale dependent dispersion length is observed experimentally. To improve our understanding and to support the use of such models on large scales, the fracture network approach represents the geometry of the flow system more directly. It is not feasible to characterise the fracture network directly, since we cannot measure the properties of all the fractures in a rock mass; rather we characterise the network stochastically, and simulate realisations of the fracture network statistics. This stochastic fracture network models cannot predict precise details of the flow field, but instead predict statistical properties of the flows. As described below, we show how the scale of a REV can be predicted, above which continuum approximations are justified. We make predictions of bulk properties of the rock, from direct measurements of the properties of individual fractures, and we make predictions of the distribution of flow and tracer recovery on more local scales. Thus our approach can explain all the qualitative features of flow and transport as seen in mines and borehole, and provide the parameters and justification required to use conventional models on larger scales.

In order to provide understanding of the flow system, we have adopted a 'forward modelling' approach. We make simple assumptions when constructing representations of the site and investigate the consequences of these straightforward models. In principle, fracture network models have many millions of adjustable parameters: the difficulty is not in representing physical complexity in the models, rather it is in characterising our models with the sometimes limited data. This was particularly the case with the disturbed zone around the validation drift, which we do not understand (although empirical changes to network properties near the drift can give good matches to the observed flows).

Throughout, we have endeavoured to incorporate directly in our models as much as possible of the experimental data gathered at Stripa.

Our approach is embodied in the NAPSAC computer code (Grindrod et al., 1992). This is a flexible fracture network code specifically designed for the Stripa project to investigate realistically large networks. It is restricted to fracture flow and currently to steady-state flow systems. The aim is to understand the effect of the geometry of the network on flow and transport: more complex physical processes can be better investigated with other tools. It can represent the detailed flow through networks of up to 50000 fractures, involving up to 200000000 finite elements when implemented on a Cray supercomputer.

Characterisation of the SCV site

A preliminary characterisation of the flow system at the SCV site was based on the data gathered in stage 2 of the SCV programme. This was used to generate the models used to predict the inflows in the SDE experiment. Subsequently, in SCV stage 4, further data was acquired and the conceptual model of the SCV site was updated. This new conceptual model formed the basis for our characterisation of the site for predictions of inflow to the Validation Drift, and for tracer transport models. We also repeated our predictions of SDE inflow with this improved model, with very consistent results for average rock flows. The main difference between these two characterisations was our ability to represent the H-zone as a fracture network in the later models, rather than approximating it as a single effective planar feature (as was done for the SDE predictions). Otherwise, the interpretation used the same approach and techniques. We describe the characterisation of fracture network models used in both these stages, and present a summary of the final network properties inferred. Detailed descriptions of the procedure can be found in Herbert and Splawski 1990, and in Herbert et al. 1992 respectively.

The conceptual model of the site presented previously divides the rock into two distinct categories: averagely fractured rock, and discrete feature or 'fracture zones'. Three distinct zones of importance cross the SCV site: in order of significance these are the H-zone, B-zone and I-zone. Of these only the H-zone intersects the drift, and this feature dominates the flow through the site. Fracture data gathered at the site was split into separate datasets for each of these types of rock. However, there was insufficient data to characterise the networks comprising the B- or I- zones, and so we had to represent these as simple planar features with a typical transmissivity. Fortunately, these did not intersect the validation drift. It does mean that in the prediction of D-hole fluxes, the flux from the end of the D-holes, where B- and I-zones intersect, is not predicted using the fracture network approach.

We characterised the average rock and H-zone fracture networks using the following approach.

First, the cores from characterisation boreholes were logged and the fracture orientation and frequencies were measured for all hydraulically active fractures. We represent all these fractures in our initial models, although, later we calculate that 30% of these fractures can be neglected as having too low a transmissivity to contribute more than 10% of the network permeability. The data was separated into average-rock and H-zone datasets, and cluster analysis used to identify a number of independent fracture sets, each characterised in terms of a distribution of dip and azimuth angles (see for example Gale et al. 1990).

Secondly, traces were measured along scanlines and area maps on tunnel walls around the SCV site. Their orientations were used to assign these traces to the fracture sets, and a log-normal distribution was fitted to the trace length datasets. In order to characterise the network models, we had to convert these trace length distributions into fracture length distributions. This is a difficult step, and in the general case there is no unique conversion. We again made a simple assumption that the fracture length distribution was also log-normal, and were then able to solve equations relating the moments of the trace length distribution to the parameters of the fracture length distribution for square fractures. We could now also infer a density of fracture centres, and chose to do this using the borehole frequency, rather than the scan line frequency.

Finally, to complete our characterisation of fracture properties, we need a distribution of hydraulic properties. Ideally one would measure a large number of single fracture transmissivities. However this was not practical for the densely fractured networks found at Stripa. The focussed hydraulic testing approach described in section 3.4 above resulted in a set of interval transmissivities together with a list of fractures intersecting each interval. It was not possible to identify separate hydraulic properties for the different fracture sets from this data, indeed there was no measurable correlation between the orientation of fractures and the transmissivity of their zones.We therefore assumed a single log-normal distribution of fracture transmissivities for all the fractures in each network. We could then use a maximum likelihood program to estimate the parameters of this distribution (estimating the likelihood numerically for the multiple fracture intervals).

The results of this characterisation are summarised in tables 1 and 2. Note that because we have chosen to take the fracture frequency from the density of hydraulically active fractures in the core logs, the traces are denser than those mapped in the mine. The network for the H-zone network in particular posed a difficult numerical problem for flow calculations. It is very intensely fractured and involves fractures with a wide range of length-scales, as illustrated in figure 1.

Modelling the Simulated Drift Experiment (SDE)

The characterisation of the SCV site used in the predictive models of this experiment during stage 3 of the SCV programme involved two simplifying approximations. First the zones were simply represented as planar features with a typical transmissivity inferred from average borehole test results. Secondly, the fracture data had not been partitioned between average rock and zone statistics: we simply truncated the distributions to avoid the influence of transmissive H-zone fractures in our average-rock fracture network models. The average rock network properties used were therefore slightly different from the final interpretation summarised above (Herbert and Splawski, 1990).

The first stage of our modelling is to establish the flow properties of the fracture networks, in particular the permeability of the fracture flow system, and the scale of the REV. We determine the REV to be that scale for which the permeability of cubes of rock is insensitive to the realisation or details of individual fractures of the network. Permeability is calculated by generating networks in a sufficiently large region, and solving for the flow induced by a unit pressure gradient in each of the three coordinate directions in turn. The flux crossing 'test planes' in the network is then calculated for planes of varying sizes and positions. By aligning the planes with the coordinate axes, we can infer the components of the full permeability tensor. The planes must be sufficiently large that the flux crossing them is insensitive to individual fracture locations, and sufficiently far from the boundaries of the network, that they do not include artificial flows induced by uniform pressure gradients on the boundary. The smallest such scale corresponds to that scale at which the flows are independent of realisation details and for scales larger than this REV, the use of continuum models is justified. If we accept an uncertainty of about 50%, then for average rock the REV size is 12m, and in the H-zone it is 7m. The corresponding permeability tensors are:

$$k_{ij} = \begin{pmatrix} 2.1 & -0.1 & 0.2 \\ -0.1 & 2.6 & -0.2 \\ 0.2 & -0.2 & 2.8 \end{pmatrix} \times 10^{-17} m^2$$

for the average rock, and

$$k_{ij} = \begin{pmatrix} 1.7 & 0.2 & -0.2 \\ 0.2 & 2.7 & 0.3 \\ -0.2 & 0.3 & 2.2 \end{pmatrix} \times 10^{-16} m^2$$

for the H-zone. The networks for the H-zone are less well characterised, since we have less measurements of H-zone fractures, and there is an order of magnitude uncertainty in this tensor (mainly due to uncertainty in the interpretation of fracture hydraulic properties).

Having established the permeability of the fracture networks, we are now able to investigate the extent to which relatively untransmissive fractures can be ignored. We showed that 30% of the least transmissive fractures could be removed with only a 10% reduction in the permeability of the networks. We regarded this as an acceptable approximation and subsequent network calculations were made on this truncated network.

The network models are very computationally intensive and the scale of the models with our interpretation of the fracture system is limited to a few tens of meters. The scale of the flow system associated with the SDE is much larger: of the scale of the SCV site. However, we have demonstrated the validity of continuum approximations on these scales, and so to model the bulk fluxes in the SDE, we adopted porous medium models using NAMMU for the SCV stage 3 predictions and then CFEST in stage 5. Note that for these models, the finite-element size is larger than our predicted REV, and the finite-element properties are given by the predicted values from fracture network models. Our modelling remains based on the discrete fracture network approach and we have not had to calibrate the models.

Once we have the continuum approximation to the pressure field of the SDE experiment, we can return to using fracture network models directly.These network models will predict the distribution of flux to the experiment, aswell as the bulk fluxes. We note that the pressure field is cylindrically symmetric about the D-hole array on a scale of 10-20m and we therefore can generate separate network models of disks of rock centred on D1. These models extend to a radius of 12m. The results are presented separately for average rock and for the H-zone since these fluxes are independent. In stage 5 we did not predict the flux from the B-zone. This had been predicted during stage 3 using a simple single plane model, and there was no additional data to improve on this model. The inflow predictions are summarised in table 3.

Note that in the stage 5 models, the boundary conditions to the models were approximately interpolated and a better boundary condition would increase flux by a factor of 2.5. With this, our results for the average rock agree with our stage 3 prediction, as expected. The flux from the H-zone network is a factor of 2-3 lower than that given by the stage 3 single plane representation. This is not surprising given the uncertainty in the H-zone network characterisation.

Modelling inflow to the Validation Drift

The differences between modelling the SDE experiment and the Validation Drift inflows are the presence of the disturbed zone, and the increased resolution of flow measurements. We again aimed to explain the flow field by using forward modelling of physical processes to predict flows. We could at this point have calibrated the SDE model to the measured SDE flows, and then incorporated a model of the disturbed zone. We decided that there was no value in doing this, since the calibration could be performed in several ways, which might interact with our representation of the disturbed zone. We therefore present our results in terms of a change in flux, as compared with models that did not represent the disturbed zone; that is as a change from SDE results.

There was no direct experimental characterisation of the disturbed zone. We therefore chose to investigate the conventional model of disturbed zone effects: changes in hydraulic properties due to stress changes. There were no direct measurements of these stress changes, and so the Stripa project commissioned a modelling study of the SCV stress field. This used a continuum elastic approximation, and provided input to our fracture network models. The stress field predicted was in good agreement with simple analytical models of stress around tunnels, and reasonable agreement with two-dimensional discrete model of the plastic deformation around the tunnel. Using this modelled stress field, we can calculate the normal and shear stress on each fracture plane in our network model, and also the change in stress due to tunnel excavation. We first attempted to see a correlation between stress and transmissivity in the original hydraulic measurements of fracture properties. There was no such correlation in the data, possibly due to the averaging of fracture properties in the multi-fracture test intervals. We therefore applied a simple normal compliance model of transmissivity change due to changes in normal stress. This is the simplest model, and one that has been suggested as a cause of skin around tunnels:

$$\frac{T}{T_0} = \left(\frac{\sigma}{\sigma_0}\right)^{-\beta}$$

where T is the fracture transmissivity, σ is the normal stress on the fracture and subscript 0 refers to the undisturbed values. The stiffness parameter β was measured for a few large core samples with a typical value of 0.2 for Stripa fractures. The result of applying this model to our predictive network models,was to increase the inflow by about 2%. This is explained by the dominant fracture set intersecting the tunnel perpendicularly and experiencing a slight decrease in normal stress due to a Poissons ratio effect in the stress models. Clearly this model of normal stress effects does not represent the disturbed zone around the Validation Drift. More detailed parameter studies investigating the influence of stress changes are given in Herbert et al (1992).

The other aspect of the inflow prediction to the Validation Drift in the detailled prediction in inflow distribution. We did not understand the disturbed zone, and this might significantly affect the pattern of inflow: our results, neglecting the disturbed zone, are shown in figure 2. Note also that the relative importance of the H-zone is underestimated here since we have not calibrated against the known SDE inflows.

Tracer transport modelling

The final stage of our work was to model the tracer transport experiments performed in the SCV site: the Radar-Saline injection the the D-hole sink (RS1); the Radar-Saline injection to the Validation Drift (RS2); and the Tracer injection experiments with recovery in the Validation Drift and borehole T1 (TR). These experiments took place within the H-zone, and the results of the flow experiments mean that we can neglect the average rock for the purposes of these models

The objective of our transport modelling is to try and build understanding of fracture network transport, by directly modelling tracer transport in a discrete model. This is the only approach able to quantify the dispersive effect of the fracture network geometry. The models are however computationally intensive, which means they have difficulty representing complex physical effects such as rock-matrix diffusion, two-phase flow and density driven flows. All of these mechanisms may be present for the Stripa experiments, but the parameters of these processes were not measured. We chose to use NAPSAC to investigate the effect of the network geometry, and to neglect the other processes.

There are two main problems to be addressed before we can use NAPSAC to model the Stripa tracer experiments. First, there is considerable uncertainty in the flow-field, and in particular in the influence of the disturbed zone. This was not understood in the flow modelling. We wish to investigate tracer transport in

226

this stage of our modelling. Therefore, to avoid confusing flow approximations with the uncertainties of our transport model, we chose to calibrate the flow solutions to the tracer experiments and to fit as good a match to experimentally measured inflows as was reasonable in our stochastic model.

Secondly, there is the problem of scale. The flow systems of the transport experiments is typically 30-40m: an order of magnitude larger than the simulations of the flow experiments. Here, in contrast to the flow modelling, we are not necessarily justified in using porous-medium approximations, since there are well known scale dependencies in the transport properties of heterogeneous rock. We therefore increased the scale of our models to represent network of more than 60000 fractures, 180000 intersections and more than 200000000 finite elements. We also further truncated the fracture network distributions, showing that if we accept a 30% loss of transmissivity, then 70% of the smallest and least transmissive fractures could be deleted. This is a significant reduction but nevertheless our results will still adequately reflect the influence of the fracture network geometry on tracer transport. The neglection of these less significant fractures will tend to result in slightly faster and less dispersed breakthrough curves. Finally, we have developed a new transport model specifically for these very large simulations. It is based on an approximation of the transport pathways on each plane, assigning a transport channel to correspond to each flux connection on the fracture plane (this approach is used by other fracture network codes as the principle transport model). This approximation was verified by comparison with the highly discretised solution to a small sparsely connected random network problem used in the Stripa verification exercise (Herbert, 1990, and Schwartz and Lee, 1991). It gave good agreement for early and peak arrivals but underestimated the tail of the recovery. Even better agreement was found when we verified the simplified model for a small region of the highly connected H-zone network.

Since we are calibrating the flow field, the most straightforward approach to supplying regional flow boundaries within the H-zone was to set up a simple two-dimensional planar model of H-zone in the SCV site and impose the observed regional pressure head gradient over the boundaries. This scoping model had a scale of 200m and included all the major sinks in the SCV site. From this model we were able to match the observed fluxes in the experiments by fitting an H-zone transmissivity, together with skins of reduced transmissivity around the Validation Drift (for RS2 and TR experiments), 360m Drift, and increasing the transmissivity of the specially chosen C2 injection interval. This was used to predict the flow fields of the RS1, RS2 and TR experiments. Pathlines in this models enabled flow divides to be located and appropriate boundaries for the network models to be identified. The requirement on the boundaries of local model regions was for them to be about 10m away from sources and sinks, and to include almost all the tracer transport pathways. In fact for RS1 we had to use a symmetry argument and model a half-region.

Once the network model region had been identified, we checked the accuracy of the regional flow solution by running a local single-plane model, interpolating boundary conditions. We could then interpolate these pressure boundary conditions onto network models. The additional characterisation required for our network models are transport apertures for the fractures, which were again represented using an equivalent-parallel-plate model, the width of the H-zone and a prescription for the disturbed zone representation. We expect the transport aperture to be smaller than the hydraulic aperture by a factor resulting in residence times being 2-10 times longer (Gale et al., 1990). To investigate this we performed a scoping study using variable aperture models, but found no difference in the equivalent-parallel-plate hydraulic and transport apertures: it seems likely that the difference is due to three-dimensional flow effects. In any case it is straightforward to apply a scale factor to travel times as a post-processing step, and for the Radar Saline experiments we present unscaled results and calibrated results using a scaling factor of three (Herbert and Lanyon, 1992). This factor is used in the predictive simulations of the tracer experiments.

The results of the RS1 experiment were known when we carried out this work, but we approached the task as a predictive modelling exercise. We obtained reasonable agreement with the experimental measurements using a factor of between 1 and 3 to scale the transport aperture. However, there were difficulties in comparing results due to changes in the experimental conditions after 100 hours and to our use of a symmetry boundary condition along an approximate flow divide.

To model RS2, we followed a similar procedure, using a regional model to derive boundary conditions for a 40m scale network model. We used the H-zone model calibrated to SDE flows, and introduced a disturbed zone, calibrated to match the observed flow field. We chose to do this by modifying the transmissivity of the regional model with 4m of the drift boundary (approximately three drift radii). Similarly in the network model we modified the aperture of all fractures within 4m of the drift. The transmissivity was reduced by a factor of 30 in the scoping calculation and by 60 in the network model (the difference being due to the use of a coarse discretisation near the drift in scoping calculations). This choice of disturbed zone influence is not unique, and different fractures are likely to experience different disturbances. This calibration gives a good match to RS2, but if there is any anisotropy to the disturbed zone, we will not represent it. The calibration will be less good for experiments that involve flows crossing other parts of the disturbed zone.

The results are shown most clearly for a pulse test, and figure 3 shows a comparison of our prediction with the measured recovery from a pre-test pulse injection of Amino-G tracer. The results from the saline injection that followed are fully consistent with this tracer pulse test. It can be seen that a better match is obtained by scaling the transport aperture to increase residence times by a factor of three, and we then have an excellent agreement. We therefore used this calibrated scaling factor to make our 'blind' predictions for the tracer experiments.

The final set of simulations concerned the tracer tests from boreholes around the drift held at near ambient pressure. We modelled 5 separate injections to the drift and 3 interhole experiments in the H-zone. The other experiments, involving tracer injection outside the H-zone and the use of a sorbed tracer were not simulated. The results of these injections are influenced by the transmissivity of the specific injection interval (in contrast to RS2 where a large flux was imposed on a transmissive interval). We simulated a large number of injections in several realisations, with particles injected in several intervals in each of the four boreholes. Thus we do not calibrate our models to match the selection procedure used to identify injection intervals, but instead predict the range of results for typical intervals within the injection boreholes. The flow disturbance due to the injection fluxes was minimal and so many transport simulations could be performed within the same flow realisation. We had shown in the verification work that 50000 particles were needed to converge the breakthrough distribution, and some of these simulations tracked swarms of 450000 particles comprising 9 different tracers from 9 injection intervals. All the particles were recovered at the drift or the model boundaries.

This results in a very large number of results to compile. This task is addressed in Herbert and Lanyon (1992), here we present a few typical results. As expected, the best results were for injections in C2, where the experiment took place in the same part of the H-zone as had been used for our calibration. Figure 4 illustrates a representative swarm of particle paths for an injection of 10.8 ml/min in an interval of C2. Figure 5 summarises the corresponding breakthrough to the drift, for all our simulations of the tracer experiment in C2. These figures show very smooth results, and our continuum approximation of the flow field is very consistent. The range of breakthrough curves for the different realisations is very small, and in this 'double-blind' exercise, they predict the experimental recovery correctly. Indeed, one of the realisations matches the experiment almost exactly.

Away from the calibration region, above the drift, the results are less impressive, and pressure field for drift inflow is in error by a factor of four. Our prediction of drift breakthroughs from these experiments are correspondingly a factor of four fast. The other experiments were the cross hole tracer tests. These took place in a flow-field that was not strongly influenced by the disturbed zone, and our flow field was quite accurate. Here again, once we have a good flow representation, our transport predictions were accurate.

Summary

Our work for the Stripa project has developed a very powerful computational tool, the NAPSAC computer code. It has proven possible to infer fracture network statistics from the experimental investigations, and to incorporate these statistics in NAPSAC models without having to recourse to unjustified simplifications. There is a clear path from data collection to numerical model to prediction. This exercise was successfully carried out for the SDE experiment. The success of this exercise has lead to a clear justification for the use of equivalent porous medium models on sufficiently large scales, and a quantification in the inherent uncertainty associated with such a conventional approach. It provides strong support to safety assessment programmes.

When predicting the inflow to the Validation Drift, the use of Napsac highlighted our lack of understanding of this region. Previous studies incorporating such a disturbed zone accounted for it empirically, with inadequate justification. The modelling has also provided a useful integration of geomechanical modelling expertise with hydrogeologic modelling. We have demonstrated a valuable tool to aid future studies that address the experimental characterisation of such disturbed zones.

Finally the modelling of tracer transport has proven feasible, even in the highly fractured H-zone. Our modelling is on scales beyond that at which continuum behaviour is shown for flux calculations: our preliminary analysis of these experiments indicates that we can accurately predict the dispersive effect of the network geometry. Future work will enable us to investigate the importance of other dispersive mechanisms, and the validity of continuum approximations.

We have succeeded in our objective of demonstrating the validity of the fracture network approach as a useful tool for understanding flow and transport in field experiments in hard fractured rocks.

References

Gale,J.E., R.MacLeod and P.LeMessurier, 1990. Site Characterisation and Validation - Measurement of Flowrate, Solute Velocities and Aperture Variation in Natural Fractures as a Function of Normal and Shear Stress, Stage 3. Stripa Project Technical Report, 90-11, SKB, Stockholm.

Gale,J.E., R.MacLeod, A.Strahle and S.Carlsten, 1990. Site characterisation and validation - drift and borehole fracture data, stage 3. Stripa Project Technical Report, 90-02, SKB, Stockholm.

Grindrod,P., A.W.Herbert, D.L.Roberts and P.C.Robinson, 1992. NAPSAC Technical Document, AEA-D&R-0270, AEA D&R, Harwell.

Herbert,A.W., 1990. Development of a tracer transport option for the NAPSAC fracture network computer code. AEA-D&R-0023, AEA D&R, Harwell.

Herbert,A.W., J.E.Gale, G.W.Lanyon and R.MacLeod, 1992. Prediction of flow into through fractured rock at Stripa. AEA-D&R-0276, AEA D&R, Harwell.

Herbert,A.W. and G.W.Lanyon, 1992. Modelling tracer transport in fractured rock at Stripa. AEA-D&R-0277, AEA D&R, Harwell.

Herbert,A.W. and B.A.Splawski, 1990. Prediction of Inflow into the D-holes at the Stripa Mine. Stripa Project Technical Report 90-14, SKB, Stockholm.

Schwartz,F.W. and G.Lee 1991. Cross-verification testing of fracture flow and mass transport codes. Stripa project technical report in press, SKB, Stockholm.

Table 1 Summary of the inferred average-rock fracture network properties

Property Distribution type	set 1	set 2	set 3	set 4	set 5	set 6
density (m^{-3}) Poisson	2.40	0.30	0.21	0.24	0.24	0.42
azimuth (°) Normal	108.6 (26.4)	210.1 (16.1)	51.7 (8.4)	143.4 (59.6)	19.6 (19.8)	247.6 (18.7)
dip (°) Normal	77.3 (26.0)	68.8 (12.6)	81.4 (10.9)	34.8 (14.4)	62.4 (15.7)	48.9 (14.8)
length (ln(m)) log-normal	-1.06 (0.92)	-0.44 (0.76)	-0.64 (0.83)	-0.68 (0.88)	-0.97 (0.99)	-1.25 (1.02)
transmissivity ($\ln(ms^{-1})$) log-normal	-24.06 (2.25)	-24.06 (2.25)	-24.06 (2.25)	-24.06 (2.25)	-24.06 (2.25)	-24.06 (2.25)

Table 2 The H-zone was modelled as a 10 m thick planar feature oriented normal to the Validation Drift

Property Distribution type	set 1	set 2	set 3	set 4
density (m^{-3}) Poisson	4.65	12.04	6.55	6.57
azimuth (°) Normal	240.2 (15.6)	168.8 (18.9)	100.6 (24.5)	159.9 (81.7)
dip (°) Normal	71.2 (10.6)	77.6 (15.5)	80.3 (25.0)	30.0 (12.5)
length (ln(m)) log-normal	-1.08 (0.60)	-1.08 (0.37)	-1.14 (0.87)	-0.80 (0.55)
transmissivity($\ln(ms^{-1})$) log-normal	-21.98 (2.07)	-21.98 (2.07)	-21.98 (2.07)	-21.98 (2.07)

Table 3 Summary of inflow predictions to the D-holes.

	D1	D2	D3	D4	D5	D6	Total
average rock (ml/m/min)	0.09	0.23	0.22	0.24	0.14	0.23	1.20
H-zone (ml/m/min)	1.67	4.65	2.88	1.69	2.25	3.48	15.75
B-zone	not modelled in SCV stage 5 (no data collected)						

Figure 1 Typical fracture network model of Stripa site

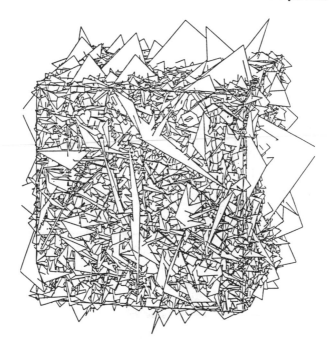

Figure 2 Predicted distribution of drift inflow

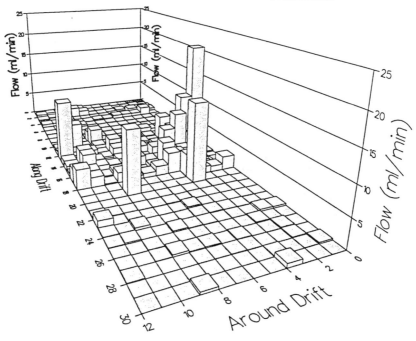

Figure 3 Simulated beakthrough for RS2

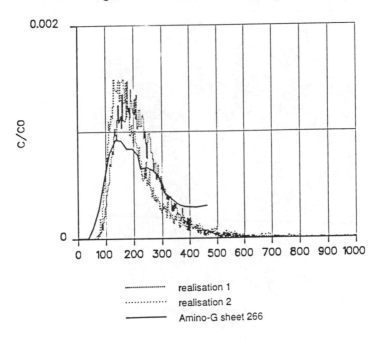

.... realisation 1
.... realisation 2
———— Amino-G sheet 266

Figure 4 Predicted plume of tracer pathways from C2

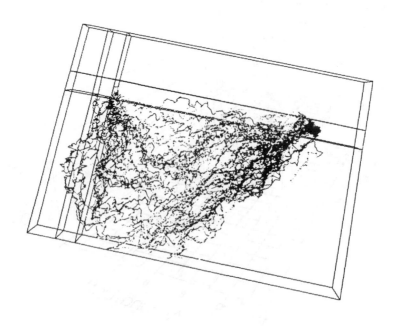

Figure 5 Prediction of tracer breakthrough from C2

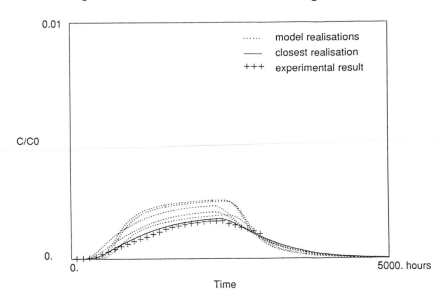

The Role of the Stripa Phase 3 Project in the Development of Practical Discrete Fracture Modelling Technology

William S. Dershowitz
Golder Associates Inc., Seattle, United States

Abstract

The Stripa Project has played a major role in developing discrete fracture analysis from a theoretical research topic to a practical repository evaluation tool. The Site Characterization and Validation (SCV) program positively answered questions regarding (1) the validation of discrete fracture models, (2) the feasibility of collecting data for discrete fracture models, (3) the ability of discrete fracture models to simulate flow in a rock volume of approximately 10^6 cubic meters using modest computing resources, and (4) the ability to model transport in discrete fractures. The SCV program also made progress on such continuing issues as the importance of in-plane fracture heterogeneity and coupled effects.

Introduction

The development of discrete fracture analysis methods for use in repository characterization and performance assessment is one of the major accomplishments of the Stripa Project. This paper's goal is to show how the Stripa Project brought about the transformation of discrete fracture analysis from a theoretical research topic to a practical hydrologic tool. The paper first reviews the Stripa Project's role in developing fracture hydrology concepts, then the paper states the key issues which were unresolved at the beginning of the Phase III program, and finally the paper shows how these issues were resolved within the Stripa Project and concurrent research activity elsewhere.

Background -- Discrete Fracture Analysis and The Stripa Project

The discrete feature analysis approach to model development recognizes that flow in heterogeneous systems, such as fracture networks, occurs within discrete conduits rather than through a continuous porous system. Discrete features are conduits which are generally geologic features of structural origin (including faults, joints, and fractures) or sedimentary origin (such as channel sands). When the features of concern are fractures, we refer to the approach as *Discrete Fracture Analysis*. Discrete Fracture Analysis uses field fracture data to build a geologically realistic model on which flow simulations can be based.

David Snow (1965) was a pioneer in this area of work, and his attempts to define the hydraulic properties of fracture networks using mapped fracture data are direct precedents for the fracture modelling that we are doing today. John Gale's original plan for hydrologic studies within the SAC (Swedish-American Cooperative) program at Stripa, was initially intended as an application of Snow's ideas of developing permeability tensors from a combination of structural geologic data and well test information (Gale and Witherspoon, 1978).

The concept of permeability tensors assumes that fracture systems behave anisotropically and homogeneously. Work on the statistics of fracture geometry through the 1970's and early 1980's showed that fracture systems were not homogeneous, and that the highly skewed distributions of spacing, size, and transmissivity (e.g. Baecher, and others, 1977) introduce significant heterogeneity to the properties of fractured rocks. The wedding of these modern statistical views of fracture geometry with numerical flow simulators resulted in the first generation of discrete fracture flow codes such as the ones at Berkeley (Long, 1983), where Snow began his work, at MIT (Dershowitz, 1983), where much of the work on geometric statistics had been performed, and AEA-Harwell

(Robinson, 1984). The initial development of discrete fracture models at Berkeley was inspired, in part, by a recognition of the strongly heterogeneous behavior of the Stripa Granite observed in well testing results (Doe, personal communication).

When the Stripa Phase III work was being planned in 1986, the discrete fracture modelling approach was still a theoretical concept, and its practical applicability was not at all clear. The experience at that time reinforced a general criticism of discrete fracture approaches that they were impractical due to the burdens of data collection among other difficulties.

Clearly, improvements in field data collection were required to make discrete fracture tools practical. Some of these improvements came with the Stripa Phase 2 crosshole program. Stripa Phase 2 had shown that only a subset of all fractures have hydrologic significance, and that fracture zones could be detected by geophysical and geohydrologic means. The availability of the fracture zone information provided significant constraints to the previously unbounded possibilities posed by purely stochastic network simulations.

Within Stripa Phase 3, the Stripa Site Characterization and Validation (SCV) program had a major goal of demonstrating the validity of an approach which integrated field methods and numerical tools to provide predictions of hydrologic performance in a repository-like setting. This integration was to occur in many ways including the development of conceptual models using a wide range of field measurements and the implementation of these conceptual models in numerical simulations (Olsson, et al, 1992). Fractures dominate the hydraulic behavior of crystalline rocks, hence the Stripa Phase III planning placed a strong emphasis on fracture data collection and on discrete fracture modelling (Carlsson, 1987). This fracture emphasis appeared in the original selection of a discrete fracture code for the SCV program (AEA-Harwell's NAPSAC), and in the subsequent decision to solicit the involvement of American codes from Lawrence Berkeley Laboratory and Golder Associates.

At the first Stripa Fracture Flow Task Force meeting in June, 1987 (Hodgkinson, 1992), there was a general consensus that radioactive waste disposal required more accurate and realistic models of fractured rock hydrology than were available using conventional continuum approaches. As mentioned above, however, it was not clear that discrete fracture models were a practical alternative. Many issues were raised concerning the data and computational requirements of discrete fracture modelling, and there was some discomfort with the stochastic approach and its lack of single "answer". At the meeting, the three discrete fracture modelling groups (AEA Harwell Laboratory, Lawrence Berkeley Laboratory, and Golder Associates) asserted confidently that they could sufficiently resolve these issues to make predictions of fluxes, but the modelling groups refused to commit to the far more difficult task of solute transport modelling. Since gross fluxes to underground openings may be predicted with reasonable confidence using simple analytical solutions, the practical contributions of discrete-fracture modelling were still in doubt.

The eight major issues which confronted the discrete fracture modelling teams were:

1. *Validation*: can discrete fracture approaches produce results which provide reasonable matches to observed behavior ?

2. *Fracture Geometry Characterization:* can a site characterization program collect sufficient data for a discrete fracture model within reasonable time and cost constraints?

3. *Hydraulic Characterization:* is there a way to obtain the hydraulic properties of individual fractures from well tests?

4. *Computational Requirements and Scale Limitations:* are reasonable computer resources available to use a discrete fracture model at repository and site scales (that is to model a rock mass containing 10^7 to 10^{12} or more fractures)?

5. *Utility of Probabilistic Predictions:* is there value in probabilistic predictions of stochastic discrete fracture models, and can discrete fracture models be calibrated and conditioned to observed hydrogeologic behavior similarly as continuum models can?

6. *In-Plane Heterogeneity:* what is the nature of fracture roughness and channeling, and how can these processes be successfully incorporated into discrete fracture analyses?

7. *Solute Transport:* what is the nature of solute transport in fracture networks, and how can the key parameters of network dispersion, mixing, and within fracture plane dispersion and retardation be derived?

8. *Coupled Processes:* how do coupled processes such as stress-strain, saturation, and geochemistry affect flow and solute transport in fracture networks, and how can coupled processes be treated within discrete fracture network models.

The following sections describe how these issues were addressed and, in many cases, resolved during the course of the Stripa Phase 3 project. Remaining issues to be addressed in future projects such as the SKB Hard Rock Laboratory at Äspö are listed in the final section.

Issue 1: Validation

The Stripa Fracture Flow Task Force developed a series of validation exercises and criteria starting with prediction of flow to the Simulated Drift Experiment, Flow to the Validation Drift, Simulation of the Saline Tracer Experiment, and finally solute transport in a flow experiment to the validation drift. For each stage of the modelling, the site characterization activities progressively produced more information and further refinements to conceptual models.

For our part, the validation activities used our FracMan system of computer codes. FracMan is a system of computer codes which covers the full scope of discrete fracture analysis from the interpretation of field data (FracSys), to fracture generation and sampling (FracWorks), to definition of boundary conditions and finite element mesh generation (MeshMonster), and finally to solution of the flow and transport equations (MAFIC). The various validation exercises focussed on different scales of simulation. At each scale we constrained the simulations to only the significantly conductive fractures rather than trying to simulate all fractures. This is an important distinction of our modelling approach which is discussed further below under the "Computational Requirements" heading.

The three main distinctions of our approach as compared with the other models are the following. First, we used a discrete fracture approach (as opposed to a stochastic continuum) at all scales of simulation. For larger regions, we reduced the number of fractures simulated to those with the most significant transmissivities. Second, we used relatively modest computing resources (i.e. workstations rather than supercomputers). And third, we integrated the field data analysis and the flow simulation into one system of computer programs.

A full review of the results of the validation exercises using FracMan is beyond the scope of this paper. The results have been presented in a series of Stripa Reports by Golder Associates' staff (Dershowitz, et al, 1992), and are summarized both in the final SCV report (Olsson, 1992) and the final report of the Fracture Flow Task Force (Hodgkinson and Cooper, 1992). As reported, the FracMan results met all of the validation criteria set out by the Fracture Flow Task Force. This apparent success does not guarantee that fracture flow models will always predict the future with high accuracy, but the fact that FracMan, and other fracture flow codes "failed to fail" some stringent validation criteria greatly increases our confidence in the fundamental validity of the discrete fracture approach. As an example of results, Figure 1 illustrates a successful prediction of the pattern and magnitude of tracer breakthrough to the Validation Drift as part of the tracer validation experiment.

Issue 2: Fracture Geometry Characterization

Discrete fracture modelling must specify the geometry and properties of the conductive features within a rock mass. There are two major parts to the problem: determination of the geometric conceptual model and evaluation of the statistical properties of the fracture system.

The geometric conceptual model provides the rules for distributing the fractures in space. Some alternate models implemented in FracMan are several random disk models (Baecher, Enhanced Baecher, and BART), a nearest neighbor model which incorporates spacial correlation, a fractal model (Levy-Lee), and a war-zone model, which generates fracture sets within fracture zones. The HeterFrac model within our FracSys program subjects digitized fracture-trace maps to a range statistical tests, which identify the most appropriate conceptual model for a fracture system. HeterFrac was developed during the Stripa Phase 3 Project. The collection of trace map data in addition to the more conventional scan-line survey is thus vital to the implementation of a discrete fracture code.

The second half of the site characterization task is the definition of geometric properties. Our ISIS module in FracSys was used to define fracture sets. The ISIS approach is somewhat unique in that various fracture characteristics, such as fillings or roughness, can be combined with orientation to define sets. FracSize provides the conversions of trace data to fracture length information. The approaches to defining conductive fractures are discussed in the following section.

In summary, during the course of the Stripa Phase 3 project we developed procedures to analyze the borehole and surface mapping data required for discrete fracture modelling (Table 1). The algorithms for these analysis procedures are described in Stripa TR 91-16, TR 91-23, and Dershowitz et al. (1992).

Issue 3: Hydraulic Characterization

Single-hole hydraulic tests measure the leakiness (transmissivity) of a section of borehole . In classic sedimentary-rock hydrology, the hydraulic conductivity or permeability is then calculated by dividing the transmissivity by the thickness of the stratum which is conducting the water. When testing fractured crystalline rock, this thickness value is never clear. There is uncertainty about how many fractures and which fractures are actually conducting fluid. Further complications arise from (1) uncertainties in distinguishing single fracture behavior from network behavior and (2) from the effects of heterogeneity within the fracture planes which are intersected.

Several analytical tools have been developed within the Stripa Project to help answer these difficult issues. These include fractional dimension well test analysis, cross-fracture versus at-borehole transmissivity, and the *Oxfilet* method of analyzing variable-spaced packer test data. These are described briefly below.

The importance of test dimension has been strongly recognized in the Stripa Project beginning with the crosshole experiments of Phase II (Noy, et al., 1988). Dimension is basically a measure of the power exponent by which conductive area (or conductivity) changes with distance from a pumping source. The literature on well testing is overwhelmingly dominated by dimension 2, radial cylindrical flow solutions, however real fracture systems may exhibit dimensions ranging from the spherical flow of three-dimensional fracture networks to the one-dimensional flow of single-channel dominated systems (Doe and Geier, 1990).

The concept of cross-fracture versus at-hole transmissivity arises from the inherent heterogeneity of fracture surfaces. A fracture network model requires the effective transmissivity between two or more line boundaries on any given fracture surface, that is, between the upgradient intersections and the downgradient intersections. Single-hole well tests, on the other hand, measure the transmissivity from what are effectively point sources on the fracture plane (Figure 2). As mentioned above, transmissivity is a measure of leakiness, and measured leakiness is strongly dependent on the geometry of the sources and sinks. Unlike a homogeneous fracture, a heterogeneous fracture does not have a single transmissivity value (Figure 3), but rather has many transmissivity values depending on what flow geometry is imposed on the system. We should not expect the transmissivity value at the point of borehole penetration to be the same as the effective transmissivity across the fracture. We will return to the issue later in the paper when we discuss roughness and channelling. For now, our approach to the at-borehole versus cross-fracture transmissivity problem has been to perform calibration studies on simulated heterogeneous surfaces to

243

develop probabilisitic relationships between the measured at-hole values and the unknown cross-fracture transmissivities.

Borehole tests provide a distribution of transmissivity values for test zones, while discrete fracture models require the spacing of conductive fractures and distribution of transmissivity among those fractures. We resolve the difference between field data and model requirement by assuming a distributional form for the single fracture transmissivities, and finding an optimal conductive fracture spacing to fit the distribution of packer test results. The Oxfilet (Osnes Fixed-Interval-Length) program within FracSys interactively performs this optimization (Dershowitz et al., 1992; see Stripa TR 91-16, TR 91-23).

Issue 4. Computational Requirements and Problem Scale

A fundamental question regarding discrete fracture approaches is the size of problem that can be addressed. The problem reflects jointly the number of conductive fractures, the scale of rock volume represented, and the computational resources available.

A model which attempts to simulate all fractures must be very large to simulate reasonable rock mass volumes. For example, a fracture frequency of only 5 fractures per meter of borehole, with a mean fracture area of 5 m^2 (approximately the values for the Stripa site), will have over a million fractures in a 100 m cube. (Dershowitz and Herda, 1992). In the FracMan/MAFIC model, each fracture is generally discretized to 20 to 100 triangular finite elements, so that the one million fractures will have $2x10^7$ to 10^8 finite elements. So large a model is well beyond the capabilities of current computers.

If it were necessary to model all of these fractures, discrete fracture modelling would be limited to scales of 1 to 10 m, and would produce behavior which can approaches that of a continuum, in direct contrast to the observed, highly heterogeneous and sparsely connected nature of fracturing at the Stripa site.

The above considerations have several implications for discrete fracture modelling. First it is not possible to model every fracture, hence a discrete fracture model must be limited to the most conductive and largest fractures except for localized regions where greater detail may be required. Second, it is not appropriate to model every fracture identified geologically as if it were a conductive feature. And third, the minimum size and transmissivity of fractures to be modelled should be based on the scale of interest . These conclusions are the key to a successful implementation of the discrete fracture approach. Table 2 presents the scales and approximate number of fractures modelled in

FracMan discrete fracture analyses to date. The Stripa Phase 3 project demonstrated that reasonable discrete fracture modelling was possible without rigorous modelling of all geologically identified fractures.

As to computer capacity, the FracMan/Mafic analyses for the Stripa SCV experiment were run in solution times of 0.03 to 3 hours on an IBM RS/6000 model 530 workstation using 10 to 200 Megabytes of real or virtual memory. With these resources we could simulate and reproduce heterogeneously connected responses, bulk flow, flow heterogeneity, and (when modelled) solute breakthrough.

Issue 5: Utility of Probabilisitic Predictions

Members of the Fracture Flow Task Force viewed the concepts of calibration and inverse modelling with suspicion, although conditioning and optimization were generally acceptable. This may be in part because the validation criteria (such as inflow to a drift or trace breakthrough at a drift) were local, as opposed to alternate criteria that might have been more global in nature, such as matching the hydraulic potential measurements over the SCV region. The data analysis procedure for the FracMan simulations were based entirely on optimization procedures which ensured that the fracture geometric and transport properties were consistent with available data. Flow predictions were produced directly from these geometric fracture network patterns without benefit of calibration of local fracture properties using observed large scale hydraulic response. In this limited sense, predictions were produced by "forward modeling".

The Stripa Fracture Flow Task Force had major difficulties in evaluating and comparing probabilistic predictions for flow and transport. Were measurements best compared against mean predictions, prediction modes, one standard deviation bounds, or 90% confidence bounds ? How far of could the actual measurement be from the mean prediction and still validate the model? During the course of the project, significant progress was made in the ability to evaluate and utilize probabilistic flow and transport predictions. By the final tracer-validation prediction, a full set of criteria were developed for evaluating probabilistic predictions. These criteria recognized the importance of evaluating individual realizations to detect patterns, together with the use of mean, mode, and range to obtain a more complete picture of the model results. However, the task force did not proceed to the use of Bayesian or statistical hypothesis testing or other more quantitative approaches.

Issue 6: In Plane Heterogeneity, Fracture Roughness, and Channeling

The FracMan predictions for the SCV experiments used fracture planes with effective homogeneous transmissivity, however, actual fracture planes clearly have heterogeneous properties over their surfaces. Channelling, which has been an object of considerable interest over the past several years, is one form of this heterogeneity.

There are important distinctions between the various conceptual models of channelling in fracture planes. These distinctions strongly affect the appropriateness of the concepts used to develop the flow simulator. Channels in one model implementation are discrete pipes, which form a network of one-dimensional conduits. In this view, most of the fracture plane is non-conductive and effectively is neglected.

An alternate, and perhaps more realistic view of channelling is that it arises from the heterogeneity within continuously conductive surfaces (Figure 3). The channels in this case are not discrete one dimensional features. Rather the channels are conductive paths that appear as geometric artifacts of a specific distribution of heterogeneities or roughnesses and a specific set of boundary conditions. In this view channels may appear or disappear with adjustments in the geometries of either boundaries or heterogeneities. This form of channelling phenomenon is closely related to the concepts of at-hole and cross-fracture transmissivity discussed above -- the apparent hydraulic properties of the fracture depend on the flow geometry within the fracture.

The Stripa channelling experiment provides some indication that the heterogeneous continuum view may be more appropriate than the pipe network model. In this experiment channel behavior was observed between two in-plane boreholes, however, the channel-flow behavior was not reversible when the source and sink holes were interchanged. In short, a change in boundary conditions affected the apparent channel behavior of the fracture.

For the discrete fracture modeler, the concept of channeling and the scale of modelling will dictate the most appropriate model for the individual fractures: (1) an equivalent homogeneous fracture, (2) a pipe network within a plane, or (3) a heterogeneous continuum within the fracture planes. The choice of channel conceptual model should have large effects on hydraulic storage and transport properties such as dispersion and diffusion.

For the scales of modelling in the Stripa SCV experiment, it should be noted that the equivalent homogeneous fracture model provided predictions within the validation criteria of the fracture flow task force, hence it possible to create a realistic model without considering channel effects and in-plane heterogeneity in detail. Nonetheless, the treatment of in-plane heterogeneity is an open research issue, particularly with regard to very near-field performance assessment, where details of the flow paths within the fractures may be important.

Issue 7. Solute Transport Dispersion, Mixing, and Retardation

Confidence in discrete fracture transport modelling was very low at the beginning of the Stripa project, primarily due to a lack of a clear understanding of the how to characterize the required fracture solute transport properties: transverse and longitudinal dispersivity, mixing at fracture intersections, and retardation effects such as matrix diffusion. During the course of the Stripa Phase 3 project, little was learned about these properties, but the principal investigators still managed a credible solute transport prediction (e.g., Figure 4). This was possible because the required solute transport parameters were derived by calibration from preliminary cross-hole tracer experiments (Stripa TR 91-23). A number of organizations are currently carrying out extensive rock block experiments which will provide much better estimates for these parameters (e.g., PNC Labrock Experiment, USGS Rock Block Experiment, AECL URL Single Fracture Experiment, SKB Aspo Rock Block Experiment).

Issue 8. Coupled Effects, Including Stress and Saturation

The Stripa Phase 3 project originally projected a 100-m long validation drift. However, from the first principal investigator's meeting in 1986, major questions were raised concerning the relative value of boreholes and excavations for evaluating discrete fracture approaches. The Stripa discrete fracture approaches did not initially include any of the excavation effects important for modelling flow: stress redistribution, stress-transmissivity coupling, blast damage zones, gas dissolution, multiphase flow, capillary pressure effects, and geochemical changes. Given the uncertainty over processes active in the drift wall, the validation drift length was to 50 meters, and an increased emphasis was placed on borehole experiments and predictions.

Considerable effort was expended by all three modelling groups in attempting to incorporate excavation effects for the predictions of flow into the validation drift. These efforts met with varying success. Golder and AEA Harwell both initially combined continuum stress solutions with simple stress-transmissivity relationships as the basis for altering transmissivity values in the vicinity of excavations. The corrections were based on stress-flow relationships such as those shown in Figure 5. This mechanism failed to explain the observed 1 to 1.5

order of magnitude reduction in flow into the drift as compared with the simulated drift boreholes. Later we simply made an empirical correction of one order of magnitude in transmissivity inside a 2-meter skin around the drift.

Although subsequent modelling and analysis has been unable to completely resolve the mechanism behind the observed skin, the Stripa Phase 3 project demonstrated that it is possible to apply skin factors in discrete fracture modelling in the same manner that they are used in continuum approaches. As a result, the issue of coupled processes such as excavation effects does not limit the usefulness of discrete fracture approaches any more than it does continuum approaches. Thus, discrete fracture flow and transport modelling can be supplemented by external modelling which provides a basis for adjustment to the properties of fractures due to concurrent changes in stress, temperature, saturation, geochemistry, or other phenomena.

Conclusions

Through the Stripa Phase 3 project discrete-fracture modelling has been established as a valuable tool for evaluation of the hydrology of fractured rocks. The use of discrete fracture modelling at scales from 3 m (packer test simulations) to 20 m (tracer simulations) to 200 m (SCV block simulations) prepares the way for the use of the approach in canister scale (10 m), repository scale (100 m) and site scale (1000 m) site characterization and performance assessment applications. The simulations indicated realistic patterns of inflow to the drift walls (Figure 6). The Stripa Phase 3 project demonstrated that the data and computational requirements of discrete fracture modelling are compatible with the level of detail required for repository characterization. Although many issues remain to be resolved, discrete fracture technology is now established as a practical tool for analysis of fractured rock hydrology.

Major issues of discrete fracture modelling which remain to be resolved by future rock laboratory facilities include:

- the nature of flow and transport within fracture planes, and at fracture intersections;

- the identification of hydraulically significant features;

- determination of fracture shape and size by borehole methods; and

248

- derivation of transport properties for fractures and fracture networks.

Acknowledgements

The support of the U.S. Department of Energy and the Stripa project, as well as the efforts of the Stripa Project Manager, Dr. Bengt Stillborg, and the members of the Fracture Flow Task Force are gratefully acknowledged. Many persons from Golder Associates deserve acknowledgement for their efforts in the Stripa modelling effort. The initial FracMan model was developed by Joel Geier of Golder's Seattle office (and later Uppsala office), and the subsequent development of the code was largely the work of Dr. Glori Lee also in Seattle. Joel Geier, Scott Kindred, and Dr. Peter Wallmann were the key people for the reductions of the data for model use and for the simulations themselves. Masahiro Uchida of the Power Reactor and Nuclear Fuel Development Corporation participated in Stripa work during his sabbatical in Seattle. Dr. Wallmann also assisted greatly with preparation of this paper and the associated oral presentation. Dr. Thomas Doe has contributed to our modelling work in many ways from his help in arranging the American modelling participation at the beginning of Phase III, to assisting with the model concepts, and finally to his extensive review and this paper contains his extensive contributions. A special credit goes to the graphics department at Golder-Seattle and to Carol Baker, in particular.

References

Baecher, G., N. Lanney, and H. Einstein, 1977. Statistical Description of Rock Properties and Sampling. Proc. 18th US. Symposium on Rock Mechanics, American Institute of Mining Engineers (AIME)

Dershowitz, W., 1983. JINX: Joints in Networks Three Dimensional Discrete Fracture Model, User Documentation. Massachusetts Institute of Technology, Cambridge, MA.

Dershowitz, W., J. Geier, and G. Lee, 1992. FracMan Version 2.3 Discrete Feature Data Analysis, Conceptual Modelling, Exploration, and Flow and Transport Modelling System. Golder Associates Inc., Redmond, WA, USA.

Dershowitz, W. and H. Herda, 1992. Measures for Fracture Intensity. Proceedings, 31st US Rock Mechanics Symposium, Santa Fe, NM.

Dershowitz, W., P. Wallmann, and S. Kindred, 1991. Discrete Fracture Modelling for the Stripa Site Characterization and Validation Drift Inflow Predictions. SKB Stripa Project Report TR 91-16, SKB, Stockholm.

Carlsson, H. (ed.), 1987. Program for the Stripa Project Phase 3, 1986-1991. Stripa Project TR 87-09.

Dershowitz, W., P. Wallmann, J. Geier, and G. Lee, 1991. Preliminary Discrete Fracture Network Modelling of Tracer Migration Experiments at the SCV Site. SKB Stripa Project Report TR 91-23, SKB, Stockholm.

Doe, T. and J. Geier, 1990. Interpretation of Fracture System Geometry from Well Test Data. Stripa Project Report TR 91-03

Gale, J.E. and P. Witherspoon. 1978. An Approach to the Fracture Hydrology at Stripa -- Preliminary Results. LBL-SAC Report 15, Lawrence Berkeley Laboratory

Hodgkinson, D., 1992. A Compilation of Minutes for the Stripa Task Force on Fracture Flow Modelling. SKB Stripa Project Report TR 92-09, SKB, Stockholm.

Hodgkinson, D. and N. Cooper, 1992. A Comparison of Measurements and Calculations for the Stripa Tracer Experiments. SKB Stripa Project Report TR 92-20, SKB, Stockholm.

Long, J., 1983. Investigation of Equivalent Porous Media Permeability in Networks of Discontinuous Fractures. PhD. Dissertation, University of California, Berkeley.

Noy, D., J. Barker, and J. Black. 1988. Crosshole Investigations -- Implementation and Fractional Dimension Interpretation of Sinusoidal Tests. Stripa Project Report TR 88-01.

Olsson, O. (ed.), 1992, Site Characterization and Validation -- Final Report. Stripa Project TR 92-22.

Robinson, P, 1984. Connectivity, Flow, and Transport in Network Models of Fractured Media. PhD. Dissertation, Oxford University, Oxford.

Snow, D.T., 1965. A Parallel Plate Model of Fractured Permeable Media. Ph.D. Thesis, University of California, Berkeley, 331p.

Table 1. Data Analysis Procedures for Discrete Fracture Modelling		
Data Requirement	Data Used	Summary of Procedure
Spatial Pattern of Fracturing	Trace Maps Geophysically Identified Zones	Statistical Confidence Tests (Chi-Squared, Kolmogorov-Smirnov) on the statistical implications of alternative geological, fractal, geostatistical, empirical, and tesselation models for fracture location.
Fracture Size	Trace Maps Geophysically Identified Zones	Simulation of the process of intersection between fractures and trace maps, with optimization of assumed fracture size to match observed trace statistics.
Fracture Shape and Connectivity	Trace Maps	Analysis of Fracture Connectivity/Intensity Relationships (e.g., termination statistics, intersection intensities)
Fracture Orientation	Borehole Logs, Formation Microscanners (FMS), Trace Maps	Discriminant analysis to determine the parameters appropriate for grouping of fractures into sets, with a probabilistic acceptance/rejection algorithm to assign fractures to sets and calculate the set orientation statistics.
Fracture Hydraulic Properties (Transmissivity, Storativity,	Single Borehole Hydraulic Tests (DST, Packer Tests)	Simulation of the process of intersection between conductive fractures and boreholes, with optimization of assumed fracture intensity and transmissivity to match observed packer test statistics and transient responses.
Fracture Transport Properties	Single Fracture Transport Experiment Breakthrough Curves, Multiple Fracture Transport Experiment Breakthrough Curves	Back analysis of single fracture breakthrough curves to derive transport aperture, and in-fracture lateral and longitudinal dispersivity, and calibration of assumed values for these parameters within fracture network models to match multiple fracture breakthrough curves.

Table 2. Discrete Fracture Modelling Scale

Site, Experiment	Model Scale	Discrete Features Modelled (approximate)
Stripa Simulated Drift Experiment	200 m x 200 m x 200 m	2000
Stripa BMT - SCV Area Interference	40 m x 40 m x 40 m	500
Stripa, Validation Drift Experiment	200 m	2000
Stripa, Transport Experiment Design	300 m	500
Stripa, Tracer Validation Experiment	200 m	1000
Stripa Saline Radar 2 Experiment	200 m	3000
SKB Aspo Hard Rock Laboratory, Block Scale	50 m	500
Nagra Sedimentstudy, USM Lower Freshwater Molasse "Marl"	100 km, 10 km, 100 m	1 to 1000 (faults only)
Nagra Sedimentstudy, OPA Opalinus Claystone	100 km, 10 km, 100 m	1 to 1000 (faults only)
Yucca Mountain Rock Block Test	0.5 m	250
Yucca Mountain, Pathways Analysis	40 m	500
PNC Kamaishi Mine, Site Scale	200 m	1000
SKB Finnsjon, Packer Test Simulation	40 m x 40 m x 40 m	500
SKB Finnsjon, Block Scale	20 m	500
SKB Finnsjon, Site Scale	300 m	2000

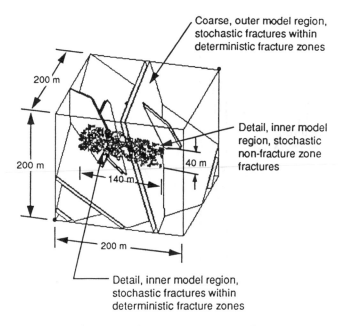

Coarse, outer model region,
stochastic fractures within
deterministic fracture zones

Detail, inner model
region, stochastic
non-fracture zone
fractures

200 m

200 m

40 m

140 m

200 m

Detail, inner model region,
stochastic fractures within
deterministic fracture zones

1a. Validation Drift Discrete Fracture Conceptual Model

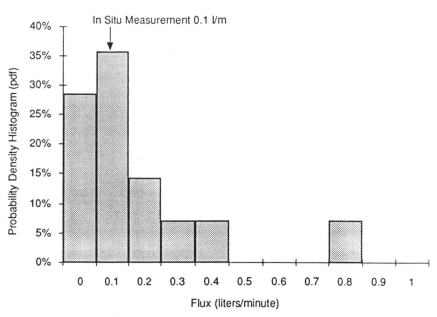

1b. Probability Drift Inflow Prediction.

Figure 1. Flow Validation Experiment.

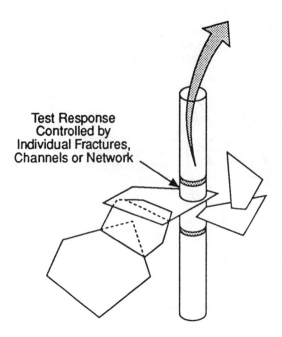

Test Response Controlled by Individual Fractures, Channels or Network

2a. Hydraulic Test in a Fractured Rock.

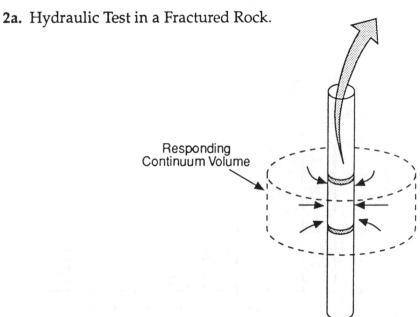

Responding Continuum Volume

2b. Continuum Assumption.

Figure 2. Discrete Feature Hydrologic Test Interpretation.

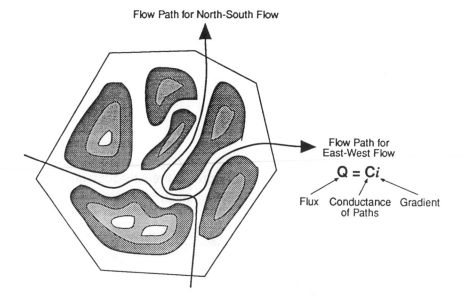

Flow Path for North-South Flow

Flow Path for
East-West Flow

$$Q = Ci$$

Flux   Conductance   Gradient
of Paths

3a. Apparent Flow Channels due to Roughness.

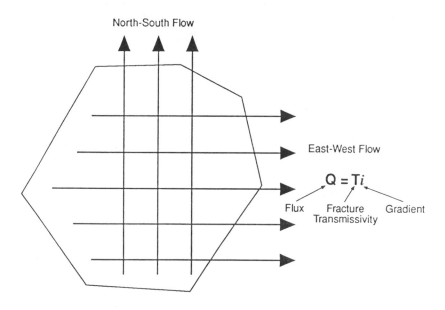

North-South Flow

East-West Flow

$$Q = Ti$$

Flux   Fracture   Gradient
Transmissivity

3b. Equivalent P "Cross Fracture" Transmissivity.

Figure 3. Cross Fracture Transmissivity.

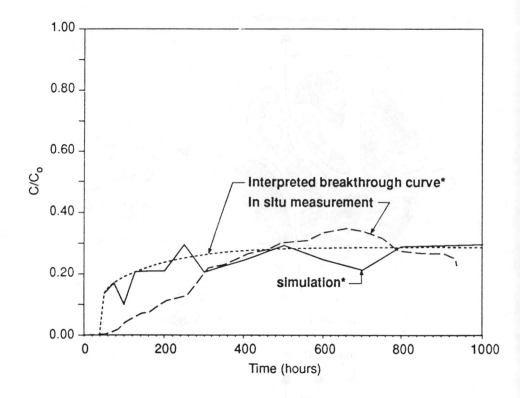

*Simulation results are coarse due to the limited number of particles used.

Figure 4. Total Breakthrough to Validation Drift.

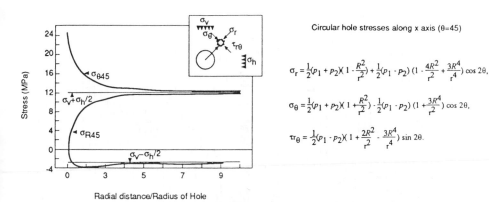

Circular hole stresses along x axis (θ=45)

$$\sigma_r = \tfrac{1}{2}(p_1 + p_2)(1 - \tfrac{R^2}{r^2}) + \tfrac{1}{2}(p_1 - p_2)(1 - \tfrac{4R^2}{r^2} + \tfrac{3R^4}{r^4})\cos 2\theta,$$

$$\sigma_\theta = \tfrac{1}{2}(p_1 + p_2)(1 + \tfrac{R^2}{r^2}) - \tfrac{1}{2}(p_1 - p_2)(1 + \tfrac{3R^4}{r^2})\cos 2\theta,$$

$$\tau_{r\theta} = \tfrac{1}{2}(p_1 - p_2)(1 + \tfrac{2R^2}{r^2} - \tfrac{3R^4}{r^4})\sin 2\theta.$$

5a. Kirsh Plane-Strain Stress Solution.

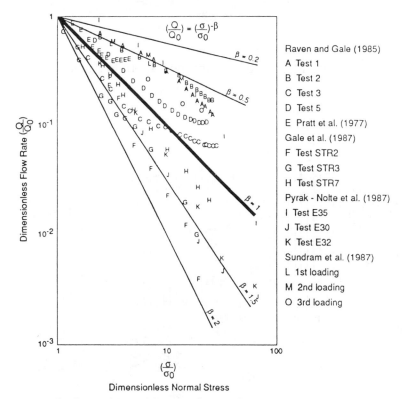

$$\left(\frac{Q}{Q_0}\right) = \left(\frac{\sigma}{\sigma_0}\right)^{-\beta}$$

Raven and Gale (1985)
A Test 1
B Test 2
C Test 3
D Test 5
E Pratt et al. (1977)
Gale et al. (1987)
F Test STR2
G Test STR3
H Test STR7
Pyrak - Nolte et al. (1987)
I Test E35
J Test E30
K Test E32
Sundram et al. (1987)
L 1st loading
M 2nd loading
O 3rd loading

5b. Stress/Transmissivity from Laboratory Experiments.

Figure 5. Stress/Transmissivity Correction.

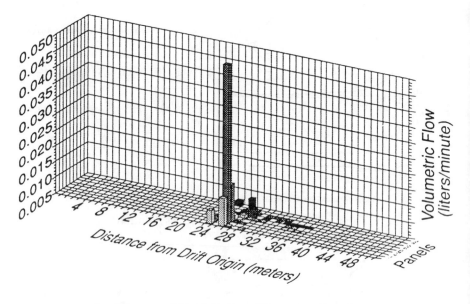

6a. Measured Drift Inflow Pattern to Panels.

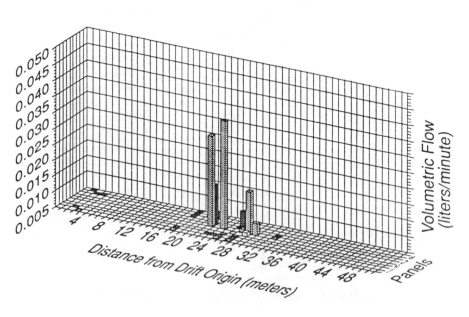

6b. Predicted Drift Inflow Pattern to Panels.

Figure 6. Validation Drift Prediction of Inflow Pattern.

Modeling Flow and Transport for Stripa Phase 3
What did we learn ?

Jane C.S. Long
Earth Science Division
Lawrence Berkeley Laboratory
United States

Abstract

The Stripa Project provided a unique opportunity to develop and apply flow and transport models for fractured rock. A complex and extensive series of measurements provided unique opportunities to plan, execute and evaluate modeling exercises. At this point, it is useful to look back on what was done and assess this work. Several key points emerge: 1) Extensive characterization efforts with geophysics and hydrologic testing resulted in a tremendous improvement in our ability to develop the conceptual model underlying our numerical analyses. 2) For this particular case, equivalent discontinuum models conditioned on well test behavior were reasonably able to predict new flow regimes. 3) The SDE in combination with modeling studies provided a unique measurement of excavation effects and important data for designing further experiments, but the excavation effects undermined the utility of the SCV as "validation" data for numerical models not designed to simulate complex physics. In fact, this example underscores the practical difficulties that plague the concept of validation in general and thus serves as a valuable object lesson. 4) We gained evidence that models conditioned on flow behavior can provide some information about the behavior of tracer transport.

The work at LBL was supported by the U.S. Department of Energy through the Office of Civilian Radioactive Waste Management, Office of Strategic Planning and International Programs, Yucca Mountain Site Characterization Project under Contract No. DE-AC03-76SF00098. The author wishes to thank Kenzi Karasaki, Amy Mauldon, Kent Nelson, Steve Martel, and Olle Olsson.

Introduction

The Stripa Project represents a major milestone in the development of technology for the characterization of fractured rock for the purpose of predicting fluid flow and transport. There is probably no other in situ block of rock of similar scale that has been probed in so many ways with such intensity over such a time scale. An examination of what was learned at Stripa is nearly equivalent to an examination of what is currently knowable about flow and transport in crystalline rock.

This paper reviews a few of the key findings which involved fluid flow and transport modeling. The analyses are based on the conceptual model described in Olsson et al. (1989, 1992) and Black (1991). This model included the definition of the major fracture zones in the Site Characterization and Validation (SCV) rock block. In particular, the 2 conceptual model identified the H-zone as a sub-vertical zone which was the only major fracture zone to intersect the Validation Drift (VD) in the SCV.

This discussion of model predictions will refer to three hydraulic experiments. The first is the set of cross-hole interference tests (Black, 1991) that took place between packed-off intervals of the SCV boreholes. In particular, we concentrate on the C1-2 test whose source was confined to interval 2 in borehole C1 that intersected fracture zone H. The second experiment was the Simulated Drift Experiment (SDE) (Black et al., 1991) which was meant to mimic the behavior of a drift from the hydrologic point of view by placing six parallel boreholes within a ring. The third is the Validation Drift inflow experiment (Olsson et al., 1992) which was conducted from a drift excavated through one half the length of the SDE boreholes. The SDE provided a unique study of the effects of excavation on flow.

The discussion will also refer to two sets of tracer experiments. The radar-saline experiments (Olsson, 1991a,b) consisted of combining a saline tracer test with radar tomography. In the first test 30 l/min of saline was injected into the C1-2 interval and collected at a sink formed by opening the D-holes in the vicinity of the H-zone. The second test was a repeat of the first but took place after the excavation of the Validation Drift. The second set of tests were tracer tests conducted after the excavation of the Validation Drift where a variety of tracers were injected at very low rates from a variety of borehole intervals intersecting the H-zone and collected at the Validation Drift, and the tracer tests (Birgersson et al., 1992)

The use of these experiments was as follows: The C1-2 test was used to develop a model of fracture flow in the SDE based on the conceptual model of Black et al., (1991). The modeling approach used by LBL was to create hierarchical fracture systems through an iterative optimization procedure, or inverse method (Long et al., 1989, Long et al., 1992a). In this approach the fracture network is optimized such that matches certain observations, in this case the C1-2 interference test. Simulated Annealing is used as the optimization technique because the inversion problem is highly non-linear and Simulated Annealing allows the optimization process to escape from local minima. A model optimized to the C1-2 test is then used to predict the SDE inflow results. Then a model is annealed to both the C1-2 and the SDE results and used to predict the Validation Drift Inflow. Finally, the ratio of flux to velocity in both of these models are calibrated such that they best predict the breakthrough curve for the first radar-saline test. These calibrated models are used to predict the second radar-saline test and the subsequent tracer tests.

The importance of the conceptual model

A conceptual model is an interpretation of all the observations of a fracture system which provides a description of how flow and transport occurs and is the basis of numerical models that can quantify flow and transport. The conceptual model is built based on interpretation of laboratory studies, field mapping, geophysical measurements and in situ hydrologic tests. Building the conceptual model is the most important part of fracture characterization because the conceptual model is critical to making predictions. For example, suppose we have in in situ measurement of transmissivity. The transport velocity we predict for this system will vary by many orders of magnitude depending on whether we use a three-dimensional porous media approximation, a two-dimensional parallel plate model for the fracture, or a one-dimensional channel model. The error associated with choosing the wrong conceptual model in this case is much more significant than the measurement error or the numerical errors.

Understanding the geometry of the fracture system and how that geometry controls flow is central to conceptual modeling. There are two types of fracture patterns that result in important flow paths: One is a well connected cluster of open fractures or a fracture zone. The other is an extension feature of significant extent. The former is apparently the case at Stripa. Nearly 100% of the flow is observed to occur in fracture zones.

The choice of an appropriate conceptual model also depends on the phenomena of interest. A simple prediction of flow rate as a function of time is not highly dependent on having a physically realistic conceptual model. Thus, the prediction of inflow was insensitive to detailed conceptual modeling. If however, one needs know where waste will migrate, a more complete understanding is critical. The purpose of the detailed conceptual modeling and development of sophisticated flow models was primarily to set the stage for the more complex transport modeling.

Some salient features of the conceptual model for the Stripa SCV can be briefly summarized as follows: 1.) The fluid flow is largely confined to fracture zones. 2.) There are approximately seven fracture zones in the block, including 3.) major zones whose location, orientation and extent have been determined. 4.) The fracture zones are not uniformly permeable.

The conceptual model formed the basis of a series of numerical models for flow and transport. These models were formed by creating a lattice of conductors on each of the planes representing the fracture zones identified in the conceptual modeling exercise. Simulated Annealing was used to conduct a random search through the elements of the lattice to find a configuration of elements that matches the hydraulic test data (Long *et al.*, 1992a). Simulated Annealing results in a solution which is constrained to behave like the hydrologic observations. The process can be repeated many times to get a series of models which all behave like the hydrologic observations. Although this was not done for the Stripa data, it is possible in this way to determine the what constraints are placed on the model by the data.

Prediction of the simulated drift experiment

Two kinds of models were created. One which looked at the H-zone alone as it was the only fracture zone to intersect the Validation Drift. The other represented all seven fracture zones (Figure 1). Each of these were annealed to the C1-2 data. Figure 2 shows the two-dimensional H-zone model annealed to the C1-2 data. Each of these annealed models was used to predict the inflow to the D-holes. The two-dimensional model predicted an inflow of 0.77 l/min and the three-dimensional model predicted 0.95 l/min. The measured inflow was 0.77 l/min. These are highly encouraging results. The models are constrained only by the conceptual modeling and a.single interference test and they predict a second test extremely well.

Issues That Came About In Modeling the Validation Drift

The next stage of the SCV project was to excavate the Validation Drift through the holes of the SDE. Prediction of inflow to the drift was to be accomplished with the models developed using the data from the SDE. The inflow measurement was meant to be used to test the predictive abilities of the models. Below we examine the reasons that this inflow measurement was inappropriate for this purpose, but was valuable as a measure of excavation effects.

The SDE was conducted by lowering the head in the boreholes in three separate steps. Lowering the head in the boreholes increases the gradient into the holes and increases the flow rate. Each step was maintained for several weeks to develop steady state conditions. Figure 3 shows the flow rate measured at the end of each step as a function of pressure head. During the last step, the pressure head in the boreholes was lowered to approximately 17m above the elevation of the boreholes.

The shape of the isopotential boundary conditions imposed by the D-holes was almost identical to that imposed by the excavation (Black *et al.*, 1991). Thus, extrapolating the flow rates observed during the three steps of the SDE to the case for atmospheric pressure gives an estimate of the Validation Drift inflow discounting any other effects due to excavation.

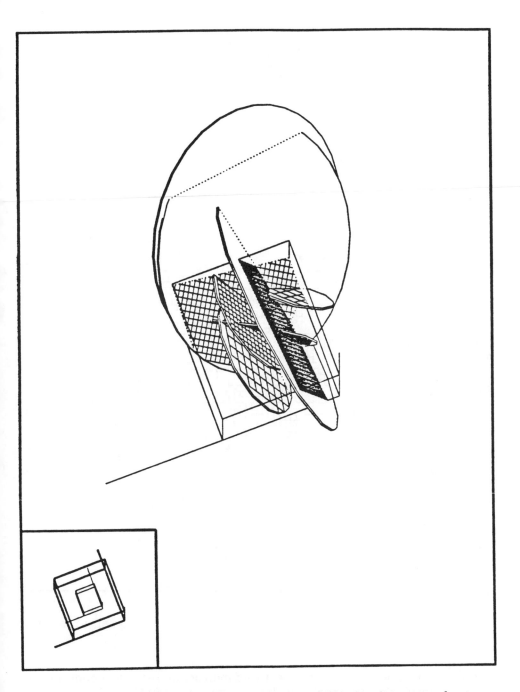

Figure 1.   The three-dimensional model of the SCV showing seven fracture zones represented by seven lattice structures. (XBL 921-5543)

263

2-D C1-2 annealed mesh
(Dead-end elements dotted)

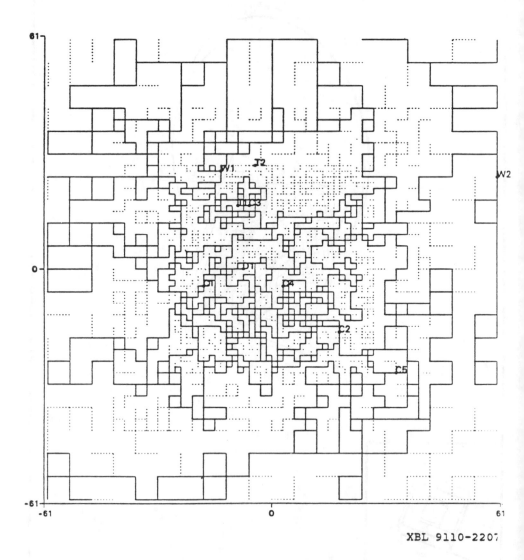

XBL 9110-2207

Figure 2. The configuration of the two-dimensional H-zone model which resulted from annealing to the C1-2 data. (XBL 9110-2207)

Flow to the D-boreholes

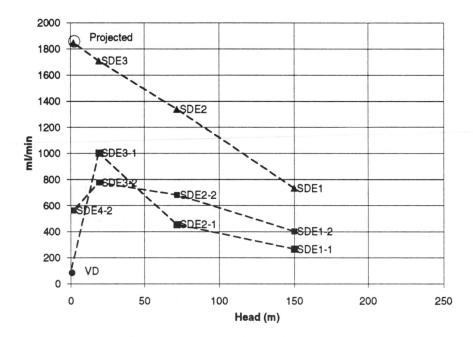

Figure 3. Inflow to the D-holes as a function of pressure head. SDEx = total flow, SDEx-1 = flow to 0 to 50 m, SDEx-2 = flow to 50 - 100m, + = flow to "average rock," i.e. rock not in the H-zone. (The x denotes the steps of the SDE.) Validation Drift inflow is shown with a • symbol.

The total inflow to the D-holes is comprised of flow to the first 50 meters of the boreholes which were later excavated plus the second 50 meters which were not excavated. Between the second and third steps, flow was redistributed among the boreholes as evidenced by the non linear flow plots for each of these two borehole sections on Figure 3. However, the total flow is linear with pressure decrease as expected. Based on extrapolating the flow rate from each of the three steps, flow into the first 50 m of the boreholes was estimated to be 0.88 l/min (880 ml/min) at atmospheric pressure.

The excavation of the Validation Drift reduced the pressure in the drift to atmospheric. Thus the flow into the excavation area should have increased and consequently the pressures in the surrounding rock should have decreased. Pressures in the surrounding rock began to rise shortly after excavation commenced and continued to rise steadily during excavation. This means that inflow must have been decreasing. After excavation, the total measured inflow was approximately 0.1 l/min, roughly a factor of 9 lower than the SDE measurement.

A series of numerical analyses were conducted, first to simulate the SDE experiment and then these simulations were modified to predict the Validation Drift inflow (Herbert *et al.*, 1991; Dershowitz *et al.*, 1991; Long *et al.*, 1992). Herbert *et al.* used an analysis of elastic, continuum stress changes due to excavation (McKinnon and Carr, 1990) to modify a discrete fracture model of the SDE to predict drift inflow. Herbert *et al.* used the computed change in normal stress acting on each fracture in the model and a relationship between stress and permeability derived from laboratory measurements (Gale *et al.*, 1990) to change the permeability of the affected fractures. The net result was a small increase in inflow over that calculated for the SDE, contrary to what was observed. Dershowitz performed similar analyses in a stochastic mode and found increases in flow for some realizations and decreases in others.

Long *et al.* (1992a) used the equivalent discontinuum models for the SDE described above which included flow only in the fracture zones in the vicinity of the D-holes. Data on pressure distribution around a drift a few hundred meters to the west of the Validation Drift (the Macropermeability Drift) was used to infer the magnitude of permeability decrease in a 5 m wide zone around the drift. These data indicate that the permeability in the first 5 meters is decreased on average by a factor of four (Wilson *et al.*, 1981). When this permeability reduction was used in predictions of the inflow to the Validation Drift the predicted inflows were high by a factor of about five. To match the drift inflow the permeability in the first five meters had to be decreased by a factor of 40. Evidently the rock around the Validation Drift experiences a significantly less permeable skin than that observed during the Macropermeability Experiment (Wilson *et al*, 1981). The maximum reduction in permeability that can be inferred from the Macropermeability Experiment is a factor of 20. So the factor of 40 for the Validation Drift may not be unreasonable.

Long *et al*, (1992b) examined possible causes for this dramatic decrease in permeability in the vicinity of the Validation Drift. These are summarized below.

The effect of additional sinks drawing off an increasing proportion of the flow after excavation was examined numerically and shown to be unlikely to explain the decrease in VD inflow.

Drifts can significantly perturb the stress field and hence the fracture conductivity within a few diameters of a drift. The stress perturbation caused by the excavation of the Validation Drift was analyzed to examine the possible effect of inflow to it along the H-zone (Long *et al.*, 1992a). The orientations of the Validation Drift and the most compressive far-field horizontal stress differ by only 4°. The stress state and the relevant geometries thus indicate that the change in stress on fractures perpendicular to the drift axis will be minimal. Therefore a 2-D plane strain analysis is useful for analyzing the stress effects on the H-zone due to excavation. Using the most recent stress measurements for the Validation Drift area, this analysis indicates the maximum increase in normal stress will occur for fractures radiating from the drift. The compressive normal stress tangential to the drift wall should increase between 50% (at the drift walls) and 133% (at the roof and floor of the drift); these effects decay to less than 10% within a few meters from the drift. Experimental work on laboratory core samples suggests that the ratio of the change in fracture hydraulic conductivity to the change in normal stress varies as $\sigma_N{}^a$, where a most likely is in the range -0.2 to -1 (Gale *et al.*, 1990b, Makurat *et al.*, 1990). Accordingly, the hydraulic conductivity along fractures oriented radially to the drift should decrease by no more than 40% at the drift perimeter and by no more than 7% as appropriately averaged over a 5-meter distance from the drift. The absolute magnitude of the normal stress changes parallel to the drift are small (<15%) and average to zero around the perimeter of the drift. There should be little direct effect on the inflow along drift-perpendicular fractures (i.e. the longest H-zone fractures).

A three-dimensional simulation of the stress field around the drift was made by Tinucci and Israelsson (1991) using the discrete fracture code, 3-DEC. The result of their work essentially confirms the conclusions based on the 2-D plain strain analysis presented above. In a two- dimensional simulation with a fully coupled model Monsen *et al.* (1991) show no significant differences with the continuum model results. It is difficult to see how changes in normal stress across fractures could decrease the overall permeability of the first 5 m of rock by a factor of 40.

On the other hand, shear displacements may occur due to excavation. These may lead either to increases or to decreases in permeability depending on the amount of shear and the accompanying normal stress acting on the fracture Although shear deformation can not be ruled out as a contributor to flow reduction, the fact that pressure heads began to rise before the main conductor was excavated does not support this theory.

Two-phase flow caused by differential drying during ventilation has been suggested as a possible reasons for a decrease in flow (K. Pruess, pers. comm., 1991). Drying could occur due to the ventilation procedure. As the ventilation rate is increased, either by raising the temperature in the drift or increasing the air flow rate, the measured inflow of water will change. At first, an increased ventilation rate will increase the measured flow as more water is essentially sucked out of the rock. Then, the rock will start to dry out and air invasion could decrease the effective permeability of the rock near the wall. During part of the 7 drift inflow measurements, the plastic sheets were removed, thus effectively increasing the ventilation rate. When the sheets were removed the measured flow rate increased. This indicates that the ventilation was not producing any significant two-phase flow effects.

Blasting can damage the rock through the creation of new fractures (i.e. increase permeability), or cause gas to be intruded into the rock mass, and shake loose fine particles which then block flow paths (i.e. decrease permeability). The effect of fracture formation may explain some of the difference between the Macropermeability Drift and the Validation Drift because the Macropermeability Drift was excavated with a smooth blasting technique and the Validation Drift was excavated with a pilot and slash technique which caused very little blast damage to the rock. It may be that in the Validation Drift careful blasting did not increase the permeability near the drift as much as blasting the Macropermeability Drift did.

The near surface waters at Stripa are rich in carbonates and the deeper waters are rich in sodium. Mixing of these waters is caused by both waters flowing towards the same sink, i.e. the Validation Drift, and results in a water that is oversaturated in calcite. Precipitation of calcite in the fractures would decrease the permeability, but this precipitation should be independent of whether the drain consists of a borehole or a drift. In addition, if calcite precipitation was significant, a permeability reduction would be expected during the SDE and this was not observed.

During the last step of the SDE, gas bubbles were observed in the outflow tubing. Gas bubbles that are constantly released as water approaches atmospheric pressure could have a very significant effect by causing two- phase flow and a significant decrease in relative permeability if the gas bubbles remain in the rock mass. Samples of the inflow water show that as much as 5 or 6% by volume of the water is nitrogen that comes out of solution at atmospheric pressure. Application of Henry's law to the measured gas content of the inflow water shows than much smaller amounts of gas should come out of solution at the 17 m pressure head used in the last step of the SDE. Thus the two-phase flow due to degassing scenario is constant with a significant flow drop between the SDE and the VD. Degassing is also consistent with the observation that pressures began to increase immediately following the start of excavation. Finsterle and Vomvoris (1991) attempted to numerically simulate degassing for the Validation Drift and found no effect on inflow rate, however, a different result might have been found if the simulations had used different parameters or geometry more like the true flow system.

In conclusion, the most plausible cause for a significant decrease in the permeability of the skin surrounding the Validation Drift is degassing as the pressure of the water is dropped to atmospheric on inflow. Excavation method, stress effects and dynamic loading may also have some importance. It should be noted that the orientation of this tunnel with respect to the maximum principal stress and the dominant fracture orientation and its small size (< 3 m diameter) may be largely responsible for the fact that stress changes appear to be relatively unimportant.

An important consequence of these observations is that hydrologic measurements in tunnels can not be used straight forwardly for characterizing the flow system due to the unknown magnitude and cause of the skin. It is certainly inappropriate to consider inflow measurements to drifts as indicative of the undisturbed hydrologic regime. In the case of the SCV modeling effort the VD measurements were dominated by physical effects that were not part of the modeling exercise. Hence, the VD inflow was not useful for examining the appropriateness of the modeling efforts.

The combination of the SDE and VD experiments were, however, very valuable for understanding excavation effects. The process of gas dissolution is eventually reversible and if a low permeability skin is caused by two-phase flow effects, refilling the tunnel is likely to cause permeability to increase. This is of importance for the storage of nuclear waste in underground repositories, for water transport tunnels and other cases where inflow to an underground excavation needs to be understood. This phenomenon may create an opportunity to study two-phase flow effects in fractured rock by controlling the pressure in boreholes drilled from an underground facility.

What was learned from modeling the tracer tests

A series of tracer simulations were compared to results of in situ tracer tests. The simulations were based on two, two-dimensional equivalent discontinuum models of the H-zone. The first, mentioned in section 1 above, was annealed to the C1-2 test and is called the "C1 model". The second, called the co-annealed case, was simultaneously annealed to the C1-2 test and the observations made at the end of the SDE.

Both models were used to simulate the first Radar/Saline (RSI) experiment, i.e. injection of tracer in C1-2 and collection of tracer in a sink caused by opening the D-holes in the vicinity of the H-zone. The actual breakthrough curve was used to calibrate the model by changing the ratio of flow to velocity in the conductive elements of the model to match the breakthrough curve. Only adjective dispersion of tracer was allowed. This calibrated model was then modified to include the low permeability skiing around the drift and used to predict the second Radar/Saline experiment (RSII), this time from C1-2 to the Validation Drift. Figure 4 shows the predicted and measured breakthrough curves for RSI and RSII respectively in the C1-2 configuration. Then a series of tracer tests from a variety of borehole intervals within the H-zone to the drift were simulated.

In the models, the change in boundary conditions from RSI to RSII had the effect of increasing the maximum C/Co. In the data, the excavation left the maximum C/Co relatively unchanged. The increase in C/Co from our simulation of RSI to RSII is explainable because the injected water is a greater proportion of the inflow to the drift than the inflow to the D- holes. However, the fact that the same trend is not observed in the data may be due to the effects of excavation and the resulting two-phase flow near the drift, or the fact that the remaining D-holes which were left open could be pulling tracer away from the drift, or that the higher injection pressures of RSII increased the local dispersion phenomenon. Thus, lack of knowledge of the boundary conditions prevents a good estimate of the maximum concentration. Both the C1 and the Co-Annealed configurations match the arrival time of the RSII data fairly well.

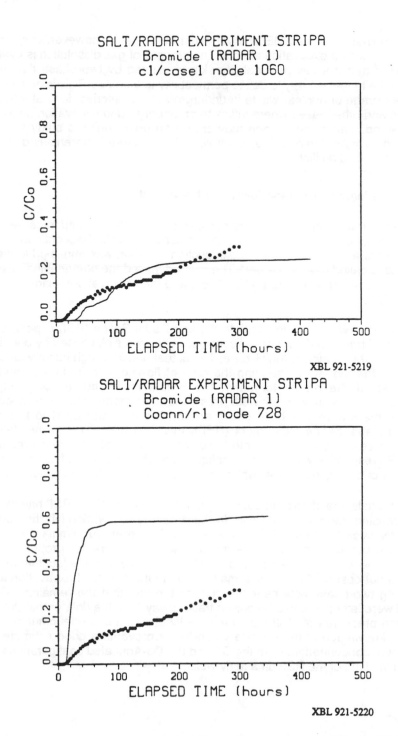

Figure 4. Predicted and measured breakthrough curves for a) RSI and b) RSII.

The tracer tests were simulated twice: once with the C1-2 configuration and once with the co-annealed configuration as described above. No dispersion coefficient was used. The breakthrough curves are all much steeper than that observed. As can be seen in Table 1 the models do extremely well in predicting the breakthrough times. It seems as if these hydraulically based models do reasonably well in predicting arrival times. However, more physical reality, e.g. more channel conductance variability, may be needed to capture more of the transport behavior. Knowledge of boundary conditions is much more critical to predicting tracer transport than it is for predictin of flow.

Table 1. Estimates of first arrival and maximum concentration for the tracer tests.

Case	Source	First arrival (hrs)		Maximum C/Co	
		Predicted	Data	Predicted	Data
1	T1:2	800 to 2800	400	0.0001 to 0.0005	
2	T2:1	200 to 1400	200	0.0001 to 0.0004	0.002
3	C2	200 to never	300	0 to 0.002	0.002

Conclusions

Some of what was gained from the Stripa Project was not planned: as in most earth science research, the results may not exactly match the original goals. Nevertheless the actual results are significant. The stated aims of the project included "validation" of fluid flow and transport codes. I would argue that this is not a possible achievement in a strict sense. Simply changing the definition of validation such that validation somehow becomes achievable trivializes and obfuscates an accurate assessment of the modeling effort. What we have learned is that the codes are a mathematical formalization of the exceedingly more important effort of "conceptual modeling". Stripa is by far the best example of conceptual modeling done to date for saturated rock.

Although I have only discussed the modeling efforts done by LBL in this paper, all of the modeling efforts showed that the key to good modeling is good characterization. Each of the codes had advantages and disadvantages and each could be applied with flexibility in meeting the challenge of predicting behavior. None of the codes would have come anywhere near making good predictions without a good understanding of the fracture system.

A recommendation for future work is that a new studiy of tracer transport in fracture networks should not be confounded with excavation effects. There are many things that we do not fully understand about the physics of the transport phenomenon in fractures. It may be possible to match breakthrough curves, but we do not yet know the best way to build predictive models. The SCV project has shown clearly that we do not understand the hydrology of excavations. Coupling these two problems makes it very difficult to interpret the experiments.

It may be very productive to conduct a series of interference tests as the basis for an iterative model development process. The idea would be to optimize the model to one test and predict the second. Then optimize the model to the first two tests and predict the third, etc. In this way it may be possible to see how the ability to predict improves with additional data.

It will be very important to study the effects of excavation in such a way that the various physical phenomena can be deconvolved. Much of the inferences about a rock mass being considered for a nuclear waste repository will be made from observations in underground excavations. We will consequently need to know how to conditions our inferences to reflect the excavation effects.

Finally, in formations rich in dissolved gasses, underground excavations in otherwise saturated rock may provide an excellent opportunity to performed controlled studies of two-phase flow in fractures.

References

Birgersson, L., H. Widen, T. Agren, I. Neretnieks, L. Moreno, 1992, characterization and validation, Tracer migration experiment in the Validation Drift, Report 2, Part 1: performed Experiments, Results and Evaluation. Stripa Project TR 92-03, SKB, Stockholm, Sweden, 1992

J. Black, O. Olsson, J. Gale, D. Holmes, Site characterization and validation, stage IV—Preliminary assessment and detail predictions. Stripa Project TR 91-08, SKB, Stockholm, Sweden, 1991.

G. G. Bursey, J. E. Gale, R. MacLeod, A. Strahle, S. Tiren, Site characterization and validation—Validation Drift fracture data, stage IV. Stripa Project TR 91-19, SKB, Stockholm, Sweden, 1991.

Davey, A., K. Karasaki, J. C. S. Long, M. Landsfeld, A. Mensch and S. Martel, 1989. Analysis of the Hydraulic Data of the MI Experiment, Report No. LBL-27864, Lawrence Berkeley Laboratory, Berkeley, CA.

W. Dershowitz, P. Wallman, S. Kindred, Discrete fracture modeling for the Stripa site characterization and validation drift inflow predictions. Stripa Project TR 91-16, SKB, Stockholm, Sweden, 1991.

S. Finsterle and S. Vomvoris, Inflow to Stripa Validation Drift Under Two-Phase Conditions: Scoping Calculations, Nagra Report 91-40, Baden, Switzerland, 1991.

J. Gale, R. MacLeod, P. LeMessurier, Site characterization and validation-Measurement of flowrate, solute velocities and aperture variation in natural fractures as a function of normal and shear stress, stage III. Stripa Project TR 90-11, SKB, Stockholm, Sweden, 1990b.

J. Gale, R. MacLeod, P. LeMessurier, Site characterization and validation-Measurement of flowrate, solute velocities and aperture variation in natural fractures as a function of normal and shear stress, stage I. Stripa Project TR 90-11, SKB, Stockholm, Sweden, 1990b.

A. W. Herbert, J. E. Gale, G. W. Lanyon, B. MacLeod, Modelling for the Stripa site characterization and Validation Drift inflow prediction of flow through fractured rock. Stripa Project TR 91-35, SKB, Stockholm, Sweden, 1992.

M. Laaksoharju, 1990. Site Characterization and Validation—Hydrochemical Investigations in Stage 3. Stripa Project TR 90-08, SKB, Stockholm, Sweden, 1990.

Long, J., K. Karasaki, A. Davey, J. Peterson, M. Landsfeld, J. Kemeny and S. Martel, 1991. An Inverse Approach to the Construction of Fracture Hydrology Models Conditioned by Geophysical Data—An Example from the Validations Exercises at the Stripa Mine, *Int. J. Rock Mech, Min. Sci and Geomech. Abstr. 28,* (2/3), 121-142.

J. C. S. Long, A. D. Maldon, K. Nelson, S. Martel, P. Fuller, K. Karasaki, Prediction of flow and draw-down for the Site Characterization and Validation site in the Stripa mine. Report No. LBL-31761. Lawrence Berkeley Laboratory, Berkeley, CA, 1992a (Also SKB 92-05)

Long, J.C.S., O. Olsson, S. Martel, J. Black, 1992b Effects of Excavation on Water Inflow to a Drift, I.S.R.M Symposium on Fractured Rock Masses, Tahoe City, California, June 1992

A. Makurat, N. Barton, L. Tunbridge, Site characterization and validation—Coupled stress-flow testing of mineralized joints of 200 mm and 140 mm length in the laboratory and in situ, stage III. Stripa Project TR 90-07, SKB, Stockholm, 12 Sweden, 1992.

S. McKinnon, P. Carr, Site characterization and validation—Stress field in the SCV block and around the Validation Drift, stage III. Stripa Project TR 90-09, SKB, Stockholm, Sweden, 1990.

K. Monsen, A. Makurat, N. Barton, Disturbed zone modeling of SCV Validation Drift using UDEC-BB, models 1 to 8 -Stripa phase 3, 1991.

O. Olsson, J. H. Black, J. Gale, D. Holmes, Site characterization and validation stage II—Preliminary predictions. Stripa Project TR 89-03, SKB, Stockholm, Sweden, 1989.

Olsson, O., P. Andersson, E. Gustafsson, 1991a. Site Characterization and Validation—Monitoring of Saline Tracer Transport by Borehole Radar Measurements—Phase 1, Swedish Nuclear Fuel and Waste Management Co., Stockholm, Sweden. Report ID No. TR 91-09.

Olsson, O., P. Andersson, E. Gustafsson, 1991b. Site Characterization and Validation—Monitoring of Saline Tracer Trasport by Borehole Radar Measurements—Final Report. Swedish Nuclear Fuel and Waste Management Co., Stockholm, Sweden. Report ID No. TR 91-18.

O. Olsson, (editor), Site characterization and validation—Final Report, Stripa Project TR 92-xx, SKB, Stockholm, Sweden, 1992.

J. P. Tinucci, J. Israelsson, Site characterization and validation, Excavation stress effects around the Validation Drift. Stripa Project TR 91-20, SKB, Stockholm, Sweden, 1991. C. R. Wilson, J. Long, R. M. Galbraith, K. Karasaki, H. K. Endo, A. O., DuBois, M. J. McPherson, G. Ramqvist, Geohydrological data from the Macropermeability Experiment at Stripa, Sweden. Lawrence Berkeley Report LBL-12520, SAC -37, 1981.

An Overview of Fracture Flow Modelling at Stripa

David P. Hodgkinson
Intera Information Technologies, Henley on Thames
United Kingdom

Abstract

This paper presents an overview of discrete fracture flow and transport modelling in Phase 3 of the Stripa Project. Alternative modelling approaches and computer codes were used by three independent groups to interpret an extensive set of characterisation data and to make blind predictions for a set of performance measures. These predictions were evaluated using a structured validation process against qualitative and quantitative criteria, in order to determine the range of validity and usefulness of the modelling approaches. The major achievement of the project was that it proved feasible to carry through all the complex and interconnected tasks associated with the gathering and interpretation of characterisation data, the development and application of complex numerical models, and the comparison of predictions and measurements. Thus a new technology has emerged from the Stripa Project which can now be applied within repository performance assessment programmes.

1 Introduction

In order to assess the long-term safety of radioactive waste disposal, it is necessary to make reliable predictions of the likely water-borne transport of radionuclides to the earth's surface. Since the time-scales of many of the migration processes are extremely long and interactions between them are complex, safety cases need to be based on extrapolations using mathematical models. These assessments will only be reliable if they are based on sufficiently realistic models of the flowpath geometry and migration processes, which are calibrated and validated using appropriate laboratory and field data.

Phase 3 of the Stripa Project had as one of its two general objectives [1]:

'To integrate various site characterisation techniques and methods of analysis for the prediction and validation of groundwater flow and nuclide transport in an unexplored volume of Stripa granite.'

The development, application and validation of computational models for the flow of water and the transport of solutes through fractured rock, plays a very important role in meeting the above objective. In tackling these problems, the Stripa Project has taken particular care to integrate the modelling and experimental work. This has proved valuable in ensuring that the models under development have realistic data requirements, and that site characterisation data is of a type, quality and quantity appropriate to the models.

In view of the importance of fracture flow modelling within the Stripa Project, a Task Force on Fracture Flow Modelling was set up with the specific objectives of:

- recommending criteria for the verification and validation of fracture flow models,

- facilitating the wider dissemination amongst countries participating in the Stripa Project of progress in the development of numerical modelling and computer codes,

- coordinating the work of the three participating modelling groups.

This paper presents a personal view of fracture flow modelling aspects of Phase 3 of the Stripa Project from the perspective of a member of the above Task Force. Section 2 presents an overview of fracture flow modelling, while the approaches and codes used within the project are discussed in Section 3. The verification of the numerical accuracy of these codes is discussed in Section 4, followed in section 5 by a discussion of the approach to validation used in the project. Sections 6-8 describe three experiments performed at Stripa, and associated blind predictions used to assess the validity of the modelling approaches. Some important conclusions of this work are summarised in Section 9 and some comments on possible future exercises of this type are presented in Section 10.

2 Overview

Groundwater flow and transport modelling plays a key role in repository performance assessment. It provides a vehicle for combining a variety of site-specific measurements with generic hydrogeological information, in order to make predictions of various quantities of importance in assessing the performance of a repository. In particular, it is used to extrapolate from measurements made in boreholes to the scales of interest for safety assessment. Besides their use in interpreting experimental measurements, such as hydraulic and tracer tests, models may be required to predict repository performance measures such as:

- average flux of water and solutes through a repository;

- variability of flux between waste container locations;

- transport of radionuclides through near-field (10m scale) and far-field (100-1000m scale) rock.

It cannot be emphasised too strongly that the type of model used and the input data requirements depend on the performance measure being evaluated, and on repository design considerations. For example, a reasonable estimate of the average water flux through a repository can be obtained straightforwardly from a porous medium model, together with average values of the hydraulic gradient and permeability. Moreover, if the design philosphy is to avoid placing waste containers in or near fracture zones then these can be excluded when calculating the average permeability, while if waste containers are to be sited regardless of the local rock properties then the fracture zones are likely to dominate the average permeability. On the other hand, if the major emphasis is on modelling radionuclide transport then a more complex model is required which includes a representation of the spatial distribution of permeability and porosity, together with transport processes such as diffusion and sorption.

The countries involved in Phase 3 of the Stripa Project (Canada, Finland, Japan, Sweden, Switzerland, UK, USA) have an interest in a variety of repository performance measures and designs, not all of which could be considered within a single project. Consequently, instead of using repository-based performance measures and designs, the project was focussed on predicting flow and transport to boreholes and an experimental drift within the previously unexplored Site Characterisation and Validation (SCV) block.

Fracture flow and transport modelling can be conveniently divided into two parts namely the fluid mechanics of water flow through the complex geometrical structure of voids in the rock, and solute migration processes such as diffusion and sorption. The work described here is concerned with the former, and therefore with the description of how permeability and porosity is distributed in space, and the solution of flow equations for these distributions. The measured quantities used to test modelling predictions are the flux of water passing through defined areas, and the distribution of inert tracers in time and space arising from injections at selected locations in the rock mass. For saturated flow and with the hydraulic head gradients in operation at Stripa, it is well accepted that within a given unit, the groundwater flux is proportional to the head gradient (Darcy's Law). Thus the key questions are through what structures does flow take place, and how are they characterised and modelled?

The most important characteristic of the hydrogeology of fractured rock masses from the perspective of repository performance assessment, is the pronounced spatial variability. Heterogeneities occur on all scales from regional fault zones through local fracture zones, small scale fractures, intra-fracture channels and microfissures. An important consequence of this hierarchy of heterogeneities is that interpreted parameters, such as permeability and dispersivity, depend on the length scale over which measurements are performed. In particular, measurements in boreholes generally need to be extrapolated in length scale to give quantities of interest to performance assessment.

The fundamental problem of hydrogeological modelling is to construct a representation of the permeability and porosity field from a finite number of measurements, which adequately reflects the spatial complexity and variability. This so-called inverse problem does not have a unique solution, and thus model predictions have associated uncertainties reflecting the range of parameters consistent with known data.

To limit the range of possibilities, additional information is used to formulate a conceptual model which restricts the ways in which the modelled permeability and porosity fields are allowed to vary in space. In mathematical terminology the imposition of a conceptual model is equivalent to restricitng the classes of functions used to represent parameter fields. In general, the restrictions are based both on additional information about the rock mass, such as the geological structure and/or the fracture pattern, and on the quantity being calculated. For example, in a fairly homogeneous aquifer where it is required to estimate the regional flow field, it is reasonable to use a continuum model with permeabilities which average over local structure. However, for interpreting hydraulic measurements in a highly variable fractured rock, and modelling transport paths from point sources, a more microscopic conceptual model is required.

The basic conceptual model used in Phase 3 of the Stripa Project derives from extensive invetigations performed in earlier phases [2,3]. On the 100 m scale of the SCV block it is convenient to divide the rock into two types. Firstly, averagely fractured rock which pervades most of the volume, with flow taking place in a sheetwork of intersecting fractures with sub-millimetre apertures, and with extents and separations of the order of metres. Secondly, the rock volume has a number of near-planar fracture zones which extend for hundreds of metres, and are a few metres to a few tens of metres wide. The fracture zones are more permeable than the average rock but derive their permeability from a similar sheetwork of sub-millimetre intersecting fracture planes.

In view of the extensive evidence for fracture flow at Stripa, together with the objective of making predictions on a scale similar to fracture separations, and the eventual desire to incorporate fracture-based retardation processes (matrix diffusion and sorption), it was decided to adopt a discrete fracture modelling approach within the project. In contrast to the conventional continuum approach, in discrete fracture models flow only takes place within fracture planes. Flow takes place across an extended region by virture of intersections between fracture planes if a certain (percolation) density is exceeded [4] as is the case at Stripa. The models have the capability of incorporating spatial variability within fracture planes, if required.

Of course, discrete fracture models are not the only class of models that are useful in describing the spatial variability of flow and transport in fractured rock. Geostatistical approaches [5] which describe the medium in terms of statistical distributions of parameters for volume elements are appropriate for some applications, although the interpretation of hydraulic and tracer experiments and the incorporation of retardation processes is complicated by there not being a direct relationship to the fracture flow paths. A particularly promising approach is to represent permeability fields in terms of statistically self-affine fractals, which allows a convenient parameterisation of observed scaling properties [6].

In a comprehensive performance assessment, it would be appropriate to use a systematic procedure to identify a set of alternative conceptual models which are consistent with known data, and to evaluate repository performance measures for all these alternatives, in order to assess the range of uncertainty.

3 Modelling Approaches

Confidence in our ability to predict groundwater flow and transport is enhanced by following a number of independent modelling approaches. This strength through diversity was achieved within the Stripa Project by involving three modelling groups from AEA/Fracflow, Golder Associates, and Lawrence Berkeley Laboratory (LBL). The groups all had access to the same extensive set of characterisation data and were able to make their own interpretations and predictions using independently developed computer codes.

At the start of the project, discrete fracture flow modelling was in its infancy, and had mainly been applied to idealised two-dimensional systems. During the project the computer codes were developed to handle realistic three-dimensional fracture systems with appropriate boundary conditions. Also, considerable progress was made in developing methodologies for using the codes to derive fracture properties from experimental measurements.

All three modelling teams used a basic discrete fracture conceptual model, as described in the last section. However, the precise implementations varied due to: different interpretations of the data; constraints imposed by computer codes and resources; the performance measure of highest priority. The most salient features of the three approaches are outlined below.

3.1 AEA/Fracflow

AEA Harwell developed the NAPSAC discrete fracture flow and transport code [7] from a prototype existing at the start of the project [8]. NAPSAC is able to include known fractures explicitly and to generate a sheetwork of fractures stochastically in parts of the rock-mass where only statistical properties of the fracture system are available. A number of realisations conforming to the observed statistics can be generated, thus allowing uncertainties associated with unobserved details of the fracture system to be quantified.

For the Stripa flow calculations [9,10], NAPSAC was used in close con-junction with a continuum finite-element code, CFEST [11] used by Fracflow Consultants. NAPSAC made use of fracture statistics and hydraulic test data to calculate permeability tensors for cubes of average rock and fracture zone rock. It was found that at a scale of about 10m with typically 10^4 fractures, the numerical rock behaved like a permeable medium with respect to groundwater flux. The resulting permeability values were used in a finite-element model of the SCV region in order to make flow predictions.

The critical assumptions made in the AEA/Fracflow analysis are that all coated mineralised fractures observed in the core are hydraulically active, and that individual fractures have a constant transmissivity derived from a log-normal distribution. It was not possible to test the first of these assumptions since the hydraulic test packer intervals encompassed many fractures. Any future project of this type should endeavour to test such hypotheses by using packer spacings smaller than typical fracture spacings on at least one section of rock [12].

For the tracer transport calculations [13], NAPSAC was used independently of CFEST, and approximated the flow system as a network of one-dimensional paths.

3.2 Golder Associates

Golder Associates have developed a suite of discrete fracture codes capable of modelling steady-state and transient flow and tracer transport problems. In particular, the FracMan package [14] was used to generate discrete fracture networks, to simulate site characterisation methods, to define boundary geometries and boundary conditions, and to generate finite-element meshes from the fracture networks. The finite-element code MAFIC [15] was used to solve the flow equations.

The FracMan/MAFIC package is built around a similar set of assumptions to NAPSAC, in that they can both model flow and tracer transport through large systems of deterministic and stochastically generated fractures. However, for the Stripa Project calculations [16-19] the FracMan/MAFIC package was used directly to model as large a fraction of the SCV site as computational limits allowed, rather than deriving permeabilities for Representative Elementary Volumes.

The key assumption which allowed this larger region to be modelled is that, in contrast to AEA/Fracflow, only a small fraction of the coated fractures observed in the core are hydraulically conductive. Evidence cited in support of this assumption is that hydraulic tests in a number of packer intervals containing fractures identified in core logs, have a negligible (less than 10^{-12} m/s) transmissivity. The controversy is connected to the question of the degree to which boreholes penetrate impermeable islands between channels in fracture planes, and thereby exposes a deeper layer of complexity for modelling and data gathering [20,21]. These issues are of crucial importance to fracture flow modelling as they lead to basic ambiguities in the modelling approach. Further detailed experimentation is warranted to resolve these questions.

In order to keep the model of the Validation Drift Experiment within computational limits, the H-zone was modelled as a set of equivalent sub-parallel fractures.

3.3 LBL

Lawrence Berkeley Laboratory (LBL) used the channel network generation code CHANGE [22] to define a regular grid of one-dimensional conductors within each fracture zone, and the three-dimensional finite-element code TRINET [23] to model the response of the zones to hydrological perturbations.

The LBL approach within the Stripa Project [24-26] was to focus on flow through the fracture zones, which carry the majority of water across the site. The intervening average rock was taken as impermeable. An equivalent discontinuum model was used for the fracture zones, which were discretised with a grid of equally conductive channels with a fraction of missing links. The pattern of heterogeneity was determined using a simulated annealing algorithm conditioned to cross-hole test data. The approach is essentially an interpolative technique for flows on the scale of fracture zones, allowing the consequences of different boundary conditions to be assessed.

4 Verification

The numerical accuracies of codes participating in the Stripa Project were tested by comparison with exact solutions for simplified problems and by inter-comparing solutions for more complex problems [27,28]. The exercise proved useful in testing the geometrical accuracy with which fracture intersections can be calculated, and the accuracy of flow and transport calculations in simple and complex fracture systems. It is hoped that the test cases will also prove useable and useful to people outside the project.

In the final analysis, the codes were judged sufficiently accurate for the purposes for which they were intended. In particular, the numerical errors are certainly less than stochastic uncertainties in any realistic application. However, perhaps the most important facet of the exercise was the lessons learned along the way to this conclusion.

The first lesson is that verification cases, including the input data, equations and output requirements, need to be specified very precisely and unambiguously, and that this is best done using the language of mathematics. The original cross-verification plan [27] contained a number of inaccuracies and ambiguities, and included some inappropriate performance measures, which caused delay in the completion of the exercise. This could have been avoided if some results had been available before the test cases were fully specified.

The verification exercise proved very valuable to the participating groups in understanding how their codes work. This was especially true when discrepancies occurred and their sources needed to be identified. In particular, it was learned that a high level of discretisation is needed for a complete match between solutions from different codes, especially for transport problems. In practical applications it is not possible to use highly discretised fracture planes, and the verification exercise proved useful in quantifying the resulting numerical error. A related benefit was the derivation of 'rules of thumb' for the number of particles needed to be tracked to obtain convergence in transport problems.

Finally, it is clear that a significant level of effort is required for a successful verification exercise. This was underestimated by the project, and led to delays and resource constraints elsewhere in the modelling programme.

5 Validation

The word validation appears in the overall objectives of the Stripa Project and of the Task Force on Fracture Flow Modelling, as discussed in Section 1. It is a word which means different things to different people, although there appears to be a general consensus on the broad definition of the IAEA [29]:

> "Validation is a process carried out by comparison of model predictions with independent field observations and experimental measurements. A model cannot be considered validated until sufficient testing has been performed to ensure an acceptable level of predictive accuracy. (Note that the acceptable level of accuracy is judgemental and will vary depending on the specific problem or question to be addressed by the model)."

In order to fulfil its objectives, the Task Force had to develop a specific approach to validation for use within the project [30]. The approach was based around the definition of a set of performance measures which were specific to the validation experiment, and which were chosen with regard to the relevance and purpose of the models being evaluated. Essentially, validation is viewed as determining whether a model is fit for the purpose for which it was intended.

The performance measures were calculated by the modelling teams independently of any knowledge of the experimental results. The evaluation of the

284

validity of the models was made by an independent peer review group (the Task Force) who compared experimental and calculated values of the performance measures according to pre-defined quantitative and qualitative criteria. Given the many uncertainties encountered in hydrogeology, the quantitative criteria for measures such as the total flow to the drift, was order-of-magnitude accuracy. The qualitative criteria were used for assessing patterns of inflow to different parts of a borehole or drift, and were thus a measure of the spatial variability.

Finally, the process was documented [31] to allow external scrutiny, together with comments from the peer review group on other pertinent considerations including the usefulness and feasibility of the modelling approach for repository performance assessment.

In general the above structured approach to validation has worked well and is considered to be an important and successful part of the project. Thus it should provide a suitable starting point for future validation exercises.

Of course, in the light of experience there will be a need to modify the approach. One problem worthy of further study is the methodology and criteria for evaluating the range of results produced by statistical models. Another concern is that to be truly effective, the peer review group needs to devote considerable time and effort to the evaluation task, and this was not always possible within the Stripa Project. Also, it should be noted that not only model predictions but also experimental results can be in error, and this has certainly occurred in the Stripa Project. Thus any future validation projects should aim to carry our repeat measurements of important quantities.

Finally, it is rather humbling to admit that the present exercise is concerned with a rather simple physical process namely Darcy flow in a saturated medium. The major complication is that the medium through which the flow takes place has a highly complex structure. It is emphasised that when more complicated physcial and chemical processes are considered, for example chemical reactions between solutes and rock minerals, different approaches to validation are likely to be required. One way forward would be to start by complementing the present work, and examining complex reactions within a simplified rock structure, so that the experimental conditions can be adequately controlled . Of course, the usual scientific method of hypothesis testing should be appropriate to such well controlled systems.

6 Simulated Drift Experiment

The Simulated Drift Experiment (SDE) measured the inflow of water to an array of six parallel 100m boreholes along the line of the future Validation Drift [32]. The original aim of the experiment was to allow comparison with the inflow following drift construction as a quantitative measure of suspected drift effects. However, the Task Force decided to use the SDE as a training ground for the modelling groups and the emerging validation process, in advance of the major validation exercise based on flow to the Validation Drift.

The input to this comparison exercise was based on data from six other boreholes including information on fracture geometry, stress, single borehole geophysical logging, crosshole and reflection radar, seismic tomography, head monitoring and single hole packer test measurements, together with fracture trace maps from adjacent drifts [33]. This information was used to develop a preliminary picture of the fracture zones crossing the SCV site and to provide statistical properties of the ubiquitous fractures in the intervening average rock.

Using the conceptual models outlined in Section 3 toegether with the above input data, the three modelling groups made blind predictions [9,16,24] for a number of performance measures defined by the Task Force and these were compared with the experimental results [34,35].

The measured total flow rate to the D-hole array was 1.71 litres/min (range 1.67 - 1.75). This is to be compared with the predictions of AEA/Fracflow, 1.45 (0.36 - 5.80), Golder Associates, 0.055 (0.001 - 0.156) and LBL, 3.1 (0.0 - 7.7). It is seen that all the predictions are within the order-of-magnitude criterion discussed in Section 5, except for the Golder Associates calculations which suffered from calibration problems (a subsequent recalibration gave a mean inflow of 1.5 litres/min).

The comparison exercise also examined the division of total inflow between the fracture zones and average rock, and the pattern of distribution along the D-holes, where the models performed satisfactorily [34,35]. It was not possible to evaluate predictions of flow patterns into the average rock because the measurements were not made with sufficient precision. It is disappointing

286

that this performance measure could not be tested because the average rock is likely to be favoured by repository designers as the location for waste containers, and flux predictions are important for assessing near-field corrosion and radionuclide release rates [36]. This highlights an important lesson, that regions with the greatest water flow are not necessarily the most important from a repository safety perspective.

7 Validation Drift Experiment

Following the completion of the SDE, a Validation Drift was excavated along the first 50 m of the previours D-hole array crossing the H-Zone approximately half way along its length. The Validation Drift Experiment (VDE) measured the spatial distribution of water inflow into a grid of 2m x 1m cells on the surface of the 5m diameter drift. On the roof and sides of the drift the water was collected in plastic sheets stuck to the rock, while in the floor it was collected from sump holes.

At the time the VDE was performed, the continuing site characterisation programme had led to an improved database of fracture statistics and fracture zone properties [37]. This information was used by the modelling groups to make predictions for the following seven performance measures defined by the Task Force, as surrogates for the repository performance meausres discussed in Section 2:

D-1:   Total rate of groundwater flow to
       Validation Drift.

D-2:   Rate of groundwater flow from H-zone and
       spatial distribution.

D-3:   Rate of groundwater flow from average
       rock and spatial distribution.

D-4:   Characteristics of fractures in the Validation Drift.

S-1:   Magnitude and spatial distribution.
        of head changes due to drift excavation.

S-2:   Magnitude and spatial distribution of
        head response due to opening borehole T-1.

S-3:   Distribution of groundwater inflow to
        remaining sections of D-boreholes.

On the basis of comparing predicted and measured values for these performance measures, the validity of each modelling approach was evaluated according to the following validation criteria [30]:

Quantitative: Do the predictive calculations adequately reflect the measured values? That is, are the predictions of the correct order of magnitude as compared to the measurements?

Qualitative: Are the predicted distribution patterns sufficiently accurate as compared to the observations?

From the viewpoint of the overall applicability of a given modelling approach, the above criteria were addressed in relation to the following two questions:

Usefulness: From the viewpoint of an assessment of the expected performance of a geologic repository, is the modelling approach useful for representing groundwater flow in a geohydrologic environment which is similar to that at the SCV site in the Stripa Mine?

Feasibility: Can the characterisation data required to fully support the modelling approach be collected in a feasible and timely manner?

A major complicating factor to the comparison of predictions and measurements was the occurrence of drift effects of unknown origin, which lowered the inflow from the H-zone to 14% of its SDE value. Also the flow from the average rock was only 1.6% of that expected from scaling the SDE results, although the SDE measurement is now considered to be an overestimate (Olsson, private communication).

A number of possible explanations for these drift effects have been suggested, including: blast-induced fracturing; blast-induced fracture deformation; chemical changes, including precipitation at the water/air interface; deposition of evaporation residue; two-phase flow effects; dissolved gases coming out of solution; plastic deformation of the rock around the drift; thermal stresses; and elastic deformation of fractures caused by the stress concentrations induced by the presence of the tunnel.

Most effort on this problem within the project has focussed on quantifying the effect of elastic deformation using stress analysis models and measured relationships between normal stress and fracture aperture. However, this effect appears to be too small to explain the observed large reduction in inflow.

A plausible explanation for the drift effects are that they are a consequence of the blasting during excavation. It is pertinent that similar effects have been observed in other drifts at Stripa, but have not occurred at the Grimsel Laboratory in Switzerland where tunnels were excavated with a boring machine. However, other conceptual models, such as chemical changes due to degassing causing mineral precipitation, cannot be ruled out with the present information base.

Thus it is probable that the drift effects problem does not imply any deficiencies in the models, but that the local rock properties were changed by the drift excavation. As no measurements of the local rock properties were made following excavation they could not be used as input to the models. This limits the usefulness of the validation exercise since important characterisation data was not available.

In order to make progress, the Golder Associates and LBL teams introducted a high permeability skin into their models with parameters estimated from other drifts at Stripa. The AEA/Fracflow team chose to omit drift effects, other than a small coupling between stress and fracture aperture, since its origins were not understood.

Viewed from another perspective, this experience casts doubt on the usefulness of experimental data obtained from drifts that were excavated by blasting. With the benefit of hindsight, experiments based on boreholes, such as the SDE, appear to be more reliable.

Measured flow into the drift occurred almost entirely from the H-zone, and within the H-zone it came mainly from a single narrow fracture, which matched the location of a large inflow to the D-holes. This is a clear demonstration of the need for discrete fracture models.

An important part of the Stripa validation process was the peer review provided by the Task Force, and the documentation trail. Despite problems associated with drift effects, the performance measures have been evaluated in detail and the process documented [38].

It was found that all the modelling approaches met the order of magnitude quantitative criterion where it could be tested. For example, Figure 1 compares the predicted and observed total inflows to the Validation Drift. Also, the major patterns, in particular the dominance of a small number of fractures in the H-zone, were reasonably predicted by both the AEA/Fracflow and Golder models. Due to the lack of steady-state data, it did not prove possible to confirm the accuracy of the predicted heads.

Regarding the question as to how useful the modelling approaches are, it is clear that for predicting the average flux of water through the rock, they are unnecessarily complicated. For the more detailed calculations considering spatial variability on a scale of metres, the AEA/Fracflow and Golder Associates models are useful, for example in estimating distributions for container lifetimes and radionuclide release rates [39].

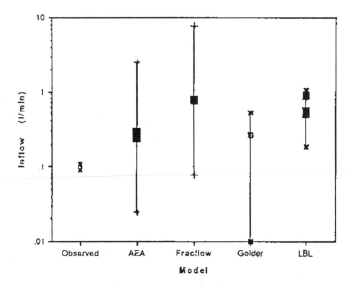

Figure 1. **Comparison of the predicted and observed total inflows into the Validation Drift.**

Regarding the feasibility of collecting the required site characterisation data, there is no doubt that this has been amply demonstrated by the Site Charac- terisation and Validation Project.

8 Tracer Experiments

It is of key importance to repository performance assessment that models of solute transport are adequately tested. As a first step in this direction, the Stripa Project performed some migration experiments with inert dissolved tracers within the H-zone, and compared these with model predictions [40]. In view of the paucity of characterisation data on the porosity field, this was not considered to be a full validation exercise, but rather a test of the feasibility of using discrete fracture models for tracer migration calculations.

8.1 Radar Saline Experiments

The first two experiments involved the injection of saline tracer (potassium bromide) into the H-zone from borehole C2. Its migration to i) the D-holes (RSTE1) and ii) the Validation Drift (RSTE2) was measured in two ways. First,

291

the flowrate and saline concentration of the water entering sections of the D-holes and the drift panels were measured as a function of time to produce a set of breakthrough curves. Secondly, radar tomography was undertaken in three tomographic planes to track the progress of the saline plume [41]. These initial experiments were aimed at providing characterisation data on the transport aperture and dispersivity for subsequent modelling of the Tracer Migration Experiment (TME), and at providing a test bed for experimental and modelling techniques.

8.2 Tracer Migration Experiment

In the Tracer Migration Experiment (TME) a mixture of two tracers, one metal complex and one dye, were injected in each of nine intervals, all but one of which were located in the H-zone [42, 43]. The injection flowrates were kept low, in order not to disturb the natural flow field significantly. The tracers were injected over a period of one to seven months, and the flowrate and concentration were measured in the array of plastic sheets and sumps in the drift. Ideally, the flow systems would have been maintained at steady state throughout the experiment. However, this was not achieved in practice, partly due to the opening and closing of boreholes in other parts of the mine. This made it more difficult to interpret the tracer breakthrough curves.

As with the water inflow, most of the tracer recovery was in a few sampling areas located along the line of a single fracture. Moreover, the largest tracer mass flowrates were found in the same sampling areas as in the Validation Drift Experiment indicating that the number of flowpaths is very limited.

There were a number of problems with the design and implementation of the tracer experiments. Of course, a major problem is whether the results are sufficiently representitive of undistrubed rock, given the well documented drift effects. Another is that it would have been preferable to have used shorter injection pulses to provide better discrimination of the flowpaths and to avoid some of the problems arising from the non-steady-state flow field. The experiments made some attempt to use a variety of different path lengths to examine the scale dependence of transport, but a more systematic design might have led to more light being thrown on this important question.

The tracer migration experiment did not succeed in advancing the state-of-the-art beyond the 3-D Migration Experiment performed at Stripa some years earlier [44]. With the benefit of hindsight, it would have been better to have performed the tracer (and flow) experiments using sections of boreholes rather than a drift. This would not only have avoided the problem of drift effects but would have allowed experiments to have been performed at a number of different flow rates by changing the water pressure in the sink holes.

In contrast to the limited experimental progress, the modelling groups performed tracer migration calculations that were considered infeasible at the start of the project. For example, the AEA group performed transport simulations in a model of the H-zone containing 60 000 fractures [13]. This was calibrated by adjusting the zone thickness, the mean fracture transmissivity in the disturbed and undisturbed rock, and the ratio of transport to hydraulic apertures. A simplified flux connection method was used for the transport calculations. This was developed from the flow solution by approximating the system by a network of one-dimensional flow paths. Pathlines were followed for a large number of particles (typically 50 000) moving through this system. Also complementary three-dimensional finite-element transport calculations were carried out by the Fracflow group [45].

Golder Associates used a fracture system model similar to that employed in the VDE predictions but with updated fracture characteristics. The model was conditioned to have approximately the measured drift inflow and RSTE2 breakthrough times. Because of time and resource constraints only about 1000 particles were tracked through the system whereas one to two orders of magnitude more were required to obtain converged breakthrough curves.

LBL performed a limited tracer migration study based on a two-dimensional discretised network of the H-zone [26]. Two simulations were performed, one of which was annealed to the C1-2 cross-hole test and one co-annealed to this and to the SDE steady-state results.

As with the flow predictions, the Task Force defined a set of performance measures to aid the comparison of predictions and observations, including breakthrough curves to interesting grid elements and the drift.

The closest agreement between predicted and observed breakthrough curves to the drift were for tracer injection in borehole C2, since the models were calibrated using saline injection data from this borehole. Viewed from the perspective of an order-of-magnitude criterion the model predictions are acceptable. This remained true for all the other AEA predictions. The pattern of inflow predicted by AEA shows slightly more panels than observed, but in general there is good agreeement.

The Golder Associates predictions show a wider spread in the predicted range of breakthrough times and relative concentrations than observed. However, the average predictions are within an order of magnitude of the measurements. The Golder Associates flow field was found to be very sensitive to perturbations in the boundary conditions, and this metastability reflects the behaviour of the actual flow system in the vicinity of the drift.

The two configurations used by LBL showed that for their simulations the conceptual model uncertainty was large, and often encompassed the observations. The predicted breakthrough curves were too steep because of coarse discretization and the use of zero dispersivity.

In general Golder Associates predicted a larger spatial variability of breakthrough curves than the observed spread between the different injections, while AEA showed less variability than observed. The difference in behaviour between these models reflects their different fracture system densities and modelling philosophies. In particular, AEA used a fracture intensity an order of magnitude larger than Golder Associates, but an average fracture transmissivity an order of magnitude less, resulting in comparable average permeabilities.

9 Conclusions

The Stripa Project has nurtured the development of a new technology for modelling the flow of water and transport of inert tracers through heterogeneous fractured rock. At the start or the project the available discrete fracture software could only handle small idealised systems of fractures and there was little understanding of how it could be used with practical characterisation data to make predictions for real rock masses. The major achievement of the work summarised here is that it has proved feasible to carry through all the complex and interconnected tasks associated with the gathering and interpretation of characterisation data, the development and application of complex numerical models, and the comparison of predictions with measured water flows and tracer movements.

In order to produce a technology that is useful and useable by the radioactive waste disposal community, the project has had to grapple with the contentious issue of validation. A pragmatic validation process and associated criteria have been defined and used by the project. This examines whether a model is fit for its purpose by asking a peer review group (in this case the Task Force on Fracture Flow Modelling) to compare blind predictions for a set of pre-defined performance measures with experimental results. The procedure has worked well and should form a suitable starting point for future validation exercises.

The idea of involving independent modelling groups has worked well, and has demonstrated that alternative conceptual models can account for the measured data. In particular, it highlighted the issue of the extent to which mineralised fractures observed in core samples are hydraulically active. Further experimentation is required if this issue is to be resolved.

Some experimental problems have been encountered which should be addressed in any future project of this type. First, it was demonstrated unambiguously and quantitatively that the excavation of the Validation Drift resulted in considerable changes to the local rock properties. This casts doubt on the degree to which flow and transport data obtained in drifts are representative of undisturbed rock. Secondly, the measurements of flow from averagely fractured rock were inadequate and so this important aspect of the models could not be properly tested. Thirdly, the tracer tests would have been more effective if pulse inputs had been used and the effects of length scaling and flow rate had been investigated more systematically.

Despite these problems, the project has demonstrated that the majority of flow at Stripa takes place in a small number of fractures, and that discrete fracture models can be used to make reasonable quantitative and qualitative predictions. However, Stripa is only the start. There will undoubtedly be further developments and refinements to the models and more particularly to the ways in which they are applied within repository performance assessment programmes.

10 Outlook

The Stripa Project has focussed attention on the uncertainties inherent in modelling flow and transport through spatially variable rock masses. Clearly, such complex systems cannot be characterised and modelled with the same degree of confidence as engineered systems.

Repository performance assessments need to quantify the effects of such uncertainties. A number of established approaches are available for handling parameter uncertainty, but further development and testing of techniques for handling conceptual model uncertainty are required.

The present project has focussed most attention on the development of discrete fracture conceptual models, and has made use of equivalent permeable medium models as an adjunct. In any future exercise of this type it would be appropriate to use a systematic procedure for the selection of a comprehensive set of conceptual models. A calibration and validation exercise of the type used in the Stripa Project could then be used to determine which models are appropriate for predicting each of a number of performance measures based on experimental data. Finally, performance measures based on the behaviour of a repository could be calculated with each appropriate model in order to quantify the extent of conceptual model uncertainty.

Acknowledgements

I would like to thank Her Majesty's Inspectorate of Pollution and the Stripa Project for supporting my participation in the Task Force on Fracture Flow Modelling. Also, thanks are due to the members of the Task Force for sharing their many insights into the fascinating topic of fracture flow modelling.

References

[1] Stripa Project, Program for the Stripa Project Phase 3: 1986-1991, Stripa Project Technical Report 87-09, SKB, Stockholm, 1987.

[2] Stripa Project, Executive Summary of Phase 1, Stripa Technical Report 86-04, SKB, Stockholm, 1986.

[3] Stripa Project, Executive Summary of Phase 2, Stripa Project Technical Report 89-01, SKB, Stockholm, 1989.

[4] P. C. Robinson, Connectivity of Fracture Systems: A Percolation Theory Approach, J. Phys. A: Math. Gen. 16, 605-614, 1983.

[5] G. de Marsily, Quantitative Hydrogeology Groundwater Hydrology for Engineers, Academic Press, 1986.

[6] M.D. Impey and P. Grindrod, Application of Fractal Geometry to Geological Site Characterization, paper presented at the British Computer Society International Conference on Applications of Fractals and Chaos, London, 1992.

[7] P. Grindrod, A. Herbert, D. Roberts and P. Robinson, NAPSAC Technical Document, Stripa Project Technical Report 91-31, SKB, Stockholm, 1991.

[8] P.C. Robinson, Flow Modelling in Three-Dimensional Fracture Networks, UKAEA Report AERE-R. 11965, 1986.

[9] A.W. Herbert and B. Splawski, Prediction of Inflow into the D-Holes at the Stripa Mine, Stripa Project Technical Report 90-14, 1990.

[10] A.W. Herbert, J. Gale, G. Lanyon and R. MacLeod, Modelling for the Stripa Site Characterisation and Validation Drift Inflow: Precition of flow through fractured rock, Stripa Project Technical Report 91-35,1991.

[11] S.L. Gupta, C.R. Cole, C.T. Kinaid and A.M. Monti, Coupled Fluid, Energy and Solute Transport (CFEST) Model: Formulation, Computer Source Listings and User's Manual, Battelle Menorial Institute Technical Report BMI/OWNI, 1986.

[12] P. J. Bourke, E.M. Durrance, M.J. Heath and D.P. Hodgkinson, Fracture Hydrology Relevant to Radionuclide Transport: Field Work in a Granite Formation in Cornwall, Commission of the European Communities Report EUR 9854EN, 1985.

[13] A.W. Herbert and G. Lanyon, Modelling Tracer Transport in Fractured Rock at Stripa, Stripa Project Technical Report 92-01, 1992.

[14] W. Dershowitz, G. Lee and J. Geier, FracMan Version Beta 2.3 Interactive feature data analysis, geometric modelling and exploration simulation: User documentation, Project Report 913-1058, Golder Associates Inc., Redmond WA, 1991.

[15] I. Miller, MAFIC Version Beta 1.2 Matrix/Fracture Interaction Code with Solute Transport - User Documentation, Golder Associates Inc., Report prepared for Battelle Memorial Institute, Office of Waste Technology Development, Willowbrook, Illinois, USA, 1990.

[16] J Geier, W. Dershowitz and G. Sharp, Prediction of Inflow into the D-Holes at the Stripa Mine, Stripa Project Technical Report 90-06, 1990.

[17] W. Dershowitz, P. Wallmann and S. Kindred, Discrete Fracture Modelling for the Stripa Site Characterisation and Valdiation Drift Inflow Predictions, Stripa Project Technical Report 91-16, 1991.

[18] W. S. Dershowitz, P. Wallmann, J.E. Geier and G. Lee, Preliminary Discrete Fracture Network Modelling of Tracer Migration Experiment at the SCV Site, Stripa Project Technical Report 91-23,1991.

[19] W. Dershowitz and P. Wallmann, Discrete Fracture Modelling for the Stripa Tracer Validation Experiment Predictions, Stripa Project Techncial Report 92-15, 1992.

[20] P. J. Bourke, Channelling of Flow Through Fractures in Rock, Proceedings of the SKI/NEA GEOVAL-87 Symposium, 167-177,Stockholm, Sweden, 1987.

[21] H. Abelin, L. Birgersson, H. Widen, T. Agren, L. Moreno and I. Neretnieks, Channelling Experiment, Stripa Project Technical Report 90-13, 1990.

[22] D. Billaux, J. Chiles, K. Hestir and J. Long, Three-dimensional statistical modelling of a fractured rock mass - an example for the Fanay-Augeres Mine, Int. J of Rock Mechanics and Mining Science and Geomech, 26, 281 - 299, 1987.

[23] K. Karasaki, A new advection-dispersion code for calculating transport in fracture networks, Report No. 22090, Earth Sciences Division, Lawrence Berkeley Laboratory, 1986.

[24] J. Long, K. Karasaki, A. Davey, J. Peterson, M. Landsfeld, J. Kemeny and S. Martel, Preliminary Prediction of Inflow into the D-Holes at the Stripa Mine, Stripa Project Technical Report 90-04, 1990.

[25] J. Long, A. Mauldon, K. Nelson, S. Martel, P. Fuller and K. Karasaki, Prediction of Flow and Drawdown for the Site Characterisation and Validation Site in the Stripa Mine, Stripa Project Technical Report 92-05, 1992.

[26] J. Long and K. Karasaki, Simulation of Tracer Transport for the Site Characterisation and Validation Site in the Stripa Mine, Stripa Project Technical Report 92-06, 1992.

[27] W. Dershowitz, A Herbert and J. Long, Fracture Flow Code Cross-Verification Plan, Stripa Project Technical Report 89-02, 1989.

[28] F.W. Schwartz and G. Lee, Cross-Verification Testing of Fracture Flow and Mass Transport Codes, Stripa Project Technical Report 91-29, 1991.

[29] IAEA, Radioactive Waste Management Glossary: Second Edition, IAEA - TECDOC-447, 1988.

[30] Paul Gnirk, Process and Criteria for Validation of the Ground-Water Flow Models in the OECD/NEA International Stripa Project, High Level Radioactive Waste Management: Proceedings of the Second Annual International Conference, Las Vegas, Nevada, 895-901, 1991.

[31] D. Hodgkinson, A Compilation of Minutes for the Stripa Task Force on Fracture Flow Modelling, Stripa Project Technical Report 92-09, 1992.

[32] D. Holmes, M. Abbott and M. Brightman, Site Characterisation and Validation - Single Borehole Hydraulic Testing of C Boreholes, Simulated Drift Experiment and Small Scale Hydraulic Testing Stage 3, Stripa Project Technical Report 90-10, 1990.

[33] O. Olsson, J. Black, J. Gale and D. Holmes, Site Characterisation and Validation Stage 2 - Preliminary Predictions, Stripa Project Technical Report 89-03, 1989.

[34] D. Hodgkinson, A Comparison of Predictions and Measurements for the Stripa Simulated Drift Experiment, Stripa Project Technical Report 91-10, 1991.

[35] A. Herbert, W. Deershowitz, J. Long and D. Hodgkinson, Validation of Fracture Flow Models in the Stripa Project, Proceedings of GEOVAL-90, Stockholm 1990.

[36] D.P. Hodgkinson and M.J. Apted, Key Scientific Issues for Near-Field Performance Assessment, High Level Radioactive Waste Management, Proceedings of the Second Annual International Conference, Las Vegas, Nevada, 874-884, 1991.

[37] J. Black, O. Olsson, J. Gale and D. Holmes, Site Characterisation and Validation - Stage 4 - Preliminary assessment and detail predictions, Stripa Project Technical Report 91-08,1990.

[38] D. Hodgkinson and N. Cooper, A Comparison of Measurements and Calculations for the Stripa Validation Drift Inflow Experiment, Stripa Project Technical Report 92-07, 1992.

[39] Swedish Nuclear Power Inspectorate, SKI Project-90, SKI Technical Report 91-23, 1991.

[40] David P. Hodgkinson and Neill S. Cooper, A Comparison of Measurements and Calculations for the Stripa Tracer Experiments, Stripa Project Technical Report 92-20, 1992.

[41] O. Olsson, P. Andersson and E. Gustafsson, Site Characterization and Validation - Monitoring of Saline Tracer Transport by Borehole Radar Measurements, Final Report, Stripa Project Technical Report 91-18, 1991.

[42] L. Birgersson and T. Agren, Site Characetrizations and Validation - Tracer Migration Experiment in the Validation Drift, Report 1: Instrumentation, Site Preparation and Tracers, Stripa Project Technical Report 92-02, 1992.

[43] L. Birgersson, H. Widen, T. Agren, I. Neretnieks and L. Moreno, Site Characterization and Validation Tracer Migration Experiment in the Validation Drift, Report 2, Part 1: Performed Experiments, Results and Evaluation. Part 2: Breakthrough Curves in the Validation Drift Appendices 5-9, Stripa Project Technical Report 92-03, 1992.

[44] H. Abelin, L. Birgersson, J Gidland, L. Moreno, I. Neretnieks, H. Widen and T. Agren, 3-D Migration Experiment Report 3; Part I Performed Experiments, Results and Evaluation; Part II Appendices 15, 16 and 17, Stripa Project Technical Report 87-21, 1987.

[45] R. MacLeod, J. Gale and G. Bursey, A Site Characterisation and Validation - Porous Media Modelling of Validation Tracer Experiments, Stripa Project Technical Report 92-10, 1992.

SESSION III

ENGINEERED BARRIERS

SÉANCE III

BARRIÈRES OUVRAGÉES

Chairmen - Présidents
R. Lieb (Switzerland)
W. Danker (United States)

The Stripa Clay Sealing Program - Intentions, Results and Afterthoughts

Roland Pusch
Clay Technology AB

Abstract

The Stripa clay sealing program involved testing of the function of smectite clay in the form of commercial bentonites for plugging and grouting purposes. Plugs were prepared by compressing clay powder to blocks which hydrated, swelled and sealed off the space they were put into, while grouting was made by use of thixotropic clay gels, which were injected into fractures in low-viscous form and transformed to stiff clay with time. The longevity of smectite clay in repository environment was investigated throughout the project. The function of smectite clay seals of high density was concluded to be very good over very long periods of time, while clay grouts may be short- or long-lived depending on temperature and geochemistry.

1   Background

The need for effective sealing of all possible passages through which radionuclides may escape from underground repositories was realized early in various countries, and embedment of waste canisters and plugging of boreholes, shafts and tunnels were hence concluded to be necessary by various authorities and governments. This was one ground for conducting the sealing part of the Stripa Project, in the course of which it also became clear that flow passages created by the excavation of shafts and tunnels in the form of mechanically disturbed zones may require sealing by grouting. The project led to the idea that a number of different sealing activities - illustrated by Fig. 1 - may combine to give a very effective isolation of highly radioactive waste in deeply located repositories in crystalline rock

2   Canister embedment - the issue of the sealing project in Stripa Phase 1, the Buffer Mass Test (1980-1985)

2.1   Buffer and backfills

Dense smectite clay was suggested to SKB as suitable embedment of waste canisters early in 1977 on the grounds that it has a well documented longevity and ability of retarding groundwater percolation and ion diffusion, as well as of serving as a suitably yielding mechanical buffer with sufficient bearing capacity to carry heavy canisters. The way dense, homogeneous smectite clay can be applied as canister embedment and the way it is expected to mature had been outlined on the basis of small-scale experiments and the sealing part of Stripa Phase 1, the Buffer Mass Test (BMT), became an approximately half-scale experiment of the Swedish KBS3 concept.

The field experiments were primarily planned to give information on the rate of wetting and associated build-up of swelling pressures of dense smectite clay in simulated canister holes and in bentonite/sand backfills applied in the overlying tunnel in differently fractured and water-supplying rock. The interest was focussed on the degree of water saturation under the influence of repository-like temperature gradients, and also on checking the validity of numerical calculations of the development of temperature fields. Important issues were also to find out if large blocks of uniform density could actually be produced on an industrial scale and applied under humid field conditions, and if homogeneous smectitic backfills could be prepared and applied with sufficiently high density. A matter of interest was finally to check if changes in chemical, mineralogical and physical properties had taken place in the course of the up to 3 years long tests at temperatures of as much as 130°C.

Fig. 1 Sealing options in a repository (1)

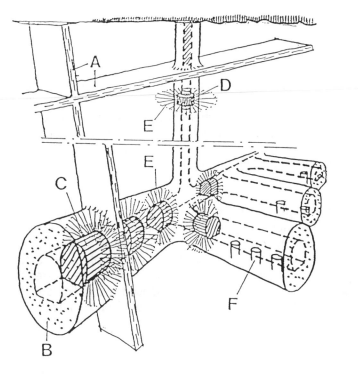

A: Water-bearing
   zones
B: Conductive,
   disturbed
   zones
C,D Tunnel and
   shaft plugs
E: Grout "cut-
   offs"
F: Deposition
   holes

As to the practicality in preparing and applying large and dense clay blocks, and in mixing and bringing in and compacting the clayey tunnel backfill, the entire experiment was very successful with one exception discussed later in the paper. The BMT was also successful with respect to the applicability of the simple models of wetting and associated build-up of swelling pressures and heat transfer, especially where there was good access to water in the heater holes. A major finding from these tests, which gave reliable field data by the use of suitable instrumentation, was that practically complete water saturation was obtained despite the counteracting temperature gradient. Still, the experiments confirmed the expected complexity of the water saturation process, a matter that is being further investigated in SKB:s research program and turns out to be of considerable practical importance. Laboratory testing of samples extracted from one of the long-term tests with almost saturated clay showed that no changes in chemical and mineralogical composition or in physical properties could be identified (2).

2.2   Interaction with rock

A separate task of the BMT was to test an idea of rational selection of suitable sites of the deposition holes with 6 m spacing in KBS3-type repositories. The plan was to drill pilot holes with 3 m spacing and to "overcore" those with small or moderate water inflow and limited number of major fractures to yield holes for canisters, or in the BMT case, electrical heaters. This technique combined with the development of simple generalized rock structure models for each 6 m distance appears to be promising for optimum use of the rock in KBS3 repositories.

The BMT also had an unexpected spin-off effect of great importance: Even at the end of the 3 year experiment only very insignificant water pressures were recorded at the interface of the rock and tunnel backfill, while successively more water was discharged to the front of the confined and largely water saturated backfill. The only logical explanation of this phenomenon was the existence of a shallow blasting-disturbed zone with strongly enhanced hydraulic conductivity. This gave the incitement of the sealing project of Stripa Phase 3, which confirmed this idea (3).

3   Plugging of boreholes, shafts and tunnels - the issue of the sealing project of Stripa Phase 2 (1985-1987)

3.1   Borehole plugging

The Borehole plugging experiment comprised field tests of the sealing function and practicality of handling and application of plugs consisting of jointed segments of perforated metal casings filled with cylindrical blocks of highly compacted sodium bentonite. Pilot tests had shown that such clay swells out through the perforation and embeds the casing.

The field experiments, which involved plugging of a subhorizontal 56 mm diameter hole with about 100 m length, and of two vertical 76 mm diameter holes of 14 m length, demonstrated that even very long holes can be effectively sealed by such plugs and that the clay becomes very homogeneous and forms a tight contact with the rock in a relatively short time provided that water is available.

Matured plugs become practically impervious and the tight contact with the rock eliminates flow along the clay/rock contact (4).

3.2   Shaft plugging

Shaft sealing by use of highly compacted sodium bentonite was investigated in an experiment conducted in a 14 m deep shaft, open at both ends. The shaft had been prepared by very careful blasting over 50% of its periphery and by slot-drilling over the rest. A reference test was first made with two concrete plugs cast on site by use of expansive cement, the plugs being separated by a chamber filled with water saturated sand. In the subsequent, main test the plugs consisted of bentonite blocks and a slot was sawed around the lower plug. The slot was filled with bentonite blocks and the entire plug construction instrumented to give swelling pressures at a few sites.

The main leakage from the sand chamber took place through steep fractures exposed in the shaft wall in both tests, but the slot effectively reduced the axial flow and led to a drop in leakage to a few percent of that of the reference test (5).

3.3   Tunnel plugging

The tunnel plugging test was planned to check how effectively a strongly water-bearing fracture zone can be isolated from an intersecting repository tunnel in the construction period by applying a lining and effective seals at its ends. Since no such zone was available in the mine, an artificial zone was made in the form of an injection chamber similar to the arrangement in the shaft plugging test but on a larger scale. Thus, two plugs with highly compacted sodium bentonite as sealing components were constructed at about 9 m distance and they were connected by a 1.5 m diameter steel tube simulating a lining. The 9 m long space between the tube and the rock in the about 4 m wide and high tunnel was filled with water saturated sand, which was pressurized up to 3 MPa in the test in order to measure the leakage along the plugs as a function of the pressure.

The leakage from the pressurized sand-filled chamber dropped from more than 200 l/hour at 100 kPa water pressure early in the test to 75 l/hour at 3 MPa after about 20 months, primarily due to maturation of the bentonite which became virtually impermeable and created a very tight contact with the rock. Hence, it was concluded that the major part of the remaining leakage was caused by flow through discontinuities in the surrounding rock (6).

This was an incitement of an additional test - financed by SKB - for investigating whether strategic grouting, i.e. injecting grout into clearly identified discrete fractures, could significantly reduce axial flow along blasted repository tunnels. This test, in which a new dynamic injection technique was tried for the first time, clearly demonstrated the possibility of sealing discrete, relatively narrow fractures. This formed the base of the sealing project in Stripa Phase 3.

4 Sealing of rock by grouting - the issue of the sealing project of Stripa Phase 3 (1987-1992)

A Task force was established for finding the most suitable grout material and identifying structures suitable for grouting in Stripa. Its work led to the conclusion that smectite clay and cement are major candidates because of their potential sealing power as demonstrated by a number of practical applications and because of their longevity potential. The most effective ways of retarding groundwater flow in repositories by grouting were concluded to be sealing of canister deposition holes, sealing of the excavation- disturbed zone if such a zone really exists, and sealing of natural, water-bearing fracture zones intersecting repository tunnels.

A comprehensive pilot test at Stripa showed that "strategic grouting", i.e. injecting discrete relatively narrow discontinuities, works if smectite clay or fine-grained cement grout are used, applying the "dynamic" injection technique (7). This first part of the study comprised development of a grout flow model taking oscillatory strain into consideration and also construction of a suitable prototype of an injection machine. The flow model is valid both for grouting at a constant, static pressure with non-Newtonian material properties and for "dynamic" injection yielding Newtonian material behavior. It was found that well defined viscometer tests yield relevant parameters.

The investigation of physical properties of candidate clay and cement grouts with respect to hydraulic conductivity, shear strength, sensitivity to changes in fracture aperture, and of the chemical stability, showed that, provided that the grouting holes hit fracture channels, effective sealing should be offered if the channel aperture exceeds about 20-30 μm, and that any rock can have its bulk conductivity reduced to about 10^{-10} m/s. Under chemically favorable conditions the operative lifetime can be millions of years, while critical conditions reduces this time to a few hundred years or even less (8).

The field tests in the main phase of the project comprised investigation of excavation-induced disturbance and attempts to seal the disturbed zone, and also grouting of deposition holes and of a natural fine-fracture zone. The first two studies were made in the BMT drift, while the latter was conducted in the northern, right arm of the 3D cross. The investigation of excavation-disturbance caused by blasting and stress changes showed that the axial conductivity of the blasting-disturbed zone increased by 100-1000 times, while that of the surrounding stress-disturbed zone, extending to about one diameter from the periphery, had increased by 10 times. The radial conductivity of the latter had dropped by about 4 times while that of the blasting-disturbed zone was the same as the one in axial direction (9). Grouting of the blasting-disturbed zone by closely spaced short holes ("hedgehog" grouting) turned out to be ineffective, the main reason probably being that debris from disintegrated fracture infillings like chlorite prevented grout from entering while water could still easily penetrate it (10).

Discrete fractures intersecting the drilled heater holes in BMT and forming passages in the natural fracture zone were effectively sealed, although it could be shown that heating reduces the sealing effect by altering the fracture apertures and activating previously closed fractures (11,12). Detailed studies of the character of the cement grout in the sealed natural fracture zone illustrated the physical state of the cement in narrow fractures (12).

5 Major outcome of the Stripa sealing activities: What is the whole thing worth? What did we learn and what should have been done differently?

Only large-scale tests like the ones in Stripa can give the huge amount of information on the usability and behavior of clay seals that is needed for designing safely operating repositories of the KBS3 type. In this respect the tests over the 15 year long research period were extremely valuable. A fact of substantial importance is that the three-phase project showed that strategic sealing in the fashion indicated in Fig. 1 is actually feasible. Thus, the combination of a very effectively isolating canister-embedment and "in-hole" grouting means that groundwater movement is displaced away from the vicinity of the canisters (Fig. 2), and local sealing of intersecting natural water-bearing zones by grouting (Fig. 3), can result in formation of stagnant groundwater regimes (13).

The Stripa sealing tests demonstrate that seals in the form of blocks of highly compacted bentonite, applied to yield canister embedment and plugs, are easily constructed and work very well and in general agreement with simple models for the saturation rate and the development of swelling pressures and temperatures. They appear to provide very good isolation of the radioactive waste in a short term perspective and results from the laboratory studies and from other research activities focussing on the chemical stability, suggest that the isolating ability is

Fig. 2   Principle of shunting off fractures (D) serving as short-circuits;
thick lines indicating parts sealed by grouting. A is a major
conductive zone, B is conductive, excavation-disturbed zone, C is
deposition hole

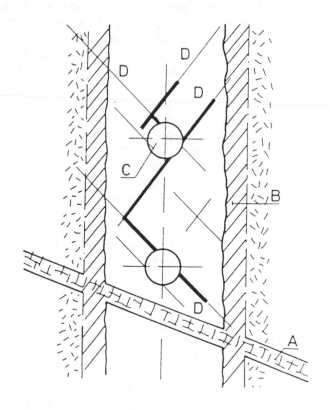

maintained for many tens or hundreds of thousands of years provided that the
temperature does not exceed 100-120°C and that the clay is not supplied with
very much potassium. In this respect, clay grouts may be less reliable since their
mass is very small and their chemical buffering capacity hence insignificant. Also,
their ability to resist high hydraulic gradients without undergoing "piping" is
limited.

The value of the field tests is somewhat limited by the fact that the electrolyte
content of the Stripa groundwater is very low and particularly that the calcium and
potassium concentrations are insignificant. A higher salt content with calcium as
dominant cation is expected to have yielded significant salt accumulation at the
wetting front and therefore in the vicinity of the heaters, a process that may have
a substantial effect on steel canisters. Also, it may have produced less effective
expansion and sealing effect of plugs and grouts and less tight contacts with the

Fig. 3 Example of KBS3 tunnel used for estimation of the influence of local grouting on the axial water flow through the disturbed zone D. Applying reasonable parameters, grouting is found to reduce the flow to less than 4% (13)

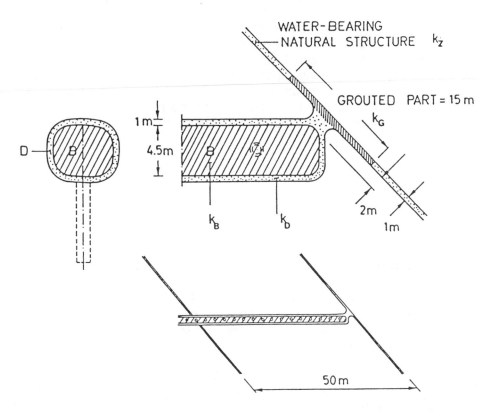

adjacent rock. Hence, corresponding testing in rock environment with higher salt contents are strongly asked for in order to fully realize the ability and limitations of good sealing performance of clay grouts.

With the limited understanding in the planning stage of many of the processes associated with the hydration and redistribution of solid clay and water that are associated with temperature gradients, the BMT could hardly have been designed and conducted in a better way. The same goes for the plugging experiments, although it should have been realized at the planning of the 76 mm borehole tests and the shaft plugging test that the absence of high water pressures would not yield complete maturation.

Still, there were disappointments, a major one being the deficiency in effective backfilling of the top part of the BMT. Pilot tests conducted by a construction company of considerable reputation, were completely misleading and the density and heterogeneity of the top backfill became unacceptably low in the Stripa test. A much improved method for application and compaction is required.

Finally, in the light of the outcome of the entire series of sealing experiments and of ongoing investigations of the behavior of "nearfield" rock, it is clear that much more detailed characterization of the rock environment would have been helpful in the planning of the grouting experiments in the BMT drift. The rock structure modelling and application of numerical methods for calculating excavation- and heat-induced rock strain that were initiated in the course of the grouting project were very helpful and form a good basis for further work.

6   REFERENCES

1.   Pusch,R., Gray,M., Huertas,F., Jorda,M., Barbreau,A. & Andre-Jehan,R. Sealing of Radioactive Waste Repositories in Crystalline Rock. Proc. of NEA/CEC Workshop in Braunschweig on Sealing of Radioactive Waste Repositories, OCDE/OECD, 1989

2.   Pusch,R., Börgesson,L. & Ramqvist,G. Final Report of the Buffer Mass Test - Volume II: test results. Stripa Project Technical Report 85-12, SKB, Stockholm, 1985

3.   Pusch,R. Alteration of the Hydraulic Conductivity of Rock by Tunnel Excavation. Rock Mechanics and Mining Sciences. Vol.26, No.1, 1989 (pp. 71-83)

4.   Pusch,R., Börgesson,L. & Ramqvist,G. Final Report of the Borehole, Shaft, and Tunnel Sealing Test - Volume I: Borehole plugging. Stripa Project Technical Report 87-01, SKB, Stockholm, 1987

5.   Pusch,R., Börgesson,L. & Ramqvist,G. Final Report of the Borehole, Shaft, and Tunnel Sealing Test - Volume II: Shaft plugging. Stripa Project Technical Report 87-02, SKB, Stockholm, 1987

Pusch,R., Börgesson,L. & Ramqvist,G. Final Report of the Borehole, Shaft, and Tunnel Sealing Test - Volume III: Tunnel plugging. Stripa Project Technical Report 87-03, SKB, Stockholm, 1987

Pusch,R. et al. Rock Sealing - Interim Report on the Rock Sealing Project (Stage I). Stripa Project Technical Report 88-11, SKB, Stockholm, 1988

Pusch,R., Karnland,O., Hökmark,H., Sanden,T. & Börgesson,L. Sealing Properties and Longevity of Smectitic Clay Grouts. Stripa Project Technical Report 91-30, 1991

Börgesson,L. et al. Final Report of the Rock Sealing Project - Identification of Zones Disturbed by Blastingand Stress Release. Stripa Project Technical Report 92-08, 1992

Börgesson,L. et al. Final Report of the Rock Sealing Project - Sealing of Zones Disturbed by Blasting and Stress Release. Stripa Project Technical Report 92-21, 1992

Börgesson,L. et al. Final Report of the Rock Sealing Project - Sealing of the Near-field Rock Around Deposition Holes by Use of Bentonite Grouts. Stripa Project Technical Report 91-34, 1991

Pusch,R., Börgesson,L., Karnland,O. & Hökmark,H. Final Report on Test 4 - Sealing of Natural Fine-fracture Zone. Stripa Project Technical Report 91-26, 1991

Pusch,R. Executive Summary and General Conclusions of the Rock Sealing Project. Stripa Project Technical Report 92-27, SKB, Stockholm, 1992

Properties and Functions of
Smectite Clays in a Repository

Lennart Börgesson
Roland Pusch

Clay Technology AB

ABSTRACT

Smectite-rich clays are suggested to be used in repositories to fill up and seal off voids in the rock with dimensions ranging from fractures with apertures smaller than 100 µm up to caverns, shafts and tunnels with volumes of thousands of cubic meters.

The composition and properties of the clay materials vary with the intended function. The three phases in the Stripa project have involved determination of the properties of these materials, the techniques and possibilities of emplacing the material, and the resulting in situ functions.

A short review of the required properties and the need for modeling the different functions of smectite clay in a repository will be given, as well as some examples of material models and predicted functions.

ripa Project has offered opportunities of testing the function of smectite
ᴄ.ᴀᵧ   fferent sealing and filling applications. It has also, at least to some
extent, entailed possibilities to test the clay in the laboratory and to model the
behavior.

Since the project has lasted for more than 10 years it is natural that the
models for predicting the behavior have developed during the years. Some of
the original simple models e.g. for the BMT test, although useful for rough
estimates, have today been replaced by more accurate models that do not only
involve better predictions but also better understanding of the processes. Some
of the models have not been worked on very intensely in later time and need
more attention while others are developed recently and are up to date e.g. the
grout flow models.

This article will briefly describe the role of smectite clays in a repository and
the need for modeling the properties and functions. Some examples will be
given as well.

THE ROLE OF SMECTITE CLAY IN A REPOSITORY

Smectite clay has been suggested to play several roles in a repository, the
main ones being to:

- protect canisters (buffer material);
- fill up large open space with material of lower hydraulic conductivity
  than the surrounding rock (backfill);
- seal boreholes and plug shafts and tunnels;
- reduce the conductivity of nearfield rock by fracture grouting using
  bentonite slurry.

THE NEED FOR MODELING THE PROPERTIES AND THE DIFFERENT FUNCTIONS

The wide range of use of smectite clays requires basic understanding of the
behavior of the clay. The need for modeling varies with the intended role of the
clay. The following properties are the most desirable and hence need to be
modeled:

Buffer material

Smectite clay as buffer material to protect the waste canisters requires several functions. The clay should be soft enough to physically protect the canister but stiff enough for limiting settlement. It should have low hydraulic conductivity and low ion-diffusivity but a high thermal conductivity. If highly compacted blocks are used they must have a high water-absorbing and swelling (self-healing) ability. The clay must survive for tens of thousands of years with acceptable properties also under severe conditions, e.g. high temperature and unfavorable ground water composition.

Backfill material

The demand on the backfill in the tunnels is less than on the buffer material. However, some functions are vital: the material should not have a higher hydraulic conductivity than the surrounding rock and it should be stiff enough to withstand the swelling from the buffer and to secure the stability of the roof. It must also have an acceptable longevity for keeping the backfill mainly unaltered during the lifetime of the repository.

Sealing material for plugging boreholes, shafts, and tunnels

This type of sealing located far away from the waste canisters is also exposed to less critical conditions than the backfill, but should still be made of highly compacted bentonite. The sealing material should be stiff enough to withstand high external pressure and erosion. It should have a swelling potential and create a swelling pressure that guarantees an excellent sealing of the rock/bentonite interface. It should of course also have an acceptable longevity.

Bentonite slurry for grouting

While the buffer, backfill, and plugs require material with high density and low water ratio ($w \approx 20\%$ at saturation) the bentonite slurry must have a water ratio $100\% < w < 700\%$ depending on the composition. The requirements for the slurry are primarily that it must be liquid during injection and plastic after injection, meaning that it should regain strength after injection, i.e. show thixotropy. Other requirements are a reasonably low hydraulic conductivity and some swelling and self-healing ability to withstand small rock displacements. If the grout is required to function after closure of the repository it must have an acceptable longevity to operate during the lifetime of the repository.

The available space in this article does not allow for presenting all properties that have been measured and/or all functions that have been considered. Instead, three short examples will be given for some of the most important applications.

Physical properties of the buffer

A physical material model of the highly compacted bentonite in the buffer or a plug after water saturation must include a number of property parameters. During phases 1 and 2 some simple geotechnical parameters were used but recently a complete material model has been developed which despite some shortcomings describes the behavior in a much more relevant way.

The material model handles stress-strain and water flow in the clay and is based on the effective stress theory using a modified Drucker-Prager Plasticity model combined with a Porous Elastic model. The model can be used in finite element calculations (ABAQUS) and includes mainly the following properties:

- Failure surface function;
- Yield function;
- Void ratio as a function of the swelling pressure;
- Poissons ratio;
- Hydraulic conductivity as a function of the void ratio;
- Properties of the pore water and the clay particles (bulk modulus, density, thermal expansion);
- Temperature dependence of all parameters.

The model has been partly validated in different laboratory tests and has also been used for calculating different repository scenarios, e.g. the effect of a 10 cm rock displacement across a deposition hole, and swelling of buffer material into a poorly compacted backfill, resulting in stress generation in the canister.

Rheological properties of the bentonite slurry used for grouting

The rheological properties of bentonite slurries have been extensively investigated in the laboratory in order to choose the right composition and to make flow models.

The material is thixotropic and the shear strength just after mixing and stirring has been proved to increase by a factor of 10. In order to further reduce the shear resistance during injection the dynamic technique was introduced and it was shown that the shear resistance can be further reduced by a factor of 10-100.

The measurements show that the shear stress and the shear rate are expressed in a relevant fashion by Eqn 1.

$$\tau = m \left(\frac{\dot{\gamma}}{\dot{\gamma}_o} \right)^n + \tau_o$$

where τ = shear stress (Pa)
$\dot{\gamma}$ = shear rate (1/s)
$\dot{\gamma}_0$ = normalized shear rate (=1.0 1/s)
m = parameter (Pa)
n = parameter

When an oscillating shear strain is applied on the slurry the parameter m is reduced according to Eqn 2.

$$m = a \left(\frac{\gamma_A}{\gamma_{AO}} \right)^b$$

where γ_A = osc. shear strain amplitude
γ_{AO} = normalized osc. shear strain amplitude ($\gamma = {}_{AO}1.0$)
a = parameter (Pa)
b = parameter (negative)

These material models form the basis of a mathematical theory of the fracture penetration of a grout exposed to a combined static and oscillating pressure. The theory has been partly validated by injection tests using an artifical fracture with 3 meter length.

LONGEVITY

The chemical stability of smectite grouts under repository conditions was considered both by applying current and new degradation models, hydrothermal experiments, and reference to geological evidence. The matter is very complex and rather crude estimates had to be made for prediction of the long term performance.

CONCLUSIONS

The Stripa Project and other separate projects have yielded a large understanding of the usefulness of smectite clay in a repository. They have yielded possibility to test some smectite clay functions and interactions with the rock. The results are also available for reevaluations in the light of new knowledge.

Research on Cement-Based Grouts for the OECD/NEA International Stripa Project

Maria Onofrei

AECL Research, Whiteshell Laboratories
Pinawa, Manitoba, Canada ROE 1L0

Abstract

This paper details some of the major findings from the laboratory research program conducted to obtain the information for a specially developed high-performance cement-based grout. The results indicate that it is possible to manufacture low water content, high-performance cement-grouts the performance of which would be acceptable for at least thousand of years and probably for much longer periods. These high-performance grouts, which were injected into fractures in the Stripa granite and diverted water flows in the rock, are shown to have negligible hydraulic conductivity, associated with very low porosity and small pore sizes, and are highly leach resistant under repository conditions. Microcracks generated in these materials from shrinkage, overstressing or thermal loads are likely to self-seal.

Neither laboratory investigations nor theoretical studies provided a definitive assessment of the longevity of the high-performance cement-based grouts when they are used to seal a repository. The results of the studies suggest that the high-performance grouts can be considered as viable materials in repository sealing applications. Further work is needed to fully justify extrapolation of the results of the laboratory and the numerical studies to timescales relevant to repository performance assessment.

INTRODUCTION

The Task Force established under the auspices of the OECD/NEA international Stripa Project recommended cement- and clay-based materials as candidate materials for use in sealing underground repositories for heat generating radioactive wastes and identified technical issues requiring detailed attention. For both of these materials, it was noted that insufficient information was available to allow for reasonable analyses of the performance of the sealing materials and the systems of which they form a part (Coons, 1987). A coordinated series of laboratory, *in situ* and theoretical studies were carried out to provide the necessary information to allow for an appraisal of the long-term performance of the selected sealing materials.

This paper focuses on the work being carried out on cement-based grouts by AECL research in Canada. Emphasis is placed on the laboratory studies into longevity of the materials. The laboratory tests were carried out to develop a conceptual model for the morphology and long-term performance parameters of high performance grout comprised of reground sulphate resisting Portland cement (SRPC), silica fume, superplasticizer and water. Test were conducted to determine the effect of superplasticizer on the long-term properties of the grouts. Specifically, the hydraulic conductivity/porosity relationships for the material was determined to improve understanding of the new high-performance material and to provide input to the numerical model. Leach test were carried out to provide details of solid/water reaction processes to improve conceptual models of the material characteristics and for comparison with predictions from numerical models.

MATERIALS

Prior to the investigations in Phase 3 of the Stripa Project, extensive research and laboratory testing had been undertaken by AECL Research and the US/DOE to investigate the detailed properties of alternative cement-based grout materials and mix proportions. This work provided the foundation for the selection of the high performance superplasticized, cement grouts containing silica fume used in Phase 3 of the Stripa Project. Admixtures such as silica fume and superplasticizer were incorporated in grout mixtures to produce desired changes in the physical and chemical properties of the grouts.

The reference grout mixture adopted for further use in field investigations and in the laboratory for longevity studies had the following composition:

* 90% by mass, Sulphate, Resistant Portland Cement (SRPC, Canadian Type 50), reground to a Blain fineness of 600 m^2/kg,

* 10% by mas, silica fume,

* 1% by mass, superplasticizer (sodium salt of sulphonated naphthalene formaldehyde condensate),

* water (*W/CM* between 0.4 and 0.6 by mass)

The rationale leading to the choice of these materials for investigation has been given elsewhere (Onofrei et al. 1992). The materials were successfully injected into an hydraulically active fracture zone at the Underground Research Laboratory in Canada (Gray and Keil 1989). The grout penetrated microfissures in granite with aperture smaller than 20 μm (Onofrei et al, 1992a)

The stripa Project, Phase 3 built on this data and experience and selected the above mix composition as the basis of its laboratory and *in situ* investigations.

SUPERPLASTICIZER

In designing cementitious grouting materials several additives may be included once the basic type of cement is selected. With regard to longevity, pozzolanic materials (e.g., fly ash and silica-fume) are used to limit the quantity of free $Ca(OH)_2$ in the hardened cement paste (Feldman 1981, Hooton 1985). Silica fume, which is the finest and most reactive of commercially available pozzolans, is included in the Canadian reference grout material. Due to very large surface area of the cement or cement/silica-fume powders, quantities of water in excess of those necessary solely to hydrate the cement are required to provide the materials with the workability characteristics needed of a grout. Conventional grouting uses water to cementitious materials ratios (*W/CM*) in excess of two.[1] When such grout hardens, the extra water, not participating in the cement hydration, is trapped in the solid structure and creates a high porosity. This increased porosity decreases the strength, increases the hydraulic conductivity and shrinkage, and decreases the longevity of hardened cement-based materials. Thus, the full benefits of using finely ground cements and, particularly of admixing silica-fume, can only be realized in combination with the use plasticizing agents or, more specifically, superplasticizers which allow for the manufacture of workable cement-based grouts at *W/CM* of 0.4 or less (Onofrei and Gray 1988, Onofrei et al. 1989).

While the implications of using superplasticizers on the engineering properties of high-performance grouts were well understood by the Task Force, several important issues related to their use in nuclear-waste disposal applications were not clear. These included the following fundamental questions:

1. What are the mechanisms by which superplasticizers allow for decreasing *W/CM* ?

2. Superplasticizers are organic materials. Will these materials enhance the mobility of radionuclides by increasing the quantity or changing the quality of organic matter in a repository ?

3. Are the engineering benefits gained by the use of superplasticizers offset by decreased longevity of the hardened cement-grout product in a repository setting ?

Attempts to address these questions were made through a series of studies which included reviews of available information, tests on unset grouts,

---

1Stoichiometrically and theoretically, portland cement can be fully hydrated at *W/CM* ~ 0.3.

microstructural investigations of hardened grouts containing superplasticizers and leaching tests on the grouts. The full description of the experimental designs and methods has been given elsewhere (Onofrei et al. 1991a).

Two theories are available for the fluidifying effects of superplasticizers on cement-water pastes; the pellicle and colloidal theory (Aitcin et al, 1989). The pellicle theory is qualitative and cannot be verified through measurements. The application of colloid theories explains some of the observed phenomena but requires other qualitative descriptions of particle-superplasticizer interactions to explain deviations of the actual from the predicted behaviour of unset cement-water pastes. It appears that superplasticizers do not significantly interfere with the long term hydration reactions. Little information could be found in the literature to describe particle -superplasticizer interactions in unset pastes containing silica fume. Moreover, with respect to the longevity of cement-based grouts in a repository, no direct information was available on the place of residence of superplasticizers in the hardened materials, the persistence of the superplasticizer at these sites and the effects of the superplasticizers on the long term properties - such as leaching and dissolution - of the hardened materials and associated releases of organic materials into the repository environment. These issues were directly relevant to the fundamental questions which studies carried out for the Stripa project were required to address.

Laboratory tests showed that superplasticizers likely interfere with the fastest hydration reactions in setting high-performance grout (Onofrei et al, 1991a). Consequently, the superplasticizer is principally located on the tricalcium silicate hydrate (C_3SH) and tricalcium aluminate hydrate (C_3AH) phases in the cement once it has hardened (Figure 1). It is likely that the superplasticizer is incorporated into these phases by its association with the hydration water and, for C_3AH, by reaction between the sulphonic group on the superplasticizer molecule with the aluminate. Not all of the superplasticizer is involved with these reactions. Some of the organic material is only physically sorbed on the surfaces of high-silica phases or is left in the capillary pore space. The methodology used to effect these observations involved the use of superplasticizer incorporating ^{35}S as a tracer (Onofrei et al, 1991a). The conclusions are drawn on the unproven, but reasonable assumption that the superplasticizer molecules did not break down during adsorption on and reaction with the cement phases.

Leaching tests on hardened high-performance grouts incorporating the radioactively labelled superplasticizer showed that the superplasticizer can be released into solution (Onofrei et al. 1991a). The cumulative fraction release for the superplasticizer are in the order of 10^{-16} to 10^{-14} kg/m^2. It was generally concluded that the release is derived from the unadsorbed fraction of the superplasticizer in the pore space and concomitant gradual dissolution of the C_3S and C_3A hydrated phases. The release rate of superplasticizer changes with time and tend to increase with increasing temperature, silica fume content, chemical composition of the leachate and the superplasticizer content.

It is now certain that the use of superplasticizers will increase the total organic load in the groundwaters around a repository. The significance of this increase can only be judged when realistic assessments of the quantity and type of naturally occurring organic materials present in a repository are available. This is likely to be case dependent. The data obtained in the studies performed for the Stripa project can provide a guide for numerical evaluation of the increase in the quantity of organic matter. The exact structure of the organic materials released from high - performance grout remains uncertain as does the influence of the material on radionuclide mobility in the groundwater. Further studies are required to clarify these issues.

LONG-TERM PERFORMANCE

Changes in the performance properties of cement-based grout materials with time may be caused either by external environment or internal causes. Environmental factors, such as the local geochemistry and stress conditions, can induce changes in the properties of cement-based grouts. During the design life of a repository the cementitious materials are expected to come in contact with the local groundwaters. Groundwaters that are only partially in equilibrium with the system. Therefore, grout may be leached in or, otherwise, react chemically with groundwater. Internal causes of degradation such as continued internal microstructural changes and phase transformations can also affect long-term changes in grout properties. However, these changes alone may not necessarily result in decreased performance. For a decrease in long-term performance to results, adverse changes in the microstructure (pore structure) of the grout must occur. It was shown by Alcorn et al (1990) and Onofrei et al (1990) that the rate of penetration of fluids across a cementitious matrix can be related to the volume of pores and size and interconnectivity of these pores.

Leaching properties

Leaching by water involves the penetration of grout by water or aqueous solutions, the dissolution of soluble constituents of the hydrated cement paste, and transportation of the dissolved species to the surrounding water. The depth of penetration of groundwater into cement matrix will be largely controlled by the permeability of the hardened cement matrix and the hydrostatic pressure of the water.

Different types of leaching test methods were applied to the study of the cement-based grout materials (Table 1). Laboratory tests were carried out to evaluate the effect of these variables on the leaching mechanisms. The full description of the experimental designs and methods has been given elsewhere (Onofrei et al, 1992). Since the grout itself is complex in structure and composition, DDW was used as a reference leachant because it provided the simplest environment in which to study the grout leaching mechanisms. However, since a major goal of this investigation was to evaluate grout leaching performance in the realistic environment, other reference leachants than the DDW were used. Canadian reference groundwaters, WN-1 and SCSSS, were selected as reference leachants for two reasons (Onofrei et al, 1992). First, their composition closely matches that of groundwaters encountered at 500 m to 1000 m depth in granitic rock and Baltic shield. Secondly, their methods of preparation are straightforward, thus facilitating their use in the laboratories.

The dependence of the leaching behaviour of grouts on the extent of interaction with the leachant and on the resulting changes in leachant chemistry were evaluated by considering two extreme cases. In one approach the grout was leached under static conditions, allowing the leachant in the grout/leachant system to approach saturation with major grout components such as Ca^{2+} and Si^{4+} leached out of the grout. The other approach involved dynamic conditions. In these conditions the accumulation in the leachant of species leached out of the grout is to small to alter the composition of the leachant in a sufficient measure to affect its reactivity toward the grout.

The analyses of leaching behaviour of grouts were based on the concentrations of calcium and silicon in solution. Leach rates for Ca^{2+} and Si^{4+}, in both static and dynamic leach tests, were calculated using the mathematical expression given in equation 1:

$$\text{Leach rate} = [\, X \,]\, (\, 1/t \,)(\, V/SA \,) \qquad (1)$$

where X is the concentration of the element (Ca^{2+}, Si^{4+}) in solution at time t. The concentration of the element in solution were calculated by subtracting the initial concentrations of element in the starting leachants from the measured concentration of element (Ca^{2+}, Si^{4+}) in leachates.

The rate at which Ca^{2+} and Si^{4+} were leached from grout mixtures of the reference grout (SRPC, Canadian Type 50) mixed at *W/CM*= 0.4 and 0.6 when reacted with DDW in static leach tests are shown in Figure 2. The results show that there is an instantaneous release of both Ca^{2+} and Si^{4+} which increases directly with temperature. The leach rates of Si^{4+} ($\sim 10^{-9}$ kg m^{-2} s^{-1}) in solution were found to be much lower than the leach rate of Ca^{2+} ($\sim 10^{-8}$ kg m^{-2} s^{-1}). The results indicate an incongruent dissolution process, i.e. more than one phase is leaching and the individual releases are not simply chemically equivalent to those in solid phases from which they are released. It is suggested that when hardened grout is in contact with water the calcium hydroxide ($Ca(OH)_2$) produced during the hydration reactions and unreacted with the added silica fume is rapidly leached. The dissolution of this $Ca(OH)_2$ is primarily responsible for the initially high rates of release of Ca^{2+} to solution. The release of Si^{4+} in solution was principally related to dissolution of unreacted amorphous silica (silica fume). The leach rates of both Ca^{2+} and Si^{4+} were found to decrease asymptotically with leaching time. It was concluded that both solution concentration effects and formation of a dissolution - limiting surface layer contributed to the observed decrease in leach rates with time (Onofrei et al, 1991b, Onofrei et al, 1992b). In addition to solution concentration and surface reaction layer, changes in the microstructural characteristics (pore volume, pore radius and pore size distribution) of the grout have been shown to contribute to the observed decreases in the leach rates. Mercury intrusion porosimeter (MIP) analyses of the grout specimens leached for various lengths of time show that the microstructural characteristics of the materials changed during leaching (Onofrei et al. 1992). Leaching was accompanied by decreases in the mean pore size and total pore volume (Figure 3). The degree of change in the microstructural characteristics was found to be a function of the initial *W/CM* ratio, temperature, initial porosity and the type of cement. The effects were attributed here to both the continued hydration of the cement phases and formation of the phases observed on the surface of the samples. The surface phases contributed to progressive densification of the grout structure as the DDW penetrated the specimen with time. SEM analyses of the leached samples provided evidence to support this hypothesis. SEM/EDX examination of the leached grout samples showed that most of the large pores formed before leaching, during curing of the grouts, contained solid infilling hydration products such as; $Ca(OH)_2$, calcium silicate hydrate (Onofrei 1992a).

The data show that for the reference grout mixed at low *W/CM* (0.4) when leached in WN-1 synthetic groundwater the variations in leach rate of Ca^{2+} with time differed considerably from those that occurred when the grouts were leached in DDW. Initially, calcium had negative leach rates (Figure 4). Calcium present in the initial leachant (4.85 $x10^{-2}$ mol/L) was removed from solution and deposited on the bottom of the leaching cell and on the surface of the leached specimen. For the reference grout mixed at high *W/CM* (0.6), calcium concentrations follow a similar trend with exception of the systems at 25 °C and 85 °C for which Ca^{2+} was initially removed from the specimens. The rate at which the calcium was removed from solutions and the contribution of the grout sample to the Ca^{2+} in solution was found to depend on the grout composition, temperature and time. However, in all leaching experiments involving the reference grouts investigated, calcium concentrations gradually increased with time approaching the initial concentration value in the original WN-1 groundwater The extent of calcium leaching was less in WN-1 groundwater than in DDW. The WN-1 solution generally decreased the rates of elemental leaching.

The initial decrease in calcium concentrations in solution were attributed to the precipitation of phases with which the groundwater is supersaturated (i.e., $CaCO_3$, $CaSO_4$, $Ca(OH)_2$). The XRD and SEM analyses of the precipitate on the base of the leaching cell and on the leached sample surfaces provided evidence to support this hypothesis. The precipitate on the base of the leaching cell was identify by XRD as a mixture of $Ca(OH)_2$ and $CaCO_3$. The SEM examination of the leached specimens revealed the formation of a crystalline surface-layer. The composition of the surface layer on the leached specimen consisted of brucite, $Mg(OH)_2$.

There was some evidence of an initial enhancement of leach rate in SCSSS groundwater (Figure 4). These results suggest that solution composition controls the leaching characteristics. However, it was expected that the higher the initial concentration of the elements (i.e., Ca^{2+}, Si^{4+}) in the leachant, the slower their leach rates will be from grouts. The concentration gradients of Ca^{2+} and Si^{4+} through the grout-solution interface should be reduced when these elements are initially present in the leachant. The decreased concentration gradients should reduce the driving forces for leaching/dissolution from the grouts. However, the grouts leached more rapidly in SCSSS (groundwater with high ionic strength) than in WN-1 (lower ionic strength). One possible explanation for this behaviour is the presence of higher concentrations of ions such as Ca, Na, Mg, Cl^-, SO_4^{2-}.

From microscopic examinations it was observed that leaching/dissolution in SCSS is also accompanied by the formation of reaction layers which include precipitations and growth of an assemblage of secondary phases (i.e., $Mg(OH)_2$, $CaCO_3$ and $Ca(OH)_2$). The presence of these phases was also predicted by the durability models being developed to estimate long-term performance of cement grouts. The thermodynamic stability of these alteration/precipitate phases may influence the long-term performance of cement-based grouts and in view of very low hydraulic conductivity, surface leaching is likely to be the major process by which bulk high-performance cement-grout will degrade. More information on the morphology, chemical composition, and the rate of formation of the reaction layers would further add insight into the long-term dissolution/leaching processes which grout would be expected to undergo in a repository and decrease uncertainties with respect to grout performance.

Based on the findings, it was concluded that the leaching behaviour of grouts at a given temperature not only strongly depends on grout composition but is also sensitive to leachant composition. The data confirmed that the initial leachant composition assumes a major role in grout leaching/dissolution by controlling the extent of leaching/dissolution required to produce saturation.

Porosity - K relationship

Laboratory investigations on the permeability of the grouts used in the laboratory and field test (Onofrei et al 1992) showed that the reference grouts are practically impermeable ($k < 10^{-16}$ m s^{-1}) under hydraulic gradients up to 35 000 m m^{-1}. Moreover, the material's porosity was found to decrease with time due to continued hydration, precipitation and associated reactions. The very low hydraulic conductivity and fine pore size distribution did not allow for pore-water exchange and associated cement dissolution effects on the grout performance to be observed in short time scales in laboratory tests. The permeability of water through water saturated cementitious materials has been shown to be related to the porosity and pore size distribution (Hooton and Wakeley, 1989; Powers, 1958). In this context, it is commonly assumed, in the durability models used to estimate the long-term performance of cement grouts, that materials will degrade through pore-water exchange and associated dissolution/leaching processes. A decrease in the performance of cement grouts is considered to be the result of increased porosity (Coons, 1989; Berner, 1987).

Laboratory investigations on the longevity of cement-based grouts was expanded to include assessments of hydraulic conductivity-porosity relationship in cement-based grout materials with high porosity and to confirm changes in pore volume and size in response to dissolution and precipitation. These studies not only provided basic information on material performance but also offered a porosity/permeability relationship for use in the numerical modelling of cement grout longevity (Alcorn et al, 1992)

The problem of assessing the hydraulic conductivity-porosity relationship in cement based grout with high porosity and to confirm changes in microstructure in response to dissolution precipitation was analyzed using both compacted granulated hardened grout and grouts with high *W/CM* ratios approaches. The full description of the experimental designs and methods has been given elsewhere (Onofrei et al, 1992).

The hydraulic conductivities, k, of granulated hardened reference grout compacted at densities, ρ, between 1.50 and 1.60 Mg m^{-3} are shown in Figures Figure 5a and 5b). The results show considerable decreases in the hydraulic conductivity with permeation time. These effects are considered to be caused mostly by the continued hydration and/or associated reaction, and therefore, progressing densification of the compacted specimen and decreases in the pore volume and mean pore diameter. An indication that the decreases in hydraulic conductivity is accompanied by changes in the pore system is given by the results from MIP (Figure 6a and 6b). The data show over the period of the test the proportion of very large pores present (>1 μm) decreased and the proportion of the smaller pores (<1 μm) increased. In all tests, it was found that the pore structure became finer as permeating time increases. Moreover, the compacted specimen with the largest decreases in hydraulic conductivity exhibited the largest changes in the pore size distribution (Figure 6, $\rho=1.60$ Mg m^{-3}). The differences in the volume of large pores in the compacted grout specimens, although difficult to relate directly with hydraulic characteristics of the compacted grout, are nevertheless reflected in the hydraulic conductivity of corresponding compacted grouts. The specimens with higher hydraulic conductivity tended to have higher volumes of large pores (>1 μm).

SEM analyses of the surfaces of fresh fractures in the specimens at the termination of the hydraulic conductivity tests revealed that a massive fibrous phase had formed on the surface of some grout grains and in the available pore spaces between the grout grains (Figure 7). EDX analyses of the fibrous phase showed it to contain Ca, Al, S and some Si. The material was identified with X-ray diffraction as ettringite

$(C_3A.3CaSO_4.32H_2O)$. The ettringite formation is a result of the reaction of C_3A with gypsum and water.

The decreased hydraulic conductivity and the densification of the microstructure (reduction of the large pores > 1 μm) can be explained on the bases of ettringite formation. However, this cannot be considered as being the sole cause. Another possibility is an increase in the quantity of hydration reaction products. As a result of crushing the hardened grout it is likely that micro-cracks are introduced in the hydrated layer which forms during the hydration reaction around the cement grain. Thus, residual unhydrated cement within the particles was exposed to the permeating water. This led to renewed development of hydrolysis and hydration processes.

The results indicate that the rate with what the hydraulic conductivity decreases with time and its final value was controlled by the particle size of the granulated grout and the compacted density. The use of more finely crushed grout material in the preparation of the compacted grout specimens (Figure 5a) resulted in faster decreases in the hydraulic conductivity and to ultimately lower hydraulic conductivities (k below 10^{-10} m s^{-1}) than those given by specimens formed from the coarser grains (k $\sim$ 10^{-8} m s^{-1}) (Figure 5b). The initial hydraulic conductivities of all the specimens was in the range $10^{-7} < k < 10^{-6}$ m s^{-1}. These results were attributed to a larger surface area of grout in contact with permeating water as well as to a higher activity of the granulated powder.

In all tests the hydraulic conductivity did not appear to approach a steady state during the tests. The long time required to achieve equilibrium in the hydraulic conductivity tests may reflect a a complex pore structure in the compacted granulated hardened grouts as well as the continuation of hydration and/or associated reactions (e.g., precipitation of ettringite).

Autogenous healing properties of high performance grouts

One of the possible functions of cement-based sealants is to seal for long time periods fractures that may contribute to the dispersal of the dissolved nuclear waste to the environment. To accomplish this objective the sealant must have acceptable low hydraulic conductivity (e.g.,

$< 10^{-10}$ m/s). It is evident that the presence of any defects (cracks, capillary pores) in grout structure have the potential to provide a more rapid transport pathway for advection or diffusion of radionuclides. The presence of these defects could impair the performance of grouts as a barrier to radionuclides migration. If self sealing can be assured, particularly for the modified cement-based grouts (containing silica fume and superplasticizer) in the likely chemical conditions within an underground disposal vault, the cracks are of much less significance.

The ability of the cement-based grouts to self seal was investigated on both bulk grouts and thin films of grouts (Onofrei et al, 1992). In both cases the self sealing capabilities of the high performance grouts were investigated with water flowing through the grout.

The autogenous sealing was studied through changes in pore structure (decrease in pore radius and volume of pores) and changes in the rate of water flow through the cement based grouts.

The results from all tests indicated that the hydraulic conductivity of the reference grout mixed at low and high W/CM and with imposed porosity decreased with increasing permeating time. Pore structure characterization by both MIP and SEM indicated that changes occurred in the grout structure during the permeability tests. The results showed decreases in the total pore volume and a shift of pore radii toward smaller values. These changes in the pore structure led to the refinement of the pore structure and densification of the microstructure.

From the hydraulic conductivity tests carried on the grout with high and low water content, analysis of the material observed to form in the pore space available revealed $Ca(OH)_2$ and amorphous CSH. These materials decreased the capillary porosity, decreased the total pore volume and increased the fraction of fine porosity. No pores larger than 0.1μm remained after the test. Thus, a mechanism by which $Ca(OH)_2$ and microporous gels, develop and grow into the larger pores, shifting the pore size distribution towards the finer pore size and completely blocking up some of the larger pores can be inferred.

The material observed to form in the pore space available and to bond the grout grains in the compacted hardened grout specimens exposed to DDW in the hydraulic conductivity tests carried out in air was identified as ettringite. The observed densification of the microstructure and the decrease in the hydraulic conductivity of the compacted granulated hardened grouts was attributed to the ettringite formation.

Experimental work indicated that self sealing also occurs in thin films of hardened grouts (Onofrei et al. 1992). The infilling material formed in the cracks was identified as $Ca(OH)_2$. The infill material may have formed as a result of either continuous hydration reaction or dissolution of $Ca(OH)_2$ in the grout and increasing the concentration of Ca^{2+} in the cracks and consequently recrystallization of $Ca(OH)_2$.

The results indicate that self sealing occurs in both thin film and bulk hardened grouts under both low and high hydraulic pressure gradients. More than one mechanism may be responsible for promoting self-sealing. These include the formation of ettringite and portlandite as well as calcite in the permeable connected porosity.

The results show that the reference grout (a modified grout) containing silica fume as pozzolanic material has the ability to self seal. The present observation obviates the concern that modified grouts, containing silica fume or fly ash, may not heal due to lower concentration of Ca^{2+} in the pore water. However, some aspects of mechanisms providing remain obscure and require further investigation. The results also indicate that the formation of calcite is not the only self sealing process.

SUMMARY AND CONCLUDING REMARKS

Without the superplasticizer, the high-performance grouts cannot be injected into the rock. Moreover, other associated work (Onofrei et al. 1990, 1991b 1992) strongly indicates that the low water content, high-performance materials are significantly more durable than normal cements and concretes with higher water contents. The durability of high-performance cement grouts is enhanced by the ability of the materials to autogenously seal. The fact that the superplasticizer does not significantly alter the phase distribution in the hardened grout allows for theoretical evaluation of grout longevity such as that carried out by Alcorn et al. 1992 with respect to nuclear fuel waste disposal.

Laboratory studies confirmed that decreasing the water content of the cement grouts by admixing superplasticizer (water reducing agent) increased strength and decreased the hydraulic conductivity of the hardened grouts. The hardened grouts have unconfined compressive strength and elastic moduli approaching those of granite and lower hydraulic conductivity (i.e., $< 10^{-14}$ m s^{-1}) than that of intact granite (10^{-12} m s^{-1}).

Evaluation of the results from leaching studies showed that the leaching behaviour of grouts at a given temperature depend on grout and leachant composition. Initial groundwater composition assumes a major role in grout leaching/dissolution by controlling the extent of leaching/dissolution required to produce solution saturation with leaching elements. The results indicated that leaching under both static and dynamic conditions is very complex involving several processes such as; incongruent dissolution, formation of alteration layers and precipitations. Leaching mechanisms and the microstructure of the leached surface strongly depend not only on the composition of the leachant and the contact time between the leachant and grout but also on the grout microstructure. Further work is needed to fully understand the mechanisms and rates of reactions between hardened cement grouts and groundwaters.

It was shown that the high performance grout, containing silica fume as a pozzolanic material and superplasticizer as a water reducer, if mechanically disrupted has the ability to self seal. The results obviate the concern that modified grouts, containing pozzolanic materials may not heal. The ability of modified grouts to heal requires further investigation. It appears that the modified grouts can heal, certainly when they are relatively young (~2 years curing). Some uncertainty remains with much older grouts.

The results indicate that it is possible to manufacture high-performance cement-grouts which can be injected into very fine fractures and divert water flow in granitic rock. These grouts are shown to have negligible hydraulic conductivity, associated with very low porosity and to be highly leach resistant under repository conditions. Microcracks generated in this materials from shrinkage, overstressing or thermal loads are likely to self-seal. The results of the laboratory and *in situ* studies suggest that the high-performance grouts can be considered as viable materials in repository sealing applications.

ACKNOWLEDGEMENTS

The work described in this report was jointly funded by International Stripa Project and the CANDU Owners Group. Particular thanks are due to the staff of the Analytical Science Branch of AECL Research, who carried out many of the chemical and microstructural analyses. The continuing support and constructive comments from the members of the Task Force on Sealing of the International Stripa Project- Phase III are greatly appreciated.

REFERENCES

Aitcin, P-C., Onofrei, M. and Gray, M. (1989). Superplasticizers. Tr-447 Fuel Waste Technology, Whiteshell Laboratories, AECL Research.

Alcorn, S.R., W.E. Coons and M.A. Gardiner, 1990. "Evolution of Longevity of Portland Cement Grout Using Chemical Modeling Techniques". Mat. Res. Soc. Symp. Proc. 176, pp. 165-173.

Alcorn, S.R., W.E. Coons., T. Christian-Frear and M. Wallace, 1992. "Theoretical Investigations of Grout Sealing-Final Report. Stripa Project, Technical Report 92-23, SKB, Stockholm Sweden.

Berner, U, 1987. "Modeling the Incongruent Dissolution of Hydrated Cement Minerals", Radiochimica Acta, Proc. Int. Conf. "Migration 87".

Coons, W.E., (ed) 1987. State of the art report on potential useful materials for sealing nuclear waste repositories. Stripa Project TR-87-12, SKB, Stockholm, Sweden.

Coons, W.E. and S.R. Alcorn, 1989. "Estimated Longevity of Performance of Portland Cement Grout Seal." Sealing of Radioactive Waste Repository, Braunschweig, Conference Proceedings, F.R. of Germany May 22-25.

Feldman, R.F. 1981. Effect of fly ash incorporation in cement and concrete. Materials Research Society Symposium Proceedings 40, pp.124.

Gray M.N. and L.D. Keil, 1989. "Field Trials of Superplasticized Grout at AECL's Underground Research Laboratory". Third International Conference on Superplasticizers and Other Additives in Concrete, Ottawa, Ontario, October 4-6.

Hooton,R.D. 1985. Permeability and pore structure of cement pastes containing fly ash, slag and silica fume. Pro. ASTM C-1 Symp. on Blended Cements, Denver, June 27, pp. 128-143.

Hooton, D.R. and L.D Wakeley, 1989. "The Influence of Test Conditions on Water Permeability of Concrete in Triaxial Cell". Mat. Res. Soc. Symp. Proc. 137 pp. 157-164.

Powers,T.C. 1958. J. Am. Ceram. Soc. 41, 1, pp. 1-5.

Onofrei, M and Gray, M.N. 1988. Cement grout longevity-Laboratory studies, 5th Meeting of The Task Force on Sealing Materials and Techniques For The OECD/NEA Stripa Project, Tokyo, Japan, October 25-27.

Onofrei, M., M.N. Gray, L.D. Keil and R. Pusch. 1989. "Studies of Cement Grouts and Grouting Techniques for Sealing a Nuclear Fuel Waste Repository". Mat. Res. Soc. Symp. Proc. 137, pp 349-358.

Onofrei, M., Gray, M.N., Breton, D. and Ballivy, G. (1990). The effect of leaching on the pore structure of cement-based grouts for use in nuclear fuel waste disposal. Mat. Res. Soc. Symp. Proc. 212, pp 417-425.

Onofrei, M., Gray, M.N., and Roe, L. (1991a). Superplasticizer function and sorption in high-performance cement based grouts. Stripa Project TR 91-21. SKB, Stockholm, Sweden.

Onofrei, M., Gray, M.N., and Roe, L. (1991b). Cement based grouts - longevity: laboratory studies of leaching behaviour. Stripa Project TR 91-33. SKB, Stockholm, Sweden.

Onofrei,M., Gray, M.N., Pusch, R., Borgesson, L., Karnland, O., Shenton, B. and Walker, B.(1992). Final Report of the sealing project - Sealing properties of cement based grout materials. Stripa Project TR 92-28, SKB, Stockholm, Sweden.

Onofrei, M., Gray, M.N, Coons W.E. and Alcorn S.R. 1992a. High performance cement-based grouts for use in a nuclear waste disposal facility. Waste Management, vol 12, pp 133-154.
Onofrei, Maria, Malcolm N. Gray and Leyton Roe, 1992b. "Cement-Based Grouts - Longevity Studies: Leaching Behaviour". Technical Report 92- , SKB, Stockholm, Sweden.

Table 1. Experimental parameters of test methods

Test Method	Leachant	Temperature (°C)	Flow-Rate (mL)	Time (day)
Static test	DDW, SCSSS, WN-1	10 to 100	-	1 to 32
Continue low flow-rate test	DDW, WN-1	25 to 85	12 mL/d	1 to 28
Continue high flow-rate test	DDW, WN-1	25 to 100	240 mL/d	1 to 28

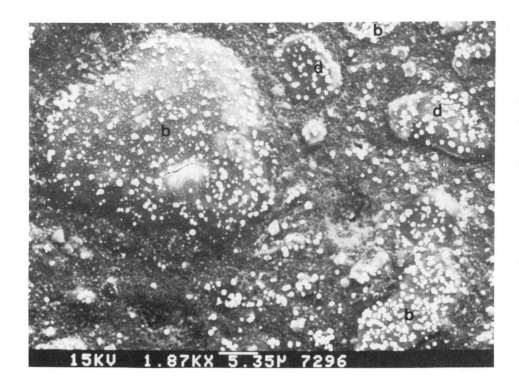

Figure 1.

Electron micrograph of polished surface of reference high-performance grout mixed at $W/CM = 0.2$, with 10% silica fume and 3% of ^{35}S - labelled superplasticizer. The bright spots indicate the the location of ^{35}S. b = calcium silicate phases, d = calcium aluminate phases.

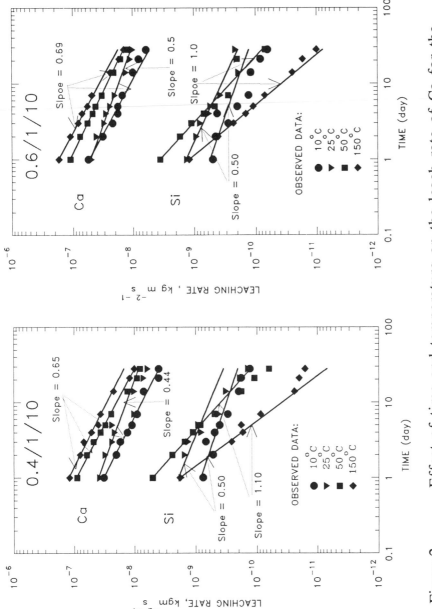

Figure 2. Effect of time and temperature on the leach rate of Ca for the reference grout based on Type 50 cement with the *W/CM* = 0.4 and 0.6 leached in DDW under static conditions.

343

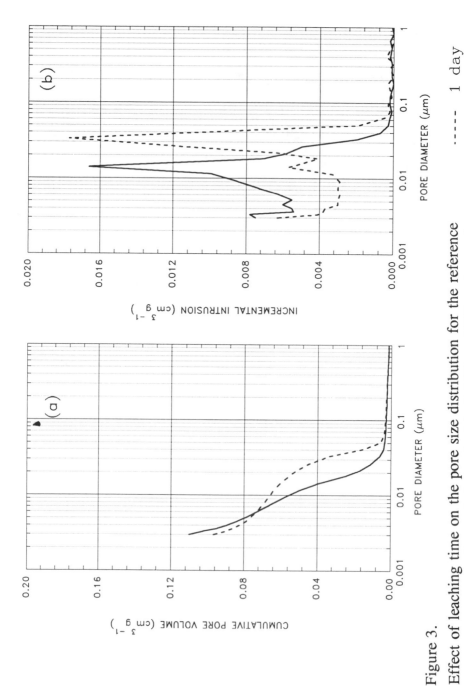

Figure 3.

Effect of leaching time on the pore size distribution for the reference grout mixed at $W/CM = 0.4$, leached in DDW at 10°C under static conditions.

    ---- 1 day
    —— 32 days

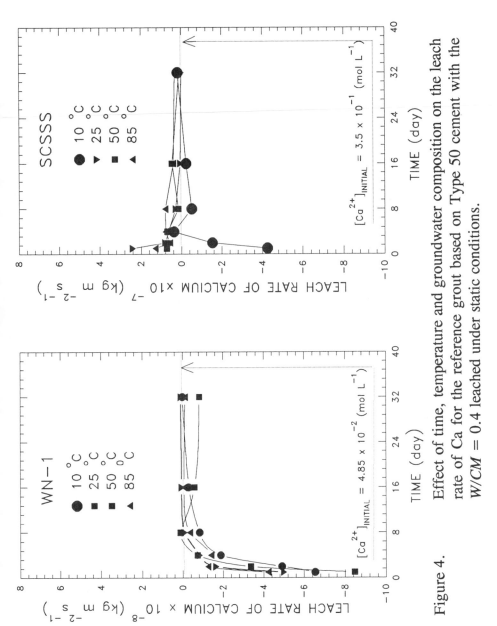

Figure 4. Effect of time, temperature and groundwater composition on the leach rate of Ca for the reference grout based on Type 50 cement with the $W/CM = 0.4$ leached under static conditions.

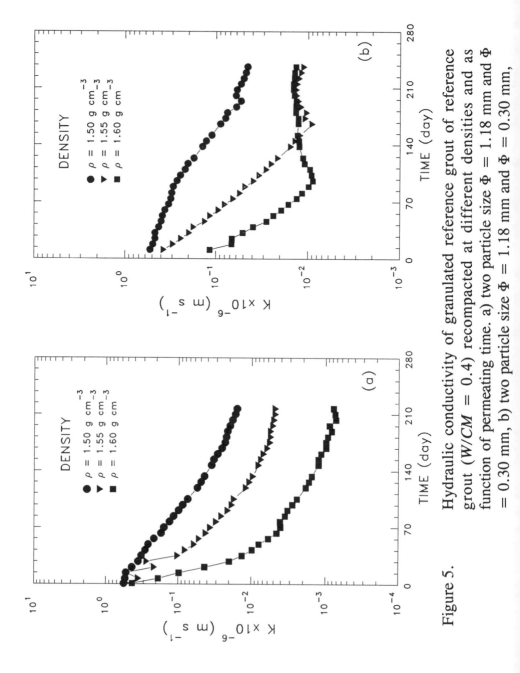

Figure 5. Hydraulic conductivity of granulated reference grout of reference grout ($W/CM = 0.4$) recompacted at different densities and as function of permeating time. a) two particle size $\Phi = 1.18$ mm and $\Phi = 0.30$ mm, b) two particle size $\Phi = 1.18$ mm and $\Phi = 0.30$ mm,

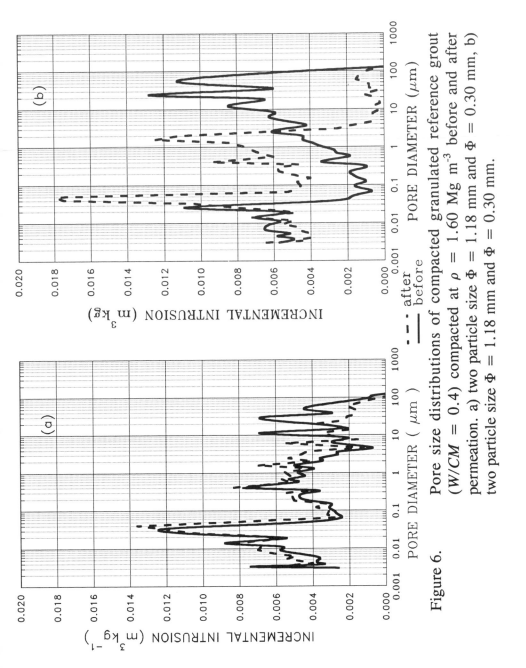

Figure 6. Pore size distributions of compacted granulated reference grout (W/CM = 0.4) compacted at ρ = 1.60 Mg m^{-3} before and after permeation. a) two particle size Φ = 1.18 mm and Φ = 0.30 mm, b) two particle size Φ = 1.18 mm and Φ = 0.30 mm.

The Buffer Mass Test (BMT)

Roland Pusch
Lennart Börgesson
Clay Technology AB

Abstract

The evaluation of the Buffer Mass Test mainly concerned heating of the bentonite/rock system that was used to simulate hot canisters in deposition holes, swelling and swelling pressures of the expanding bentonite in the holes, and water uptake of the bentonite in the holes as well as in the tunnel backfill.

The recorded temperatures of the bentonite and surrounding rock were found to be below the maximum temperature that had been set, but higher than the expected values in the initial period of testing. The heater surface temperature dropped from the rock even in the "driest" hole, which was located in almost fracture-free rock.

The water uptake in the highly compacted bentonite in the heater holes was manifested by a successively increased swelling pressure at the bentonite/rock interface. It was rather uniformly distributed over this interface and reached a maximum valued of about 10 MPa. In the wettest holes the saturation became almost complete and a high degree of saturation was also observed in the tunnel backfill. Both in the heater holes and the tunnel, the moistening was found to be very uniform along the periphery.

1   Purpose of test, test arrangement

The general objective of the Buffer Mass Test, was to check the suitability and predicted functions of certain bentonite-based buffer materials under real conditions on site, the major interest being focused on:

- the development of temperature fields in the highly compacted bentonite serving as canister envelopes

- the time-dependent distribution of pore-water in the canister envelopes

- the development of swelling pressures at the rock/bentonite interface in the deposition holes

The geometry and general features of the test arrangement were similar to those of the Swedish KBS3 concept, although on a somewhat smaller scale, but the canisters did not contain radioactive waste but electrical heaters, powered by 600-1800 W, so as to simulate the heat produced at the radio-active decay.

The BMT also served to illustrate various practical means of preparing, handling and applying soil type backfills at large depths. The test site was a 30 m long drift at 340 m depth in the Stripa mine in which the Lawrence Berkeley Laboratory, USA, had previously made extensive investigations of the gross permeability of the rock. Their work was focused on determining the frequency and distribution of joints and fractures and of estimating the amount and distribution of inflowing water as well as of the water heads in the rock, which turned out to be 1 MPa at 3-5 m distance from the periphery of the drift and very low within about 2 m distance from the periphery. The average hydraulic conductivity of the rock was found to be about 10^{-11} m/s.

Six "deposition" holes, 0.76 m in diameter and about 3 m in depth, were core-drilled for hosting electrical heaters, which were surrounded by tightly fitting blocks of highly compacted sodium bentonite. The holes were covered by sand/bentonite backfill that was compacted on site. In the inner 12 m long part of the drift, which was separated from the outer part by a rigid bulwark, the entire tunnel section was backfilled. Two of the heater holes were located in this inner part, while the outer four were situated below a 1.6 m thick concrete slab with boxing-outs that were backfilled with sand/bentonite.

The location of the holes was selected so as to find rather "wet" rock for two of them (Holes No. 1 and 2 in the inner part) and very "dry" rock for one hole (No. 6) and intermediate conditions for the other three. Fig. 1 illustrates the arrangement.

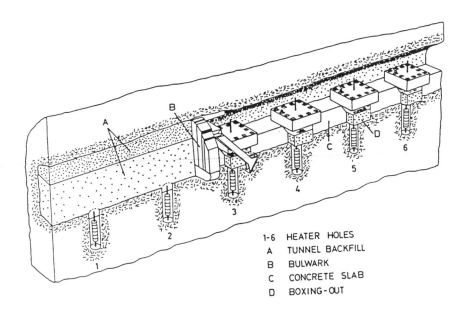

1-6 HEATER HOLES
A   TUNNEL BACKFILL
B   BULWARK
C   CONCRETE SLAB
D   BOXING-OUT

Fig.1 The BMT heater hole and tunnel arrangement

2   Sealing materials

2.1   *Bentonite in the "deposition holes"*

The required high density of the bentonite envelope of waste canisters in actual repositories can be obtained by compacting air-dry granulated bentonite powder under high pressure to form blocks, and this technique was also applied for the BMT, although it was not clear at the start of the compaction process if it would be possible to produce large, fracture-free blocks. A suitable density of the highly compacted bentonite in the final, water saturated and swollen state in the deposition holes of repositories is ρ_m=2.0-2.1 t/m^3, which implies an initial bulk density of the only partly water saturated, compressed bentonite blocks of 2.1-2.2 t/m^3 to arrive at this condition after completion of the water uptake from the rock and after lateral swelling to fill up the holes completely. Two block series were produced with a bulk density that turned out to be 2.09-2.14 t/m^3 for bentonite powder with an average water content of 10%.

351

In order to fit the heater holes as closely as possible and to form a tight contact with the heaters, the bentonite had to be prepared in the form of specially shaped blocks that could be arranged to form a tightly fitting brick-work. (Fig. 2).

The different initial bulk density of the two block series would yield different final densities and physical properties if the geometrical conditions were the same in all the heater holes. It was therefore decided to use different starting conditions chosen so as to yield the same ultimate bulk density and water content in all the holes, namely ρ_m=2.10 t/m^3 and about 20%, respectively. This was achieved by manufacturing the blocks so that a 30 mm slot between the rock and the bentonite annulus was formed for the low initial block density ρ=2.07-2.11 t/m^3, and filling the slot with loose bentonite powder with an average bulk density of 1.2 t/m^3. For the high initial block density ρ=2.09-2.14 t/m^3 at 10 mm open slot was chosen to arrive at the same ultimate bulk density. The latter slot width was in fact considered to be at minimum for bringing the bentonite/heater unit down into its hole.

The rate of water inflow into the holes had to be considered in deciding in which holes the wide slots with the loose powder should be located, since rapid inflow would make the application of the powder difficult. Such conditions were expected in holes No. 1, 2 and 5 so the following arrangement was chosen:

- Heater holes Nos. 1, 2 and 5 with open, 10 mm wide slots

- Heater holes Nos. 3, 4 and 6 with backfilled, 30 mm wide slots

2.2   Bentonite/sand backfill in the tunnel

Pilot tests had indicated that it should be possible to backfill the upper part of tunnels in repositories of the KBS3-type by blowing the material in place, while the lower two thirds are preferably filled by layer-wide application and compaction with vibrating plates or rollers. Since these two techniques give different bulk densities, the bentonite content should be higher in the upper part to yield approximately the same permeability as the lower part and to make the backfill expandable. For this purpose the upper part of the backfill of the BMT has a 20% bentonite content, while the corresponding fraction of the lower part was 10%, the ballast material being glacial-type, suitably (Fuller-type) graded.

Large concrete mixers were used for preparation of the tunnel back-fill. They operated well with as much as 500 kg of soil material and water at a time, and gave reasonably homogeneous mixtures. Certain trends of the bentonite to form coatings and large granules were occasionally noticed, especially when the components were successively added in the mixer. The best procedure was to fill the loading box of the mixer with bentonite, "filler" and sand, bring it over to the mixer and add water successively (1).

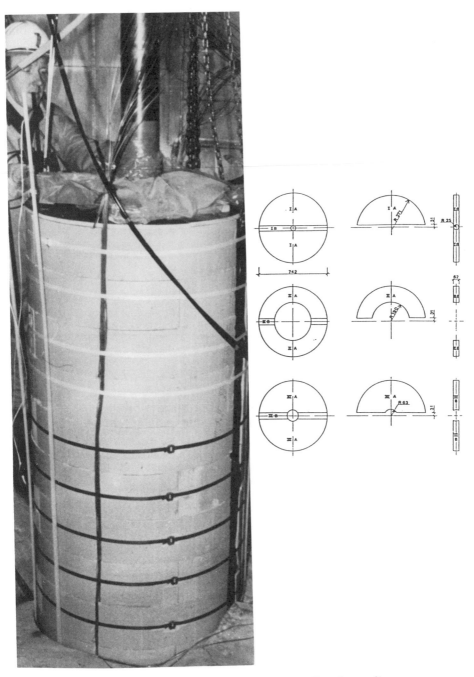

Fig. 2 Lowering of bentonite/heater unit.
Small picture shows drawing for block manufacturing

The backfill consisting of 10% bentonite mixture was applied in 0.15 to 0.30 m thick layers in the tunnel and boxing-outs and was compacted by 10-15 runs with a 400 kg electrically operated Dynapac plate vibrator. The compaction reduced the thickness of each layer to about 65% of its original height. The robot device shown in Fig. 3 was used to apply the 20% bentonite mixture in the upper part of the tunnel. No compaction was made of this upper part.

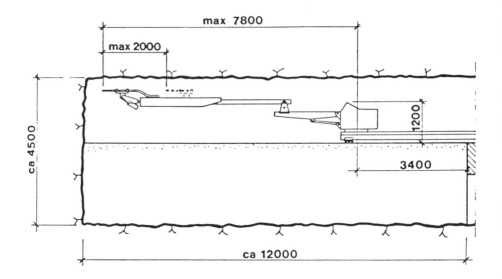

Fig.3 Schematic picture of robot for shotcreting the 20% bentonite backfill. Dimensions in mm

The arrangement of the sealing materials in the backfilled part of the drift is shown in Fig. 4. The dry density of the various zones are given in Table 1.

Table 1. Dry density of the in-situ backfillings

Zone No.	Dry density, t/m^3
1	1.81
2 and 2A	1.77
3A	1.25

3   Prediction of major data; recordings and results

All the holes and the tunnel backfill as well as the nearfield rock were richly instrumented with various back-up systems for recording the maturation processes, the temperature evolution and the rock reactions in order to check the predictions. Basic data are summarized below.

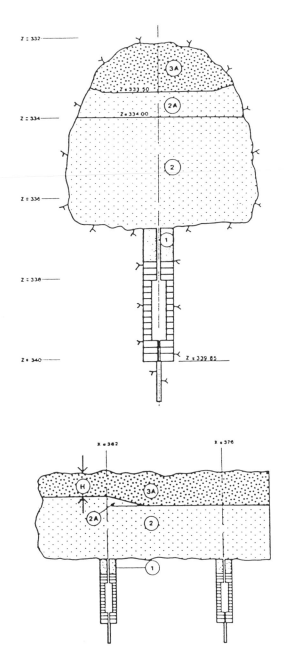

Fig. 4 Cross section through backfilled tunnel and deposition hole. Highly compacted blocks of bentonite surround the heater in the hole. 1), 2) and 2A) represent 10% bentonite mixture and 3A shotcreted 20% bentonite mixture

3.1 *Heater holes*

3.1.1 Heat evolution

Gauges: 1200 thermocouples

Prediction: FEM based on lab data ("effective" λ of 0.75 W/m, K), cf. Fig.
5 (2). Accuracy $\pm$ 10%

Outcome: Prediction $\pm$ 10%; Refined model with "zonation" of annuli (3)

3.1.2 Swelling pressure

Gauges: 60 Gloetzl cells

Prediction: Max. pressure $\geq$ 12-15 MPa (Fig. 6)
Accuracy $\pm$ 10%

Outcome: Max. pressure $\sim$ 11 MPa (3) (Fig. 7)

3.1.3 Water uptake

Gauges: Moisture sensors (LuH); H_2S gas

Prediction: Diffusion-type hydration with complete saturation in 2-450
years depending on rock structure

Outcome: 1) Corrected diffusion-type model confirmed tentatively:
Saturation rate depends on access to water (Fig. 8)
2) Evaporation/condensation at the wetting front (3)

3.1.4 Buffer/backfill interaction (displacement)

Gauges: "Coins" and tape

Prediction: Max. 5-12 cm (Fig. 9)
Outcome: 5-7 cm (3)

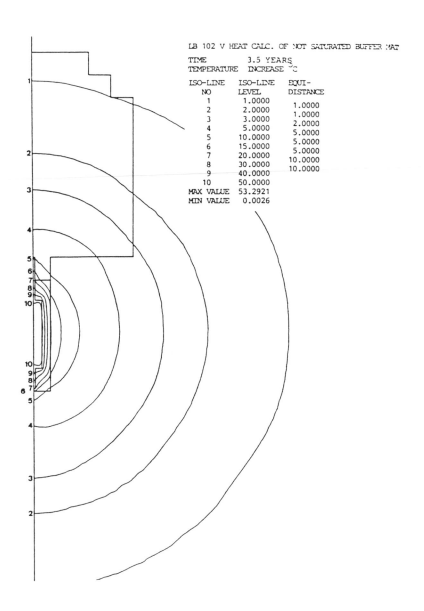

Fig. 5 Predicted temperature increase around heaters
Nos. 1 and 2 after 3.5 years (600 W)

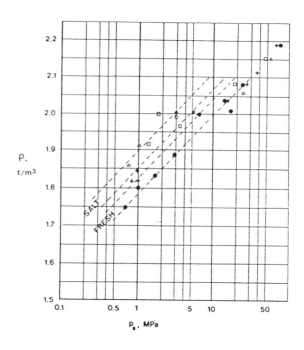

**Fig. 6 Recorded swelling pressure of MX-80 bentonite at 20°C.
Notes: + artificial groundwater, o distilled water, Δ 0.6 NaCl
solution, ▢ 0.3 M Ca Cl$_2$ solution**

3.1.5   Salt accumulation

Gauges: Chemical analysis of samples

Prediction: Some

Outcome: No clear trend (3)

3.2   *Tunnel*

3.2.1   Heat evolution

Gauges: Thermocouples

Prediction: FEM based on lab data (2). Accuracy ± 10%

Outcome: Predicted values ± 10% after 3 years (3)

358

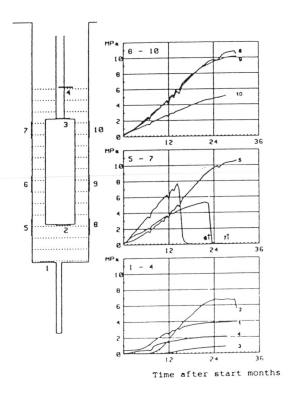

**Fig. 7 Swelling pressures in the "wet" heater hole No. 2
Cross means failed sensor**

3.2.2 Swelling pressure

Gauges: Gloetzl cells

Prediction: Max. pressure 200 kPa after 3 years. Accuracy ± 10%

Outcome: 50 kPa at max. (3)

359

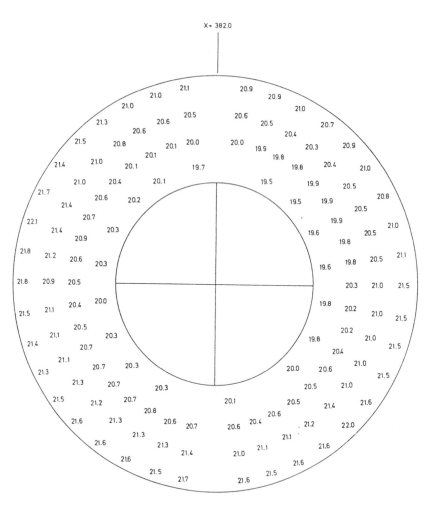

Fig. 8 Distribution of water contents in the highly compacted bentonite at the lower end of the heater in hole No.1 (1 400 W)

3.2.3 Water uptake

Gauges: Moisture sensors (LuH)

Prediction: Diffusion-type hydration with complete saturation in 200 days to 10 years

Outcome: 85-90% saturation after 3 years, indicating fair applicability of the diffusion model with very good access to water (Fig. 10)

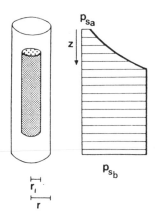

**Fig. 9 Schematic picture of the influence of
wall friction on load transfer to the
overlying backfill**

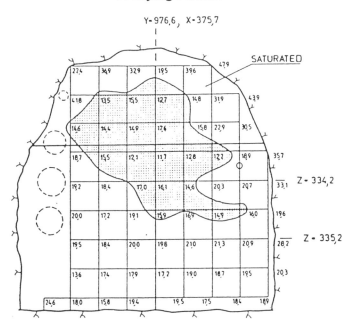

**Fig. 10 Distribution of water in the tunnel backfill,
figures denoting the water content. The central
dotted area is not fully saturated, possibly with
compressed air partly filling the voids**

3.2.4 Water pressure

Gauges: BAT and Gloetzl piezometers

Prediction: 1-1.5 MPa soon after start

Outcome: Pressure heads in interval 15-40 kPa after 3 years (Fig. 11)

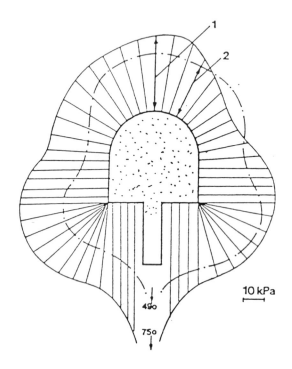

**Fig. 11 Water pressure contours at the rock/backfill
interface at sections through heater holes No. 1 and 2
at a late stage. The full line represents the section through
hole No. 2. Figures at the base of the diagrams
represent pressures at the base of hole No. 1
(750 kPa) and No. 2 (490 kPa)**

3.3 *Rock*

3.3.1 Rock structure and variation in water inflow

Gauges: Simple structure model inflow measurements in heater holes (collector) and tunnel (ventilation). Estimate of inflow distribution in BMT tunnel from inspection of wetting rate after drying

Prediction Inflow into heater holes and tunnel strongly heterogeneous

Outcome: The saturation rate in the heater holes is not controlled by the inflow except initially in the tunnel (Fig. 12). Very few water-bearing natural fractures (orthogonal-type) appear in the heater holes except in Hole 1 and in the strongly blasting-disturbed floor (Fig. 13)

3.3.2 Tunnel floor movements by expanding compacted bentonite

Gauges: Kovari installation ($\pm$ 5 μm)

Prediction: Very slight vertical heave

Outcome: No measurable heave (3)

4 Conclusions

The following main conclusions can be drawn with respect to the sealing function of the highly compacted bentonite:

1. The thermal gradient initially induced redistribution of the porewater in the bentonite blocks and the free water in the slot at the periphery ("wet" holes). Thereby, the clay swelled and established contact with the rock, while drying took place close to the heaters. The outer, expanded part of the bentonite consolidated in the course of the homogenization process.

2. Where water was richly available, i.e. in the "wet" holes where the rock water pressures was high, the wetting rate proceeded at approximately the rate predicted by the basic diffusion-type model and led to almost complete saturation despite the counteracting thermal gradient. The saturation caused development of predictable swelling pressures at the clay/rock and clay/heater contacts through which a perfect seal of the holes was created.

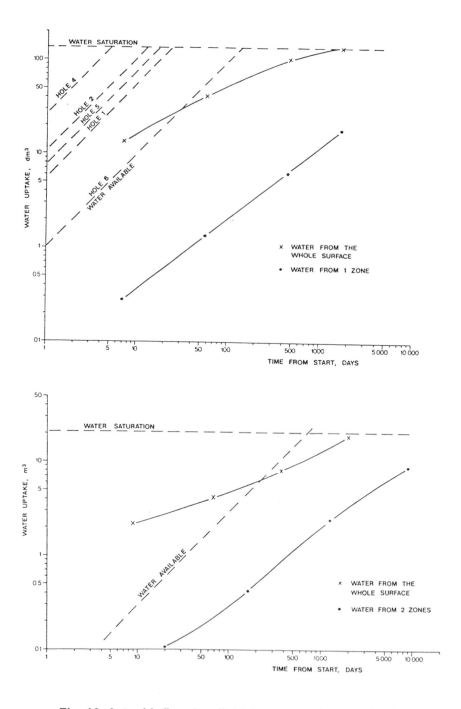

Fig. 12 Actual inflow (available) compared to required access for the validity of the diffusive hydration model

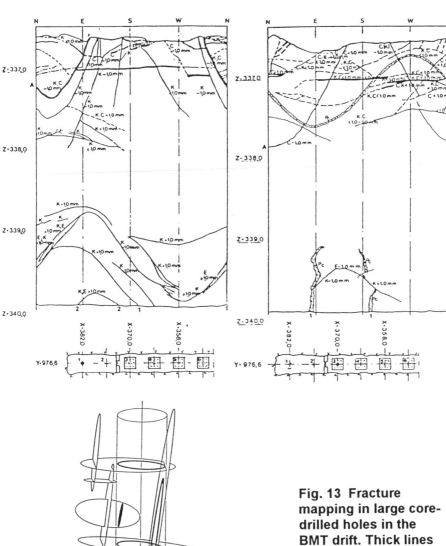

Fig. 13 Fracture mapping in large core-drilled holes in the BMT drift. Thick lines in the graphs, which show the entire periphery of two holes, represent strongly water-bearing fractures. Lower picture shows typical orthogonal pattern of two major fractures deeper than 1 m below the tunnel floor

THICK LINES INDICATE INTER-SECTION OF ELLIPSOIDS AND DEPOSITION HOLE

3. Where very limited amounts of water could be transferred to the clay a characteristic ratio of the water content gradient and the thermal gradient was developed in the initial stage. However, also in these "dry" holes water was successively taken up from the rock through a slow, apparently diffusion-type process.

4. Clay penetrated into visible open fractures and sealed them to a depth of a few millimeters. Thereby, water migrating under pressure gradients from the rock to the clay was directed through "second order" fissures, which in turn become sealed. This forced water in the rock to move through the crystal matrix and thus to enter the clay very uniformly over the entire interface, which was manifested by rather evenly distributed swelling pressures.

5. The major sealing processes were:

   - expansion made the dense bentonite fill the holes with low-permeable clay

   - development of tight contact at the clay/rock interface

   - penetration of clay into fractures

   - development of swelling pressures on the rock surface possibly causing certain fractures to be compressed

As to the sealing effect of the tunnel backfill, the following conclusions were drawn:

1. The process of saturation appeared to fit the one predicted on the basis of uniform access of water at the tunnel periphery, assuming the migration mechanism to be one of diffusion

2. With the low bentonite content of the effectively compacted backfill and with the unexpectedly low bulk density of the shotcreted 20% bentonite mixture, the swelling pressures were insignificant and without any sealing effect of fractures in the rock. However, the expansion was sufficient to make the mass occupy the entire space, which was thus effectively sealed by a backfill that had a hydraulic conductivity that was considerably lower than that of the rock, except in the uppermost part.

5   References

1. Pusch, R. & Nilsson. Buffer Mass Test - Rock Drilling and Civil Engineering. Stripa Project Technical Report 82-07, SKB, Stockholm 1982

2. Börgesson, L. Predictions of the Behaviour of the Bentonite-based Buffer Materials. Stripa Project Technical Report 82-08, SKB, Stockholm 1982

3. Pusch, R. & Börgesson, L. Final Report of the Buffer Mass Test - Volume II: test results. Stripa Project Technical Report 85-12, SKB, Stockholm 1985

Sealing of Boreholes, Shafts and Tunnels

R. Pusch
L. Börgesson

Clay Technology AB

ABSTRACT

The Borehole Plugging Experiment comprised field tests of the sealing function and the practicality in handling and application of plugs of perforated metal casings filled with cylindrical blocks of highly compacted sodium bentonite. The field tests demonstrated that even very long holes can be effectively sealed by such plugs and that the clay becomes very homogeneous and forms a tight contact with the rock in a relatively short time.

Shaft sealing was investigated in a 14 m long shaft in which two plugs were constructed with a central sand-filled central space for injecting water. A reference test with concrete plugs was followed by a main test with plugs of highly compacted bentonite. In the latter test, the outflow from the injection chamber was only a few percent of that with concrete plugs.

The Tunnel test gave evidence of very effective sealing power of Na bentonite, forming "O-ring" sealings at the plug/rock interface. Flow tests showed that the plugs were practically tight even at 3 MPa water pressure. The major sealing process appears to be the establishment of a very tight bentonite/rock interface.

1   BOREHOLE PLUGGING TEST

1.1   Purpose of test, test arrangement

The Borehole Plugging Experiment comprised three field tests of plugs with highly compacted sodium bentonite as sealing component. The function and practicality in handling and application of such plugs were tested under real conditions, the basic design principle being that cylindrical blocks of compacted clay powder were contained in perforated casings. After insertion in the boreholes, which were water-filled from the start, the clay was expected to absorb water and swell out through the perforation. Laboratory tests and pilot field tests had indicated that the expansion of the clay leads to complete embedding of the casing if the access towater is sufficient, and the main purpose of the field tests was to investigate the rate and uniformity of the water uptake under different piezometric conditions. The major components of the technical system are shown schematically in Fig. 1.

Three separate experiments were made with the following main features:

I   *The DbH2 plugging test*

The main purpose was to test if a clay plug could be inserted in a 100 m long almost horizontal, 56 mm diameter borehole, and to determine the rate of maturation by measuring the required hydraulic gradient to produce piping at various periods of time after the application of the plug (1,2).

IIa   *Test of the standard plugging technique in ∅76 mm borehole*

This test involved insertion of a 4 m long clay plug in a 14 m long 76 mm diameter borehole. The plug was equipped pressure transducers to measure the total and effective pressures in the course of the maturation process. Water overpressure was initially applied at the upper and later at the lower end to determine the critical pressure at which piping occurred. At a late stage the bond strength was determined by applying an axial force that extruded the plug (1,2).

IIb   *Test of plug with highly porous casing in ∅76 mm borehole*

The perforated casing of the standard plug was replaced by a highly porous metal net in a test which was otherwise identical to the IIa test. The aim was to investigate whether the higher porosity of the casing would yield a significant increase in maturation rate (1,2).

Fig. 1 Schematic picture of components and technique of the investigated sealing technique

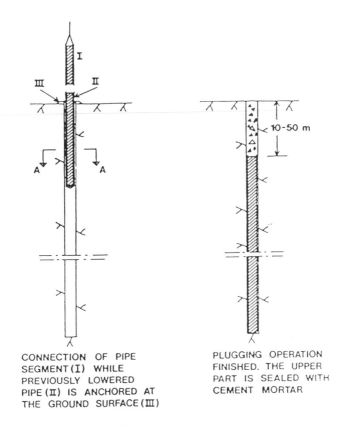

CONNECTION OF PIPE
SEGMENT (I) WHILE
PREVIOUSLY LOWERED
PIPE (II) IS ANCHORED AT
THE GROUND SURFACE (III)

PLUGGING OPERATION
FINISHED. THE UPPER
PART IS SEALED WITH
CEMENT MORTAR

10-50 m

SECTION A - A

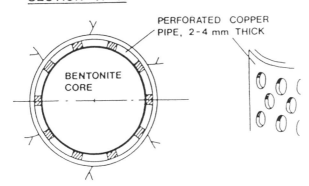

PERFORATED COPPER
PIPE, 2-4 mm THICK

BENTONITE
CORE

1.2   The DbH2 plug

Test data

The small diameter of the borehole and the significant space occupied by the central piping hosting tubings for applying and measuring water pressures in the filters required a rather thin and large diameter casing in order to obtain a reasonably high density of the clay after expansion. Still, the diameter of the casing had to be selected so that it could be inserted without difficulties.

The perforated casing consisted of 39 2.5 m long copper segments with 54 mm outer diameter and 50 mm inner diameter. The perforation, covering about 50% of the surface, was made by drilling 11 mm holes. The segments were connected at their ends, the jointing operation being made in the course of the application of the plug.

The clay blocks, which were produced by compacting Volclay MX-80 bentonite powder with an initial water content of 11% to a bulk density of 2.11 t/m^3, had an outer diameter of 48.7 mm. They were given annular shape, the inner diameter being 18.3 mm, to host a central pipe through which tubings from the filters passed to the outer end of the hole where they were connected to a system for injecting water and measuring water pressures. The compaction pressure to obtain the required density of the blocks was 120 MPa. The theoretical net density after complete homogenization was estimated at 1.9 t/m^3, corresponding to a water content of about 32%.

Filters were applied for testing the maturation rate (Fig. 2). This was made by applying water pressure in one filter at a time leaving the others open to allow for piping etc.

Results (2)

The total time to plug the hole with all the instrumentation was 2.5 hours which suggests that several hundred meter long, horizontal or slightly inclined boreholes can be plugged by the technique applied in the Stripa test. Vertical holes that can be plugged from above may be as long as 1 kilometer or more before practical problems arise. Vertical or steeply inclined holes extending upwards from tunnels may offer difficulties at considerably shorter lengths than that.

Two days after the grouting, water appeared in all the tubings and they were then connected to the injection/recording system for reading the natural water pressures that were expected to be built up in the filters. The pressure quickly rose to 1200 kPa in Filter 1 at the inner end of the plug and to 200 kPa in Filter 4, while no overpressures were recorded in Filters 2 and 3. These were still the conditions 8 days later when the first injection test took place. The fact that the natural high water pressures could be resisted by the clay plug so soon after the closure of the drainage, shows that considerable maturation had taken place very rapidly.

Filter 2 was pressurized in 30 kPa steps up to 800 kPa, each lasting for one minute. The pressure steps were then increased to 100 kPa, each with a duration of 2 minutes. When the pressure was raised to 1400 kPa, critical conditions appeared and a sudden inflow of a few milliliters took place with a simultaneous pressure drop to about 800 kPa. After releasing the pressure for 0.5 hours it could be increased to 2000 kPa without breakthrough.

Fig. 2   Schematic picture of the clay plug showing the part that was instrumented and located close to the BMT tunnel. Filter No. 1 was at the inner end of the about 100 m long plug. The distance between the hole and the periphery of the tunnel was about 1.5 m

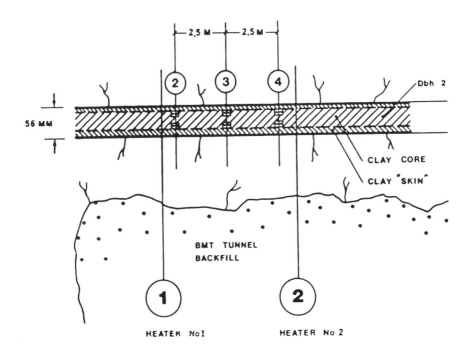

Two and a half years after the application of the plug a 0.5 m long section was extracted by slot-drilling technique. The plug, which gave the impression of being very homogeneous, was cut into two halves and 10 samples of the intact clay core were taken for determination of the water content and bulk density. Practically all the values were in the interval 32-34%, which thus demonstrates that the investigated part of the plug was completely water saturated and fully matured. The bulk density of almost all the samples was in the range of 1.8-1.9 t/m^3.

Assuming that the basic "diffusion-type" model for water uptake that was derived in the BMT is applicable also in this case, we find the time to reach complete saturation to be much shorter than two years, provided that water was available over the entire rock/clay interface. A simple calculation of the inflow of water to the hole through the crystal matrix ($k=5x10^{-13}$ m/s) showed that access to water for saturation of the clay was not a limiting factor.

1.3   The 76 mm borehole plug

Test data

The general appearance of the plugs, which were 0.5 m apart, is shown in Fig. 3. The pressure gauges were mounted in conical copper housings, which were connected to a central pipe with the cables, and which served to joint the two 2 m long plug segments in the respective hole, as well as to form the outer ends of the plugs. The gauges had the form of a few milliliter large cells equipped with pressure transducers.

The perforated copper casing of hole IIa had an outer diameter of 68.6 mm and an inner diameter of 65 mm. The perforation, covering about 50% of the surface, was made by drilling 11 mm holes. The net-type casing had of course a much higher perforation ratio, which was expected to yield a fast maturation rate but also a mechanically weak plug.

The ring-shaped clay blocks were produced by compacting Volclay MX-80 bentonite powder with an initial content of 10% to a bulk density of 2.1 t/m^3. They were given an outer diameter of 65 mm while the inner diameter was 20 mm, which yields a theoretical net bulk density of 1.82 t/m^3 after complete expansion and homogenization in those parts which were located between the pressure gauges. The corresponding water content is about 40%.

Fig. 3 Schematic picture of 76 mm plugs IIa and IIb. P_1 denotes upper chamber which was pressurized to determine the maturation rate of the clay. 1-6 are gauges for determination of the total pressure and the water pressure

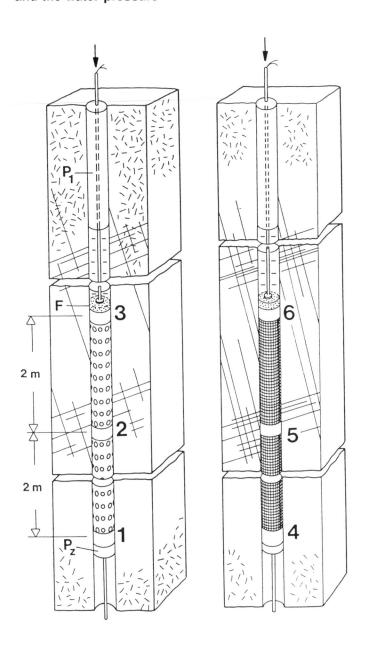

Results (2)

The main purpose of the test was to check the maturation rate of the clay by injecting water in the upper chamber as in the DbH2 test, and to record the build-up of swelling pressures as a measure of the maturation rate. Since the tests were conducted in virtually "dry" rock, very slow and incomplete maturation of the plug was foreseen.

Four days after the application of the plugs a 50 kPa injection pressure was applied with no sign of piping or clay displacements. The pressure was increased to 100 kPa after another week, and to 200 and 300 kPa after several months still with no sign of disruption of the clay. Injection pressures exceeding 200-300 kPa tended to open subhorizontal fractures in the rock and could therefore not be applied.

Theoretical considerations indicated that although the holes were not fed effectively with water because of the low water pressure in the rock, suction - which was actually recorded by the pressure gauges (Fig. 4) - would tend to bring a sufficent amount of water into the clay to make it largely saturated. The rather low total pressures that were recorded are explained by the fact that the clay density was lower at the gauges. Complete homogenization and thus equal density at the gauges and in the rest of the plugs would require many more years under the prevailing conditions.

After extrusion of the plugs, comprehensive sampling for water content and density determinations showed that the plugs were remarkably homogeneous inside the casings with an average water content of 35% in hole IIa and 37% in hole IIb. This is only slightly less than the predicted value for complete saturation and corresponds to a bulk density of about 1.85 t/m^3. The clay "skin" outside the casings appeared to be homogeneous but slightly wetter than the inner clay core. This was assumed to be due to exposure to water in the course of the extrusion experiment.

The two 4 m long plugs required about 9 t axial force each to be extruded and this corresponds to a shear stress ("bond strength") of 100-120 kPa at large strain. Almost the same value was arrived at in preceeding laboratory tests.

1.4   Conclusions

The following main conclusions can be drawn from the borehole plugging tests:

1   The plugging of a 100 m long, 56 mm diameter, almost horizontal borehole demonstrated the practicality of this plugging technique also in very long

Fig. 4 Recorded pressures in the 76 mm plugs. 3, 5 and 6 refer to the gauges in Fig. 3

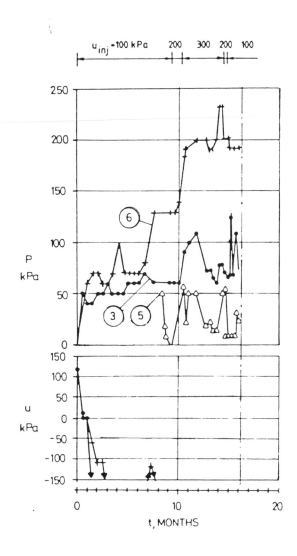

holes and this test also showed that the maturation of the plugs was sufficiently fast to resist piping or distortion by high hydraulic gradients already after about one week. The uniformity of the water content of a recovered section was tested after about 2.5 years and found to be very high. The clay was completely water saturated despite a large variation in fracture frequency of the rock, which indicates that water had passed through frequent fine fissures in the apparently fracture-free rock, leading to uniform uptake over the entire clay/rock interface and complete maturation in about 0.5 year.

2   Clay plugs located in rock with a very low water pressure but still with saturated fractures, behave as the compacted bentonite in the "dry" BMT deposition holes. Thus the bentonite expands by taking up available water in the holes and then sucks water from the rock. This yields a slower water uptake than when the holes are effectively fed with water as in rock with high water pressures. Also in the absence of a thermal gradient the conditions with limited inflow of water tend to result in a fairly uniformly distributed water content in the course of the slow saturation process.

3   The very uniform distribution of water in the axial direction of the clay plugs verified the validity of the general model for water uptake from the rock: water initially flows from wider fractures, which get sealed by penetrating clay, whereafter "second order" fissures serve as main inflow passages until they get clogged etc.

4   The bond strength measurements verified that the clay plugs were homogeneous although not completely saturated. They showed that the physical interaction between clay and rock was effective and that no wet and soft film of water or clay had been formed at the interface.

2   SHAFT PLUGGING TEST

2.1   Purpose of the test, test arrangement

The test was conducted in the same 14 m thick rock unit as the 76 mm borehole sealing tests and had the form of measuring the outflow of water from a central, sand-filled injection chamber, which was located between two plugs that were tied together and anchored to the shaft walls (1). The sealing effect of bentonite was determined by comparing the outflow in a reference test using plugs made of concrete with expansive cement with that in the main test in which bentonite plugs were used. The plugs were equipped with pressure cells for recording the successive build-up of swelling pressures which were directly related to the water uptake.

The shaft had been excavated a number of years earlier to test different techniques for making deep cylindrical holes with a fairly large diameter and a relatively smooth and regular perimeter. This yielded a 14 m deep shaft with a diameter of about 1.0 m at the upper end and 1.3 m at the lower end. The eastern part of the shaft perimeter was produced by slot-drilling, i.e. by percussion drilling of slightly overlapping, parallel 50 mm holes, while the western part was made by drilling Ø50 mm holes 20 cm apart for blasting. Both perimeter surfaces had a relatively constant radius of curvature but were naturally far from smooth. The main geometrical features of the shaft are given in Fig. 5.

As in the case of the 76 mm boreholes, the construction of the lower drift some 25 years before the plugging tests must have led to very low piezometric heads in the test site area. However, it was clear already by ocular inspection that the fractures were water saturated before as well as during the test and that the plugs had access to water over their entire length although it was under low pressure.

The hydraulic characterization comprised identification of the major water-bearing structures by ocular inspection of the shaft and by determining their flow properties by use of 16 56 mm boreholes close to the shaft. The main water passages, as identified by watching where water appeared after drying the shallow rock by use of an ordinary contractor's air-drying device, were found to be the ones listed below and illustrated in Fig. 6 (3).

Fig. 5   Geometrical shape of the shaft

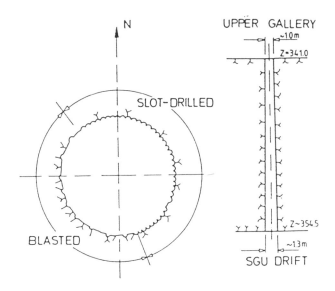

Fig. 6 **Major water-bearing structures in the part of the shaft where the plugs were located**

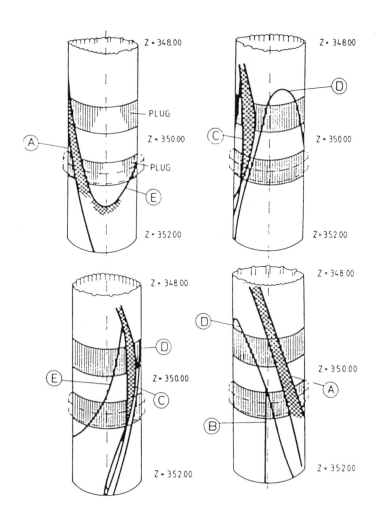

	Strike	Dip, degrees	Character
A	WNW/ESE	60	Crushed rock zone
B	W/E	90	Large, open joint
C	NW/SE	70-90	Crushed rock zone
D	NW/SE	65	Plane joint
E	WNW/ESE	60	Plane joint

The arrangement of the main test, which was preceded by the reference test with plugs cast in situ with expansive concrete and with no slot at the lower plug, is shown in Fig. 7. The observation holes were sealed over their entire length except for short segments which were intersected by the major structures. Water leaking into the shaft from these structures could be collected at various defined levels.

Fig. 7 **General view of the arrangement of the main test. hcb denotes plugs of highly compacted bentonite. ic is the sand-filled injection chamber. I, II etc are water collectors**

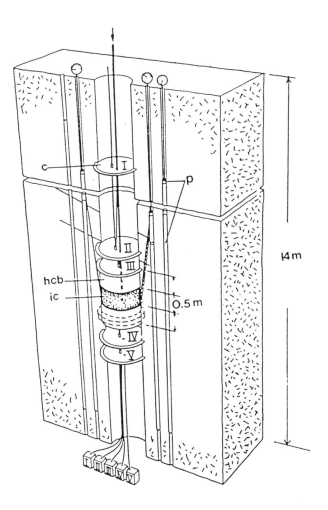

2.2 Test data

Great care was taken to saturate the 0.5-2.0 mm sand in the injection chamber between the plugs. The test program comprised stepwise increase in the injection pressure up to 200 kPa with concommittant recording of the outflow from the chamber. Different tracers were used in the two test series to identify the flow paths.

The slot surrounding the lower plug in the main test was cut by means of a rotating diamond-coated steel disc. The machine was mounted in the shaft so that a number of parallel 250 mm deep slots could be cut, the upper and lower ones having a distance of 254 mm. The rock between the cuts was then removed by use of a steel wedge that was driven into the narrow openings. This yielded a relatively even inner wall of the slot, the height of which made it possible to bring in prefabricated 250 mm thick blocks of bentonite.

The bentonite blocks were prepared from the set of larger blocks that was left over from the Buffer Mass Test. They had been produced by compacting Volclay MX-80 bentonite powder with an initial water content of 10-13%, which corresponded to a bulk density of 2.07-2.14 t/m^3. The blocks were applied as densely as possible in a regular pattern, and the space between the irregular shaft wall and the central block system was filled with smaller pieces of compacted bentonite and bentonite powder. It was estimated that the net bulk density after complete saturation would be about 2.05 t/m^3, taking into account a compression of the 0.5 m high injection chamber fill by 10% i.e. 5 cm. This density corresponds to a water content of 22-23% at complete saturation. The total pore volume of the clay plugs was 0.45 m^3 of which 0.22 m^3, i.e. 220 liters, represented air-filled voids.

The rate of maturation of the bentonite was illustrated by recording the pressure by use of the Gloetzl cells. The distribution of water and the general physical state of the bentonite at the termination of the test were determined by excavation and sampling at the end of the test.

2.3 Results (3)

Reference test

The outflow from the injection chamber was expected to take place through the five discrete structures A-E. A rough prediction of the flow was made by assuming that they were hydrologically equivalent to plane slots with an average aperture of 0.01 to 0.05 mm, which gave a predicted outflow of 0.6 to 25 l/hours at an injection pressure of 100 kPa, disregarding leakage along the rock/concrete. Since the blasted rock forming the western perimeter of the shaft would also let some water through, the total outflow was assumed to be at least a couple of liters per hour under steady state flow condition. The latter contribution was estimated by assuming vertical flow from the injection chamber in a 10 cm wide "porous" zone with a hydraulic conductivity of 10^{-6} to 10^{-8} m/s along half the periphery of the shaft. The hydraulic gradient over the two zones, each of 0.5 m axial length, would be roughly 20, thus yielding a flow on the order of 0.15 to 15 liters per hour.

The actually measured flow was intially more than 10 l/h but dropped to about 5 l/h after 4 days and to about 2 l/h after 10 days. A further drop in flow was recorded after a few weeks, probably due to clogging of the filter of the injection pipe, but the initial flow rate was again observed when water was injected through the second pipe. The average outflow rate under stationary conditions was thus concluded to be 8-9 l/h in the reference test, indicating that the larger part of the leakage took place through the identified, larger structures or along the rock/concrete interface.

The following conclusions were drawn from the study:

* No flow occurred from the injection chamber to the upper part of the shaft

* Structure B, i.e. a steep, wide fracture extending from the injection chamber and downwards, was the major water passage as indicated by the appearance of the tracer substance

* Significant leakage took place through the "crushed" rock zone A and the fracture E along the lower plug as demonstrated by the appearance of tracer substance

* No uniform flow through the blasted rock along the plugs upwards or downwards had taken place as concluded from the absence of tracers. This demonstrates that practically all the outflowing water from the injection chamber had been discharged through structures A, B and E where they contacted the concrete

* No water had flown to the observation holes as demonstrated by the complete absence of tracer substance in water samples from these holes

* Effective reduction of the flow was expected in the main test by truncating the B-structure with the clay-filled slot. The A- and E-structures were likely to be sealed by penetrating clay.

Main test

The outflow from the injection chamber in the main test was expected to be much smaller than in the reference test already at the onset of the first-mentioned test because of the blocking of the shallow, major flow passage (B), but this was assumed to be largely compensated by the much better fit between the plug material and the rock in the reference test. The initial flow rate was therefore assumed to be approximately the same in both tests, but a significant reduction was expected in the course of the main test because of various sealing effects of the maturing bentonite. The outflow from the chamber under steady state conditions in the main test at 100 kPa pressure was therefore predicted to be in the range of 0.1-1 l/h, and the lower value was actually approached after a couple of months (Fig. 8). The outflow in the main test was approximately 50% of that in the reference test in the first ten days, which is in reasonable agreement with the predictions. Hence, the sealing effect was moderate to begin with but became very significant after a couple of months.

Fig. 8   Leakage from injection chamber in reference and main tests

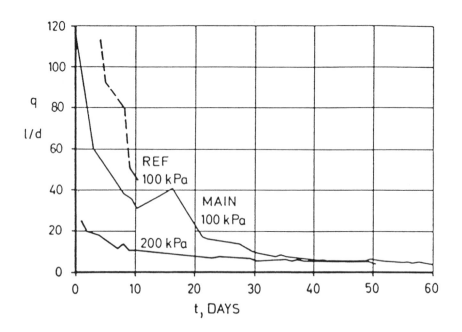

The total outflow from the injection chamber in the main test with 100 kPa pressure was approximately 1500 liters, i.e. 7 times the available empty pore space in the bentonite. Thus, the large majority of the water that left the injection chamber was discharged through rock structures which were in contact with the bentonite but which also served as drainage passages. Only a very small fraction of the discharged water could have been absorbed by the bentonite.

No flow into the collectors took place during the main test, which certified that leakage did not occur along the rock/bentonite interface or through the adjacent rock. Tracers were not found on the shaft wall or in the observation holes but appeared in the central tube, which turned out to be caused by a slight leakage at the lead-through of the injection pipe. This leakage was estimated at a few tenths of a liter per hour, meaning that the net outflow of water from the injection chamber to the rock had almost ceased after about one month.

The injection pressure was raised to 200 kPa at the end of the test but the flow did not increase substantially as shown by the diagram in Fig. 8. We see that it was less than 50% of the recorded values at 100 kPa pressure at the corresponding times after onset of the respective tests, which demonstrates that the sealing effect of the bentonite had increased considerably in the course of the test. Careful inspection of the shaft verified that no leakage had taken place along the interface between the rock and the bentonite plugs but tracers were identified at structures A and E which were obviously reactivated by the pressure increase.

The actual development of swelling pressure (Fig. 9) was in reasonable agreement with the predicted ones for cells Nos. 2 and 4, which recorded pressures that were developed by the uptake of water from the sand. Cell No. 3 gave the maximum pressure 3 MPa for which the tierods of the twin plug had been designed. The rapid pressure build-up at this cell, which was located in the northern part of the slot, and the much slower pressure development recorded by cell No. 1 on the southern side illustrate differences in the fitting of blocks and to a certain spread in density of the backfilled bentonite powder.

At the end of the test, comprehensive sampling was made to determine the distribution of the water content. At the excavation it was observed that the bentonite had established a tight contact with the rock and that it had no open joints or fractures at any of the sampling levels. It turned out to be perfectly homogeneous with no visible joints within 10 cm distance from the rock wall and the sand chamber, respectively, while closed joints were seen in the rest of the plugs.

Fig. 9 Development of swelling pressures in the main test

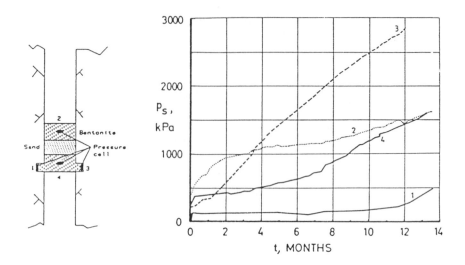

As predicted, the central part of the upper half of the top plug and lower/central part of the bottom plug had retained their original water content (Fig. 10). The presence of observed local, relatively dry peripheral parts seems to be related to the lack of clearly water-bearing rock fractures. Still, also these fracture-poor parts should have been effectively wetted if the water pressure had been sufficient to feed the fine fissures with water. When comparing the state of the clay at the blasted and slot-drilled parts no significant differences were found except for a tendency for slightly more uniform wetting at the blasted rock.

2.4 Conclusions

The drop in outflow from about 8-9 liters per hour at the start of the reference test with 100 kPa injection pressure to about 0.3 liter per hour at the end of the main test with 200 kPa injection pressure demonstrates the excellent sealing power of the bentonite plugs. We conclude from the present test that this can be ascribed to several functions which will be discussed here in some detail. They are:

1   The bentonite plugs are almost non-permeable
2   The plugs block flow passages along the rock/plug interface
3   The clay-filled slot truncates shallow flow passages in the rock and increases the length of flow paths
4   The swelling pressure tends to compress certain fractures in the rock
5   Clay tends to expand into and block certain fractures and shallow openings

Fig. 10 Typical isomoisture patterns in plugs. I, II etc denote sampling levels. Figures denote water content values

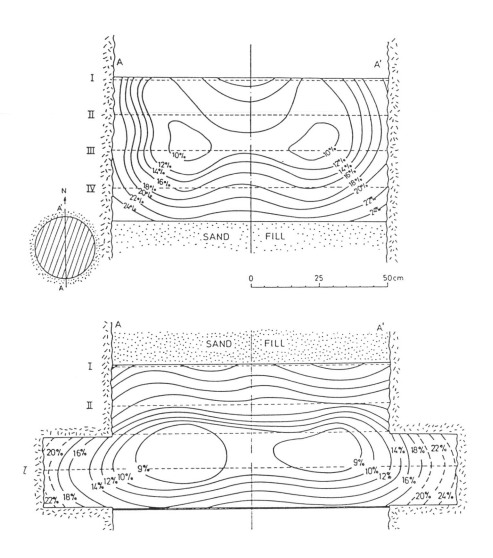

Although all these functions contributed to form an effective plug, the major effect was the blocking of flow along the rock/plug interface by the clay-filled slot and by the penetration of clay into fractures. The latter process largely reduced the flow of water through the major hydraulically active structures along the clay/rock interface and forced the water to flow at a larger distance from this interface.

Also, the tendency of the clay to penetrate any aperture in the rock surface that was exposed to the expansive clay created a perfectly integrated and tight interface.

3   TUNNEL TEST

3.1   Purpose of the test, test arrangement

General

The experiment aimed at testing the function of large plugs, intended for temporary sealing of tunnels which intersect richly water-bearing rock zones and which must be kept open for traffic and transportation purposes during several decades.

The test arrangement in Stripa consisted of a 9 m long and 1.5 m diameter steel tube, surrounded by sand and cast in concrete plugs at each end. These plugs hosted bentonite blocks arranged in the form of "O-ring" sealings at the rock/concrete interface. This simulated temporary sealing of a water-bearing rock zone penetrated by a tunnel in a repository, allowing for transports through the plug construction while minimizing the water inflow into the tunnel. The water pressure in the sand fill could be raised to 3 MPa and the associated leakage was accurately measured by flow meters and by collecting water that leaked from the plug. The swelling pressure exerted on the rock and on the sand fill by the expanding bentonite was measured by Gloetzl cells, and the deformation and displacement of the plug components were recorded by use of extensometers (4).

Hydraulic conditions

As in the shaft plugging test the hydraulic characterization of the rock was of great importance, the intention being to locate the plug so that the sealing function of the bentonite could really be challenged. This was made by constructing the outer plug where the rock contained some large coarse-crystalline and water-leaking pegmatite seams and placing the inner plug in contact with a wide diabase dike with a presumably non-perfect contact with the host rock.

Test setup

The general arrangement of the test is illustrated by Fig. 11. It should be noticed that the construction cost and time were considerable due to the difficult formwork and to the comprehensive instrumentation that had to be applied with all the requirements for effectively sealed passages for tubings and cables.

The water pressure in the injection chamber could be as high as 3 MPa in certain test phases and the expected swelling pressure was assumed to rise to about 3 MPa in the approximately 30 months long testing period. This would yield pressures producing a total axial force of about 7000 tons tending to separate the two plug members. In order to minimize the axial displacement of the concrete plugs during the pressure build-up, prestressed tierods were used to take the major part of the load. The calculated maximum displacement of the concrete/bentonite plugs in the entire tests was 4.5 mm assuming easy slip at the plug/rock contact. The actual movement was less than 1.5 mm indicating that the interaction between the plugs and the rock was very effective.

Fig. 11   Schematic view of the test arrangement. B is the "O-ring"-type sealing of blocks of highly compacted bentonite

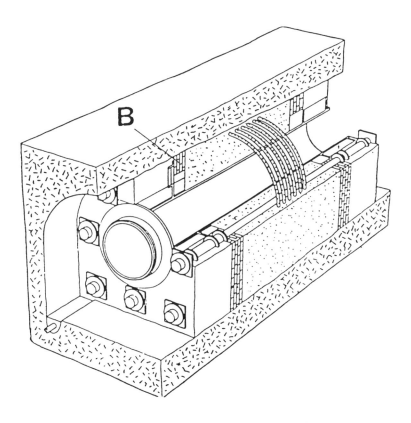

Water injection

Pressurized water was pumped at a constant rate through the circuit to which the injection chamber was connected. The pumps were fed with ordinary mine water that was stored in a small concrete pool, the excess water being discharged back to the pool, which was continuously percolated by fresh mine water to keep the temperature low and constant (10-12°C).

3.2   Test data

Instrumentation

The essential parameter was the outflow or leakage from the injection chamber. At injection pressures below 1 MPa in the initial test phase when the flow was expected to be very high, a simple turbine flow meter was used. The current recording of the flow of injected water at high pressures was made by use of a precision flow meter with an accuracy of + 1%.

An arrangement for measuring the outflow of water at the ends of the plug was also made but this arrangement could not catch all the water that leached from the interior of the plug, nor did it prevent water that could move in from other sources from being recorded. Since most of the leaking water was expected to follow the shortest paths from the injection chamber along and around the concrete units, it was assumed to reach the pumps and the measurements should therefore give approximately the same figure for the outflowing as for the injected water. This also turned out to be the case.

Gloetzl cells were used to measure the total pressure at the sand/bentonite and rock/bentonite interfaces and also to record the swelling pressure transferred to the concrete as well as to the sand fill at the bentonite surface.

The water uptake by the highly compacted bentonite from the pressurized sandfill was expected to yield fast saturation of the bentonite close to the sand interface. This would in turn yield displacement of this interface which could be observed in plexiglas tubes. The displacements could be directly viewed by use of ordinary borehole optics and TV inspection equipment. The tubes were located in steel casings equipped with flanges extending into the concrete walls. The displacement of the concrete plugs and the strain of the steel casing were measured by use of extensometers.

Bentonite

Bentonite blocks with cubical shape and 200 mm side length were applied to form a relatively tight pattern in the recesses of the concrete plugs. Small bentonite fragments from crushed blocks and bentonite powder were used to fill the space between the regular sets of blocks and the irregular rock surface. This led to local variations in homogeneity.

The total amount of bentonite blocks at the outer plug was 5515 kg and 5710 kg at the inner plug, while the amount of bentonite block fragments and powder was 1140 kg at the outer and 1055 kg at the inner plug. The total amount of bentonite was about 13400 kg.

It was estimated that the average net bulk density after complete water saturation would be about 2.0 t/m^3 in the larger part of the bentonite sealing. This bulk density corresponds to a water content of 26%.

Sand

Sand with a grain size ranging between 4 and 8 mm was used to embed the injection pipe gallery (14000 kg), while the rest of the chamber was filled with 47800 kg sand with a grain sand of 0.5-2 mm.

The sand was applied layer-wise after flooding the already applied material, so that the sand became completely saturated. Light compaction was made of the sand but due to practical difficulties in doing so, the average dry density could not be raised above about 1.2 t/m^3, yielding a bulk density of about 1.75 t/m^3 after saturation.

3.3   Results (4)

Leakage

The water pressure in the injection chamber was raised stepwise and the corresponding leakage is shown in Fig. 12. The major conclusions with respect to the flow measurements can be summarized in the following way:

*   At each pressure step a high peak value was obtained indicating rapid temporary inflow. The major reason for this is concluded to be compression of air-containing pores in the sand at the first pressure steps, and penetration of water into fractures that were expanded by the injection pressure when it reached higher values. At the lowest pressure steps strong leakage took place at the rock/bentonite interface for about 1 day.

Fig. 12   Recorded flow of injected water (a) and collected water (b)

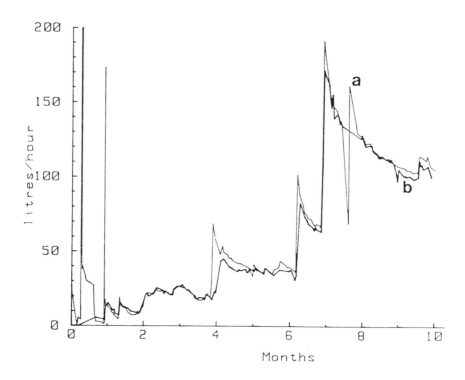

*   The flux tended to stabilize at values that were approximately proportional to the pressure up to 1.25 MPa, while it increased considerably at higher pressures.   This suggests that the natural piezometric head at the rock/sandfill interface was on the order of 1-1.5 MPa.

*   At the highest pressure, the flux tended to drop at a decreased rate from 200 l/hour to about 75 l/hour in the course of the test, and the larger part of this reduction is concluded to be caused by various sealing effects of the bentonite. The outflow decreased very much from a pegmatite zone and steep fractures in the rock walls.

*   While the bentonite sealing strongly reduced the leakage in the course of the test, water rapidly passed through the joint between bentonite and rock for a few hours at the first pressure steps. The fact that no clay was washed out suggests that the bentonite initially formed a coherent but non-homogeneous clay gel at the rock/bentonite interface and that this gel soon became consolidated and integrated in the expanding clay mass.

One possible sealing mechanism may have been that eroded clay aggregates from this gel were forced into the joint between the rock and the concrete and sealed this space.

*   Plug displacements generated by pressure changes had probably some disturbing effect on the fractures in the rock and were partly responsible for the strong, irregularly occurring peaks in the flow diagram.

Swelling pressure

The Gloetzl cell reactions were expected to reveal the maturation rate of the bentonite. The readings of particular interest were those at the rock/bentonite interface and they are illustrated by Fig. 13. We can see that there was a considerable spread in pressure but that constant values appeared to be approached in the course of the test. The major conclusions are specified below.

*   The recorded pressures at the rock/bentonite interface were much lower than the predicted minimum values for 5 months testing time, while there was fair agreement between measured pressures and predicted minimum pressures for 15 months testing time. The values measured after 20 months generally had a wide range (1-5 MPa), the maximum value being close to the predicted ones.

*   Except for the cell, which was located at the wet pegmatite zone (No. 7), all cells in the upper part of the plugs reacted slower and gave lower pressures than most of the cells in the lower part of the plug. The reason was probably the difficulties in applying the bentonite blocks in a tightly fitting pattern, and to fill in bentonite powder in the upper parts. The associated higher porosity would explain the discrepancy in the rate of maturation and the low net swelling pressure.

*   The cells at the sand/bentonite interface all reacted at approximately the same rate. The slowest pressure build-up and the lowest pressures were recorded in the upper part of the plugs where the sand was loosely layered. Here, the pressure did not exceed about 0.1 MPa, while the maximum value deeper down in the sand fill was 0.5 MPa at the end of the test.

*   A close analysis of the recorded values at the rock/bentonite interface showed that the pressure build-up was very irregular. This clearly demonstrated that the wetting and expansion was associated with fracturing and displacement of the bentonite blocks.

*   The fact that the pressure build-up appeared to have stagnated after about 18 months indicates that the degree of saturation was rather high.

Fig. 13   **Total pressure (water pressure + swelling pressure) recorded by the Gloetzl cells at the outer plug. The irregular pattern after 17 months is due to a deliberate variation in injection pressure**

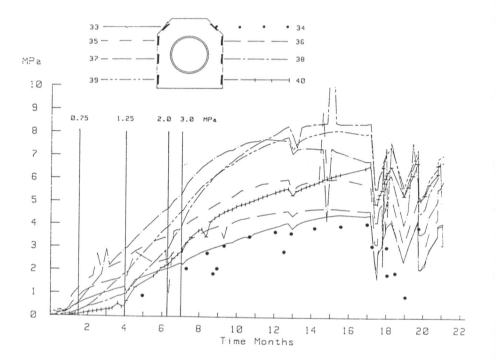

Water content distribution

Comprehensive sampling for water content determination in the bentonite was made after cutting openings in the casing at the end of the test. While the predictions based on the general diffusion-type water uptake model suggested rather slow wetting and a maximum water content of about 20% under completely confined conditions, the actual values were higher and indicated that wetting had taken place not only from the sand and the rock, but also from the concrete.

The major conclusions were the following:

*   The degree of filling of the recess with bentonite blocks and fragments was the major determinant of the swelling pressure because it controlled the bulk density. Thus, a complex geometry makes it difficult to fit in blocks in a tight pattern and this yields a high porosity and low net bulk density, while a simple space with parallel boundaries allows for a high degree of filling. Full-face drilled tunnels would be ideal.

*   It appears that water migration was considerably delayed in bentonite block fillings that contained irregular, large voids and wide joints. In well fitting block systems the water uptake is governed by the diffusion model, while the large swelling of loosely arranged bentonite blocks or fragments that is required to yield continuity in the system retards the wetting.

*   The frequency of fractures in the rock is not a major factor for the rate of wetting when the rock water pressure is high and when the clay is effectively confined by the rock. However, for porous bentonite fillings a faster wetting by richly inflowing water from fractures will increase the maturation rate.

*   The experimentally determined high degree of saturation of the bentonite is in good agreement with the earlier conclusion that the swelling pressures appeared to approach a maximum value.

3.4   Comments on the major sealing mechanism

Predictions indicated that the leakage from the plug construction had been about 1000 l/hour if no bentonite sealings had been applied, and that these sealings would reduce the flow to 60-600 l/hour at 3 MPa water pressure at the end of the test.

The leakage turned out to be about 200 l/hour at the application of 100 kPa water pressure early in the test but it dropped considerably in the course of the test and became 75 l/hour at 3 MPa pressure at the end of the about 20 months long test. During the 3 MPa pressure period, which lasted for about 10 months, the leakage dropped from about 200 l/hour to 75 l/hour and this very significant reduction must have been due to the sealing effects that we know from the shaft plugging experiment. The major effect is concluded to be the establishment of a very tight contact between the rock surface and the bentonite.

The detailed process is that shown in Fig. 14. The first stage is the imbibition of water and growth of a soft clay gel that fills up the space between the rock and the yet not wetted, dense bentonite. The second stage is the migration of water from the rock through the soft gel into the dense bentonite in which a high pore water tension is set up. The bentonite expands successively and exerts an increased pressure on the soft gel which consolidates. Ultimately, a homogeneous and fully saturated dense clay mass is formed.

When the initial rock/bentonite fitting is relatively good, as in the case of borehole plugs and and plugs in raise-drilled shafts, the initial stage lasts for a short time only. This is illustrated by the Buffer Mass Test which indicated that swelling pressures built up by initiation of the consolidation process were significant already a few weeks after the onset of the wetting, the free space between the rock and clay blocks being about 10 mm. The fit between the rock and the bentonite blocks and fragments was not at all as good in the Tunnel Plugging Test and the consolidation of the soft boundary zone was therefore slower.

4   GENERAL CONCLUSIONS FROM THE PLUGGING EXPERIMENTS

* Smectite clay plugs of high density mature quickly and can withstand high water pressures and pressure gradients soon after application.

* Where access to water is limited, i.a. where the piezometric pressure is low, the maturation rate is strongly retarded

* One of the mechanisms that give dense smectite plugs their eminent sealing effect is the formation of a very tight contact with the rock matrix.

5   REFERENCES

1.   Pusch, R., Nilsson, J & Ramqvist, G., Borehole and Shaft Sealing - Site documentation. Stripa Project Technical Report 85-01, SKB, Stockholm, 1985

2.   Pusch, R., Börgesson, L. & Ramqvist, G., Final Report of the Borehole, Shaft and Tunnel Sealing Test - Volume I: Borehole Plugging. Stripa Project Technical Report 87-01, SKB, Stockholm, 1987

3.   Pusch, R., Börgesson, L. & Ramqvist, G., Final Report on the Borehole, Shaft and Tunnel Sealing Test - Volume II: Shaft Plugging. Stripa Project Technical Report 87-02, SKB, Stockholm, 1987

4.   Pusch, R., Börgesson, L. & Ramqvist, G., Final Report of the Borehole, Shaft and Tunnel Sealing Test - Volume III: Tunnel Plugging. Stripa Project Technical Report 87-03, SKB, Stockholm, 1987

Fig. 14 Schematic picture of clay penetration into the rock matrix and formation of a tight clay/rock contact. a) Clay blocks in position, starting point. b) Early stage of expansion. c) Final stage with integrated clay (c) and crystal matrix (R). W represents water, C bentonite and R rock

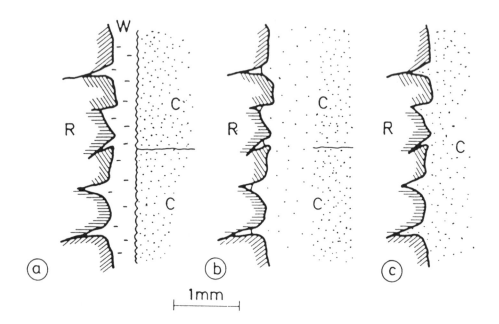

Sealing of Fractured Rock

Lennart Börgesson
Roland Pusch
Clay Technology AB

Abstract

The Stripa Phase 3 project Sealing of Fractured Rock logically followed the two first phases. While the two first phases dealt with sealing of deposition holes, tunnels, shafts, and boreholes, the third phase comprised fracture sealing.

The project involved laboratory testing and theoretical modeling of grout flow and penetration as well as grout longevity. It started with a pilot test, which showed that fine-grained grouts can be effectively injected in relatively fine fractures and that the fluidity of bentonite as well as cement grouts can be effectively increased with the dynamic technique and different additives.

The field tests comprised investigation of excavation-induced disturbance and attempts to seal disturbed rock, and, in separate tests, grouting of deposition holes and a natural fine-fracture zone. Considerable disturbance of nearfield rock by blasting and stress changes, yielding an increase in axial hydraulic conductivity by 3 and 1 orders of magnitude, respectively, was documented but various factors, primarily debris in the fractures, made grouting of blasted rock ineffective. Narrow fractures in deposition holes and in a natural fracture zone were sealed rather effectively.

Introduction

The Sealing Project of Phase 3 has involved several different sub projects. It started with Stage 1; a pilot project in which the grouting technique was developed and some in situ grouting tests made. After Stage 1 it was decided to continue with Stage 2, which included laboratory investigation of grout longevity and injection properties and development of calculation tools as well as several full scale grouting tests and large scale in situ investigations of the near field rock hydraulic properties.

Grout technique and grout materials

The following two main types of grouting material and two main types of technique were tested:

> Technique: Dynamic injection
> Static injection
>
> Material: Cement slurry
> Bentonite slurry

The properties of the grout materials are described in separate articles. A schematic diagram of the grouting system is shown in Fig 1. The water and bentonite, or the water, cement and superplasticizer were mixed in the colloid mixer for about 15 minutes. Then the grout was filled in the screw pump where it, in the case of cement, was allowed to circulate for no more than 15 minutes. In the case of dynamic injection the borehole was connected and completely filled via the injection machine, while in the case of static injection the borehole could be connected and filled directly from the screw pump. The average time from start of the cement mixing procedure to the grouting was 30 minutes, the upper limit being 90 minutes.

Fig 2 shows a picture of the injection machine. It is composed of a percussion machine, which hammers at the frequency 100 Hz against a piston that moves in the grout-filled cylinder. The variable static "back pressure" is produced by a pneumatic cylinder pushing the movable percussion machine.

Slot injection tests

The technique was carefully tested in an artificial fracture which was 5 cm wide and 3 m long, formed between two very stiff steel plates that were bolted together to assure that no deformations took place during the injection. The

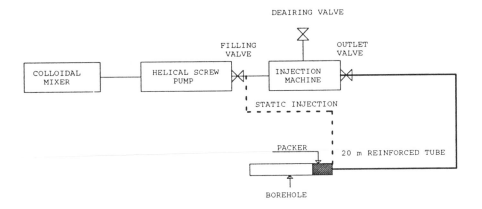

Figure 1   Schematic drawing of the components in the grouting system

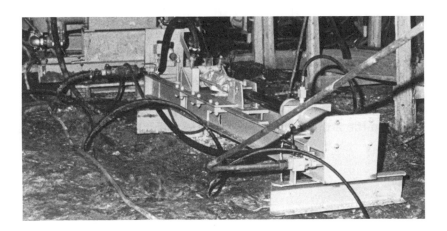

Figure 2   Picture of the dynamic injection machine

surfaces forming the fracture ware roughened. The grout was let in in one end while the other end was kept open. Since the fracture was equipped with small windows every 20 cm and pressure transducers every 30 cm the penetration as well as the pressure waves could be carefully observed. Numerous tests were performed with cement grouts as well as bentonite grouts, with the fracture water-filled or empty. The cross section represented a parallel plane fracture with an aperture of 100-300 μm or a triangular fracture with a cross section 0-100 μm.

An example of pressure measurements in the fracture is given in Fig 3. The recorded pressure in three spots at different distances from the inlet during 0.25 seconds is shown during injection of a bentonite grout. The figure shows a typical response. Every third peak with a pressure of up to 8-9 MPa comes from the strike of the percussion machine and the two following peaks with lower maximum values come from reflexions in the cylinder. The applied frequency 40 Hz is thus in reality 120 Hz. The decrease in maximum value with distance from the inlet is also shown.

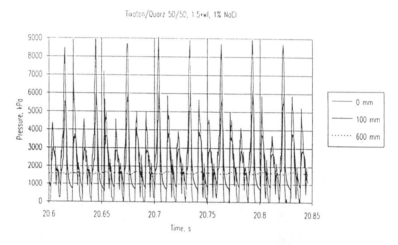

Figure 3   Measured pressure pulses at the entrance and at two locations in the artificial fracture as a function of the time from start

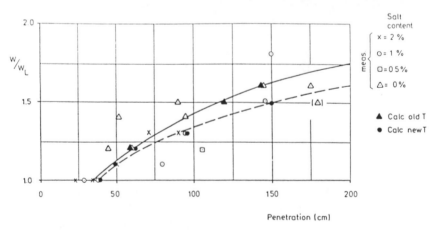

Figure 4   Comparison of measured and calculated penetration in the artificial fracture as a function of the relation between the water ratio and the liquid limit

Using the developed grout flow theory and rheological models, the penetration into the fracture can be calculated. Fig 4 shows a comparison between the calculated penetration and the measured when different stiff gels were used (expressed as the relation between the water ratio w and the liquid limit w_L). Different types of bentonite and different salt contents in the added water were used. The figure shows acceptable agreement, considering the complex nature of the grout flow, although the measured penetration has some scatter.

Pilot field tests

Pilot grouting tests were made in 8 core drilled holes in the Stripa mine. Cement grout as well as bentonite grout were used. Six holes were vertical with the depth 1.5 m (4) and 7 m (2) while 2 holes were horizontal with 40 m depth. Measurements in the holes before and after the grouting showed that the hydraulic conductivity was decreased to about $k=10^{-10}$ m/s irrespective of the initial conductivity. Excavation of the floor around the vertical holes showed that the penetration was up to several meters in the largest fracture intersecting the floor close to the grout holes.

Sealing of the near-field rock around deposition holes

The purpose of this test was to investigate the possibility to seal the nearfield rock around large-diameter deposition holes by grouting intersecting fractures. Bentonite slurry was used as grout material partly for practical reasons and partly for longevity reasons.

The field tests were conducted in the two inner heater holes of the Buffer Mass Test drift, their diameter and depth being 76 cm and 3-3.5 m, respectively. Careful mapping and hydraulic testing of the holes preceded the groutings. The hydraulic tests were made by pressurizing water over the entire periphery of 55 cm long sections in the holes by use of the Large borehole Injection Device (LID). It is a 55 cm high injection cylinder with a diameter that is only 4 mm smaller than that of the holes, the cylinder being surrounded by two large packers. This gave the average hydraulic conductivity of the rock surrounding the holes at different levels, typical ranges being $3 \cdot 10^{-7}$ m/s in the upper part of hole No. 2 and $3 \cdot 10^{-10}$ m/s in the lower part of the same hole. Very close to the floor in hole No. 1 it was higher than 10^{-6} m/s.

The injections were also made by use of the LID equipment (Fig 5). After the injections, a heater was installed in each hole and power applied to yield a rock temperature of just below 100°C. After about two months the power was turned off and the rock allowed to cool.

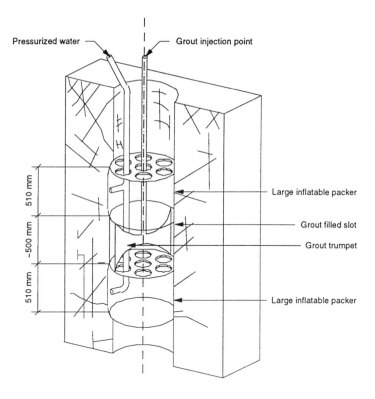

Pressurized water — — Grout injection point

510 mm

~500 mm

510 mm

Large inflatable packer

Grout filled slot

Grout trumpet

Large inflatable packer

Figure 5   Large Borehole Injection Device (LID) used for hydraulic testing as well as grouting in deposition holes

The hydraulic conductivity measurements were repeated after the grouting and after the heat pulse. The floor was leveled before and after the grouting as well as after the heat pulse. These measurements showed:

-   that the average hydraulic conductivity had decreased to $7 \cdot 10^{-10}$ m/s in hole No. 1 and $2 \cdot 10^{-10}$ m/s in hole No. 2 after the grouting

-   that the average hydraulic conductivity had increased considerably due to the heat pulse although it was still considerably lower than before the grouting

-   The residual heave of the floor around the holes was 200 µm as an average

In order to check the grout penetration into the fractures, the rock around one of the holes was excavated. The excavation showed that grout had penetrated as

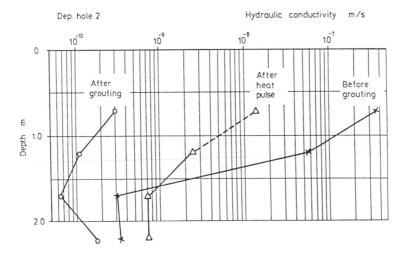

Figure 6   Measured hydraulic conductivity at four levels in hole No. 2 at three different occasions

deep as 2 m into flat-lying fractures that were free from debris of disintegrated fracture coatings like chlorite, while it also showed that steep fractures with such debris were poorly penetrated.

The results of the hydraulic conductivity measurements in hole No. 2 is shown in Fig 6 and the results of the excavation of the floor on the northern side of this hole is shown in Fig 7.

Identification and sealing of the disturbed zones around a blasted tunnel

The purpose of these tests was to measure the axial hydraulic conductivity of the zones disturbed by blasting and stress release around an excavated tunnel or drift in granite rock and to investigate the possibility to reduce the hydraulic conductivity by sealing the very fine fractures in these zones by cement grouting.

The test for identification of the disturbed zones, called the Macro-Flow Test, was performed in the innermost part of the drift used for the Buffer Mass Test and LBL's Macropermeability Experiment. Fig 8 shows the arrangements for the test. A 70 cm deep slot was cut around the inner end of the drift and a borehole curtain, consisting of 72 boreholes and extending 6.3 m into the rock from the slot, was drilled perpendicular to the drift axis. A similar slot and borehole curtain was made at the bulk head, 13 m from the inner curtain.

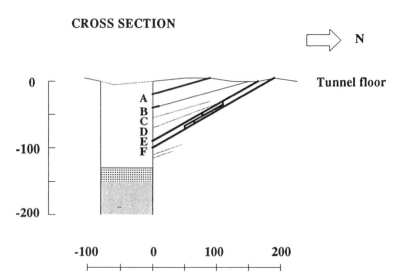

Figure 7   Cross section of hole No. 2 with fractures intersecting the hole
            on the nothern side. Thick lines denote where the bentonite
            grout was found

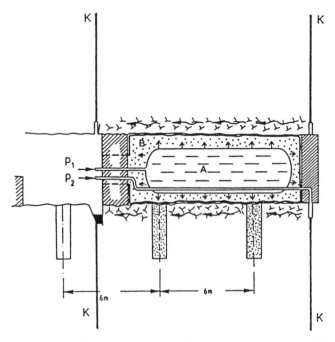

Figure 8   Arrangements for the Macro Flow Test. A bentonite slurry (B)
            was pressurized through a bladder filled with water (A)

The idea of the Macro Flow Test was to pressurize the water-filled inner curtain and measure the flow of water to the outer curtain. The surface of the rock in the drift was sealed by filling the drift with 120 m³ bentonite slurry, which was pressurized by a 100 m³ water-filled bladder. By keeping the slurry pressure higher than the curtain pressure, water was prevented from entering the drift.

The water pressure was measured at about 30 locations in the surrounding rock and the water flow was measured in 8 sections of the inner curtain and in 4 sections of the outer curtain, before as well as during the test.

The Macro-Flow Test was conducted in more than 10 steps, with different relations between the slurry pressure and the curtain pressure, the maximum pressure being just below 1 MPa. These steps showed that the relation between applied pressure in the inner curtain and measured flow was linear and independent of the pressure on the rock surface from the slurry.

The test was primarily evaluated by finite element calculations of the flow and pressure, at which the rock was simulated as an equivalent porous medium. The following four different tests were simulated:

- Before drilling of curtains (Macropermeability Test)
- After drilling of curtains (Curtain Inflow Test)
- After pressurization of the slurry with zero pressure in the curtains (Slurry Leakage Test)
- Macro-Flow Test

The calculations showed that two different rock models could be defined, which both fitted the measured water flow and pressure in the rock. Both models imply a highly permeable zone with a depth of at least 0.8 m and an average hydraulic conductivity of $1 \cdot 10^{-8}$ m/s which is more than 100 times higher than the conductivity of the virgin rock and which mainly originates from blast damage. The hydraulic conductivity is highest in the floor and lowest in the roof. Both models also imply a stress-disturbed zone reaching 3 m into the rock with a decreased radial hydraulic conductivity by about 4 times. The element mesh used in the calculations and the derived hydraulic conductivities according to model A are shown in Fig 9.

The main difference between the models concern the axial hydraulic conductivity of the stress-disturbed zone and the depth of the blast-disturbed zone as follows:

Model A

Blast-disturbed zone extending 0.8 m from surface. Stress-disturbed zone extending 3.0 m from surface with an increased axial conductivity by a factor 10.

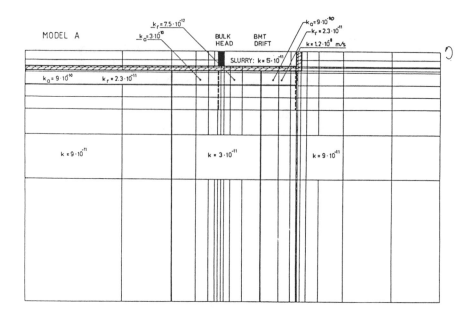

Figure 9 Element mesh and hydraulic conductivity of the different rock zones at the final rock model (Model A). The mesh is axi-symmetric around the upper boundary

Model B

Blast-disturbed zone extending more than 0.8 m from surface. Stress-disturbed zone with an axial conductivity that may range from that of Model A, i.e. 10 times higher than the conductivity of virgin rock, to a figure corresponding to the radial conductivity, i.e. 4 times lower than the conductivity of the virgin rock.

Individual measurements in the curtain holes and rock mechanical calculations support model A but the matter should be further studied.

Laboratory tests, slot injection tests and calculations of grout penetration resulted in a decision to use Alofix cement with w/c=0.45 and SP=1.4% for dynamic injections and with w/c=0.5-0.7 and SP=1.5-3.0% for static injections.

As shown in Table 1, 345 holes were percussion-drilled to a depth of 1.2 m in the floor and 1.0 m in the walls and the roof. 4 square-shaped test areas with altogether 80 holes were individually tested before the grouting, 5 holes being ungrouted and tested after grouting. As an average, no reduction in hydraulic conductivity was found in these holes.

Table 1. Borehole plan for the hedgehog holes

Location	Spacing (m)	Depth (m)	Number of holes
Inner part (2m)			Σ=90
Floor	0.4	1.2	33
E-wall	0.5	1.0	24
Roof	0.7	1.0	14
W-wall	0.7	1.0	19
Outer part			Σ=255
Floor	0.7	1.2	98
E-wall	0.8	1.0	63
Roof	1.15	1.0	34
W-wall	1.15	1.0	60
Grand total			$\Sigma\Sigma$=345

After completing the grouting, the drift was prepared for repeated flow testing by refilling the drift with bentonite slurry and installing a new water-filled rubber bladder. The slurry was pressurized and the Macro Flow Test with flow measurements repeated at different slurry pressure in the drift and different pressures in the inner curtain in the rock. A compilation of the flow measurements are shown in Fig 10.

The tests were also evaluated using the same 3D finite element calculations as for the case before grouting. These calculations showed that the effect of the grouting on the gross permeability was insignificant.

After the flow tests, the drift was emptied and the rock excavated in parts of the floor and walls for visualizing the extension and condition of the grout. The excavation showed that only fractures that had been neoformed contained grout. No grout was found in chlorite-coated fractures.

The main conclusions were that efficient grouting can be made and that penetration of grout can be seen in fractures with no coating ("infilling"), while the

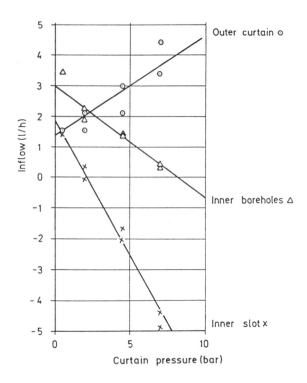

Figure 10 **Measured inflow into the curtains as a function of the curtain pressure at the Macro Flow Test after grouting. The difference in slurry leakage at different slurry pressure is compensated for**

occurrence of debris in coated fractures hindered the grout penetration into those fractures. The high frequency of chlorite-coated fractures in Stripa mine and the complex nature of the blast-induced fractures were the main reason for the poor grout penetration and the insignificant reduction in gross permeability.

Sealing of a natural fine-fracture zone

The purpose of this test was to characterize a rather richly water-bearing natural fracture zone and find out whether it could be sealed by cement grouting and whether this would yield shunting of groundwater. Observations in the preceding 3D Migration Test indicated that the zone had the form of a few steeply oriented fractures striking NW/SE and intersecting the drift. However, detailed hydrological characterization and analysis of rock structure data suggested that most of the water flowing into the test drift originated from a N/S-oriented zone that intersected the innermost end of the drift. Drilling of 75 boreholes for further characterization and for a first sealing attempt, altered this picture and indicated that a NW/SE-striking, steep zone located close to the drift was instead the major water supply, and a small part of it was grouted in a second sealing attempt (Fig 11 and Fig 12). Inflow tests before and after grouting, piezometer measurements, evaporation measurements, and hydraulic characterization of boreholes penetrating the grouted zone, all showed that the inflow from the grouted part had decreased and become redirected westwards, i.e. towards the ungrouted surrounding rock. Predictive computer calculations showed that grouting of the small part of the zone could yield a reduction in inflow from the zone by around 30%, while the total inflow into the test drift would only be reduced by about 5%. This was in agreement with the actual observations although the accuracy of the ventilation experiment used for the purpose was not sufficiently high to give a very safe value. Stronger evidence was offered by piezometric measurements, tracer testing and evaporation measurements, which all showed that the inflow of water had been reduced and displaced westwards. It was estimated that the net hydraulic conductivity of the grouted part of the zone had dropped from around 10^{-8} to 10^{-9} m/s.

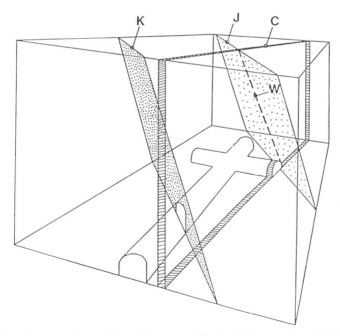

Figure 11   Perspective of the 3D area with tentative major steep structures indicated. "W" denotes the intersection of the "J" (or "D") and "C" structures

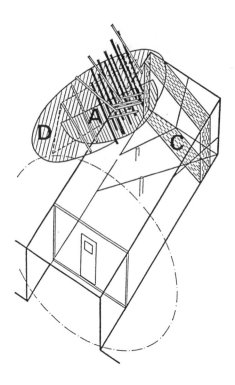

Figure 12   Schematic picture of the "D"-zone, i.e. the presumed major water-bearing structure. It is represented by the large full ellipse intersected by the flatlying "A"-zone, and contacted by the steep "C"-zone at the inner end of the drift. White boreholes represent the N-holes that intersected the "D"-zone while the black ones denote the ten core-drilled holes

Analysis of drillings through the grouted rock showed that Alofix cement grout had entered fractures with an aperture of down to 10-20 µm to several decimeters depth. Close examination of the cement grout showed that it was not homogeneous. Thus, while wider parts of the channels were filled with homogeneous, dense cement matrix, narrow parts contained very soft, permeable gel. This suggests that the average hydraulic conductivity cannot be reduced beyond 10^{-10} to 10^{-9} m/s using cement since even though a large number of fine channels can be partly filled with dense cement grout, considerable parts remain unfilled or filled with permeable cement. Still, the outcome of the study was very good in the sense that fractures with apertures that are usually considered to be far from groutable were actually at least partly filled with cement.

Conclusions and major achievements

The Stripa Rock Sealing Project is concluded to have yielded a large amount of very valuable information on the excavation-induced disturbance of nearfield rock, and on the possibility of obtaining effective sealing of finely fractured rock by clay and cement grouting, using presently available, advanced technique. The major findings of the field tests can be summarized as follows:

* Water flux through the nearfield rock in the BMT drift could be accurately described by considering the rock mass to behave as a porous medium

* Blasting produces mechanical damage that increases the hydraulic conductivity of the most shallow rock by 2 to 3 orders of magnitude. The effect over longer axial distances in a drift or tunnel is not known

* Stress-induced increase in hydraulic conductivity does not seem to extend beyond about 3 m in drifts of the Stripa BMT type. The effect is very much dependent on the orientation of the drift with respect to that of major fracture sets. In the BMT drift the axial conductivity is concluded to be 10 times higher than the average conductivity of the virgin rock, while the radial conductivity appears to be about 4 times lower than that of virgin rock

* The nearfield of large boreholes undergoes some increase in axial conductivity but it is very much dependent on the orientation of the holes with respect to that of major fractures. Heating may activate "latent" fracture channels that have not been grouted, by which some of the sealing effect of grouting may be lost

* Narrow as well as wide fractures intersecting large boreholes drilled through the disturbed shallow zone in blasted drifts, can be effectively sealed with clay by use of "megapackers" and dynamic injection technique

* Natural fracture zones of 3rd and lower orders can be rather effectively sealed by using cement injected under high pressure applying dynamic injection technique. The spacing of the injection holes determines the effectiveness of the grouting

* Shallow rock with fine fractures coated with fracture minerals like chlorite is not effectively groutable, probably because of debris produced by blasting-induced disintegration of the fracture minerals

* The groutability of rock can be predicted by hydraulic testing and evaluation of the rock structure provided that debris is not present in the fractures

References

Börgesson L, Pusch R, Fredrikssom A, Hökmark H, Karnland O and Sandén T 1991a. Final Report of the Rock Sealing Project - Sealing of the Near-Field Rock Around Deposition Holes by Use of Bentonite Grouts. Stripa Project Technical Report 91-34.

Börgesson L, Pusch R, Fredrikssom A, Hökmark H, Karnland O and Sandén T 1991b. Final Report of the Rock Sealing Project - Identification of Zones Disturbed by Blasting and Stress Release. Stripa Project Technical Report 92-08.

Börgesson L, Pusch R, Fredrikssom A, Hökmark H, Karnland O and Sandén T 1992. Final Report of the Rock Sealing Project - Sealing of Zones Disturbed by Blasting and Stress Release. Stripa Project Technical Report 92-21.

Pusch R 1988. Rock Sealing - Large Scale Field Tests and Accessory Investigations. Stripa Project Technical Report 88-04.

Pusch R, Börgesson L, Fredriksson A, Markström I, Erlström M, Ramqvist G, Gray M and Coons W 1988. Rock Sealing - Interim Report on the Rock Sealing Project (Stage 1). Stripa Project Technical Report 88-11.

Pusch R, Börgesson L, Karnland O and Hökmark H 1991. Final Report on Test 4 - Sealing of Natural Fine - Fracture Zone. Stripa Project Technical Report 91-26.

SESSION IV

OVERVIEW REPORTING

SÉANCE IV

RAPPORTS DE SYNTHÈSE

Chairman - Président
P.-E. Ahlström (Sweden)

Engineered Barriers Investigations within the Stripa Project

Malcolm Gray
AECL Research, Pinawa, Manitoba, Canada

Ferruccio Gera
ISMES S.p.A., Rome, Italy

Abstract

During the three phases of the Stripa Project the engineered barriers research group has investigated several aspects of significant importance in relation to sealing radioactive waste repositories in crystalline rocks. In Phase 1 the emplacement of bentonite-based buffer around the waste packages and the behaviour of buffer, backfill and host rock were investigated. In Phase 2 the plugging of boreholes, shafts and tunnels was demonstrated. In Phase 3 the studies addressed two main aspects: the grouting of fractured rock and the longevity of sealing materials.

1. Introduction

The broad objective of the engineered barriers studies was to demonstrate and qualify the use of different materials and techniques for sealing water flow paths in the Stripa granite, the mine excavations and the excavation disturbed zones. The engineered barriers studies were carried out through the full period over which the agreements that formed the Stripa Project were in effect. The division of the programme into phases and the overlap in the timing of each of the three phases, Phase 1 (1980 to 1985), Phase 2 (1983 to 1988), and Phase 3 (1986 to 1992), allowed for developments achieved during the project and concerns arising from independent national programmes to be incorporated in the investigations.

During Phase 1, the engineered barriers investigations focussed on the heat affected zone of the repository. Specifically, the response of clay buffers and the interactions between waste containers, clay buffer materials and the rock were studied. The Phase 2 investigations examined the feasibility of sealing boreholes, shafts and tunnels with clay sealants. In Phase 3, studies of the ability to grout and seal fractured granite, including the excavation disturbed zone, were effected and specific studies into the longevity of cement- and clay-based sealing materials were undertaken.

2. Phase 1 - The Buffer Mass Test

The Buffer Mass Test (BMT) was a half-scale mock-up of a waste deposition concept proposed by KBS. Electrically powered heaters, simulating heat-generating waste containers, were embedded in highly compacted bentonite (HCB) buffer material and placed in six large diameter boreholes drilled in the floor of a room at the 340 m level of the Stripa mine. The room above two of the deposition holes was backfilled with bentonite-sand mixtures. The layout of the BMT is shown in Figure 1.

The buffer, backfills and the rock were heated for periods of up to about four years and observations were made on the transient processes of heat and water transfer in the buffer and the effects of these processes on hydro-mechanical interactions between the rock and the buffer were studied. It was not possible, given the short duration of the test, to examine the solute transport properties of the buffer, other than by implication.

The HCB used as the buffer was prefabricated by statically compacting the bentonite to a dry density of not less than 1,88 Mg/m³. The sand-bentonite tunnel backfill materials were compacted *in situ* to minimum dry clay densities of approximately 0,43 Mg/m³ (lower backfill) and 0,37 Mg/m³ (upper backfill). At these densities the buffer probably has an osmotic efficiency approaching 100 per cent; the backfill materials probably have lesser efficiencies.

During water uptake, the buffer and backfill materials were predicted to swell and develop pressures acting against the other components of the test system. The buffer was expected to exert high ultimate pressures, 10 MPa or more, against the rock and the backfills. Pressures of hundreds of kPa were expected from the backfill materials. Deformations of decimetres at the buffer/backfill boundary were predicted. Moreover the swelling pressures could result in clay being extruded into open rock fractures that intersected the excavations in which the experiment was to be carried out.

The hydro-mechanical interactions, due to their largely unknown effects on the hydraulic boundary conditions acting at the buffer/backfill/rock interfaces, further decreased confidence in the ability to predict numerically the heater/buffer/backfill/rock performance and interactions. Observations to be made in the BMT were needed to qualify and refine conceptual models for performance and to provide an indication of the extent to which enhanced numerical prediction capabilities were required or, indeed, possible.

The work for the macropermeability experiment carried out under the Swedish-American Cooperative agreement, along with observations made in the large diameter holes ($\Phi = 760$ mm) that were diamond drilled to a depth of 3 m for the BMT, provided information used to bound the initial hydraulic conditions around the BMT room. Excavating the room at right angles to the horizontal major principal stress resulted in an excavation disturbed zone with a radial hydraulic conductivity that was less than the value of 10^{-10} m/s estimated for the surrounding rock mass. Measurable water flows occurred principally in the fractures in the rock mass although there was evidence of flow in permeable "fracture free" rock. Only a small fraction of the observed fractures visibly carried water. A large number of the natural fractures were infilled with minerals or otherwise blocked to the transmission of water. The eastern wall and the back of the room appeared likely to provide greater access to water to the backfills than the floor, the western wall and the end of the room. Natural water inflows into the deposition boreholes varied between holes. Most of the water entered the open holes through discrete fractures which, reflecting excavation disturbance, were concentrated in the upper third of the holes and sub-parallel to the floor of the room. Three holes were classified as "wet", the other three were classified as "dry".

Instruments were installed in the buffer and backfills to monitor changes in temperature, total pressure, pore water pressure, moisture content and displacement during the progress of the tests. Commercially available instrumentation was used wherever possible; much of this instrumentation required modification and calibration for the harsh environmental conditions of temperature, pressure and water salinity expected in the test. Special moisture sensors were developed to monitor transients in the clay masses.

The heaters used in the six emplacement holes were specifically designed to facilitate single point measurements of moisture content in the HCB during careful decommissioning at the and of the test.

The response of the buffer and backfills to heating and to water supplied through the bounding rock mass depended on the original hydraulic boundary conditions, the test configuration and the interactions between the clays and the rock mass. The temperature distribution, final moisture content distributions and swelling pressures developed by the HCB buffer material were largely controlled by the rate at which water was supplied at the rock/buffer interface. The buffer in wet holes became saturated within the period of the test; the buffer in dry holes showed increasing water content from the heater to the buffer/rock interface with drying having occurred near the heater. Correspondingly, swelling pressure were higher in the wet holes than in the dry ones; temperatures were generally lower in the wet holes.

Under the force of the swelling pressure, HCB was extruded into fractures intersected by the excavations. This prevented these fractures from acting as local water sources. Water uptake by the buffer took place through a thin layer of sealed rock which acted as a porous medium. The swelling pressures also ensured that the buffer mass, which originally contained construction joints, self sealed. In accordance with expectations, this self sealing was more pronounced in wet holes than in dry ones.

The results tended to confirm that an isothermal moisture transfer model could be applied to moisture transfer in the backfill materials. The model did not account for moisture transfer that occurred in the HCB buffer in which it was likely that an evaporation/condensation cycle was established down the temperature gradient developed in the unsaturated buffer. The temperature data, swelling pressure results, moisture redistribution data, results from tracer tests and retrospective analyses all support this hypothesis. This process in known to occur in loose clays and sands. Through the conduct of the BMT it is now known to be significant in dense bentonite materials under repository conditions and, depending on repository design, may need to be accommodated in models of near-field performance.

The heat conduction model used tended to overestimate the temperatures to be expected in saturated systems - the ultimate condition expected in a repository.

The mechanical performance of the backfill met all expectations by exhibiting more than adequate resistance to the uplift forces from the buffer. The magnitude of heave at the buffer/backfill interface was well predicted by a simple mathematical model. The effects of swelling pressures from the buffer on the rock mass were too low to be measured. However, the floor of the emplacement room exhibited heave, which was principally attributable to increases in the temperature of the rock and in accordance with understanding. The effects of this movement on water flow in the near-field rock mass could not be established.

Pore water pressures in the rock mass within 1 m of the tunnel faces appear to have been controlled by an excavation disturbed zone (EDZ). Adding to the results from the SAC macropermeability experiment, the BMT results indicated that the hydraulic properties of the EDZ were anisotropic; hydraulic conductivity parallel with the tunnel axis appeared higher than radial hydraulic conductivity. The high axial conductivity fed groundwater to the top of the emplacement boreholes and to the bottom of the tunnel backfills. These data provided significant understanding of the near-field rock mass for the design of grouting experiments carried out during Phase 3 of the project.

3. Phase 2 - Borehole, Shaft and Tunnel Plugging

3.1 Borehole plugging

Three borehole plugging/sealing tests were carried out. Each test was configured to allow the investigation of different aspects of borehole plugging with HCB.

In all tests HCB was introduced into smooth walled, diamond drilled boreholes and observations made on the rate of water uptake and swelling of the HCB, the resistance of the maturing bentonite to piping under hydraulic gradients and the mechanical resistance between bentonite plugs and the borehole wall. The differences between the three tests lay in the orientation of the borehole - one horizontal borehole and two vertical boreholes were sealed - and in the type of plug used - one vertical borehole was plugged using techniques which were virtually identical to those used in the horizontal borehole, the sealing system used for the second vertical borehole differed.

The sealing system in all cases consisted of hollow cylinders of HCB encased in an exoskeleton of perforated metal. The exoskeleton was needed to provide rigidity to the system as it was introduced into the borehole. The perforations were required to allow water to access the HCB, causing the material to swell and seal unfilled sections of the boreholes. The axial central hole in the HCB allowed access for instrumentation leads and hydraulic tubing and will not be present in repository seals.

The horizontal borehole plugging test was carried out in a 96.6 m long, 56 mm diameter borehole that, in part, ran approximately parallel to the drift used for the BMT.

The vertical borehole plugging tests were carried out in two 14 m long, 76 mm diameter boreholes that were specially drilled for the purpose of the tests between two vertically separated, parallel tunnels near the BMT area. The scheme of the vertical borehole plugging tests is shown in Figure 2.

The arrangement of the borehole plugging tests included mechanisms that allowed for the plugs to be subjected to high water pressure gradients so that the hydraulic properties of the sealed borehole could be examined. The resistance of the maturing bentonite to piping was specifically investigated. Moreover, total and pore water pressure sensors were included at strategic locations along the length of the plugs to assist with the interpretation of the flow measurements and to confirm aspects of bentonite behaviour and properties. After the *in situ* hydraulic testing of the borehole plugs had been completed the vertical borehole plugs were extruded and measurements were made to evaluate the water uptake and swelling behaviour of the HCB. At the end of the horizontal borehole plugging test a small volume of rock through which the sealed borehole passed was carefully excavated and the bentonite that it contained was examined.

The ease with which the borehole plugs were emplaced demonstrated the practicality of the design of the HCB sealing system for both horizontal and vertical borehole plugging operations.

Hydraulic testing of the horizontal plug as little as 14 days after plug installation proved that the bentonite plugs could sustain hydraulic gradients as high as approximately 450 without piping. A very effective seal at the HCB/rock interface was demonstrated by the pressure measurements taken in the vertical borehole plugging tests.

Examination of the recovered HCB plugs indicated that the clay was virtually fully water saturated and had expanded into the annular space between the plug and the borehole wall that was needed as a working clearance during plugging operations. This outer film of bentonite was less dense than the inner core indicating either incomplete consolidation of the bentonite or an ability for the bentonite to sustain significant stress gradients.

It can be concluded that bentonite plugs like the ones tested at Stripa will fail hydraulically by piping before they fail mechanically by extrusion. Within the pressure range in which they are effective they can be estimated to seal the borehole to an effective hydraulic conductivity in the range 10^{-12} m/s to 10^{-13} m/s.

3.2 Shaft and tunnel plugging

Shaft and tunnel plugging tests were carried out to determine the efficiency with which HCB could limit flow at the interfaces between bulkheads, backfills and excavated rock surfaces. The shaft plugging experiment was conducted near the vertical borehole plugging experiments in a 14 m deep vertical, tapered shaft that had been excavated to a diameter of 1 m (top) and 1.3 m (bottom) between two vertically separated virtually horizontal tunnels. The tunnel plugging experiment was carried out in a specially excavated, 35 m long, dead-end tunnel. The tunnel was excavated by careful blasting to a cross-sectional area of about 11 m2. The two tests are shown schematically in Figures 3 and 4.

In both the shaft and tunnel plugging tests two bulkheads were constructed within the excavations to form a test cell. The inner surfaces of the bulkheads were lined with HCB and the enclosed volume was filled with sand. The outer bulkheads, the inner, sand-filled part of the test cell and the rock acted as constraints to resist bentonite swelling. The inner sand-filled chambers were filled with water and pressurised and the resulting water flows through and around the HCB gaskets were measured.

Shaped HCB blocks with properties similar to the buffer material used in the BMT were used and tested. In the shaft plugging experiment HCB filled the entire cross-section of the excavation; in the tunnel plugging experiment the HCB was used to form a gasket (or "O" ring) on the perimeter of the inner surfaces of concrete bulkheads at the bulkhead/rock interface. The swelling and water pressures exerted on the bulkhead structures and at the HCB/rock and HCB/sand interfaces were measured along with the resulting deformations of the structures and the softer interfaces. Observations were made of the water uptake properties of the clays and, like the deformations, compared with pre-test predictions that were made using simple conceptual and numerical models of system performance based on developments made in Phase 1.

The clay plugs were readily emplaced. This demonstrated the practicality of the design of the plugging system. Moreover, the hydraulic testing showed that the HCB effectively stopped water flow through the shaft and tunnel excavations. The clay seals were less permeable than the surrounding Stripa granite through which, in both the shaft and tunnel plugging experiments, most of the water was lost from the test cells.

The Stripa rock was hydraulically variable on the scale of metres. The paths in the rock around the shaft plug through which water was lost were not defined. However, for the tunnel plugging test it was clear that a pegmatite zone and a series of discrete, steeply dipping, connected fractures in the floor of the tunnel allowed water to circumvent the plug.

Measured water content distributions in the HCB showed that at the end of the tests the inner cores of the HCB was not saturated. The measured water content distributions were compared with values calculated using a simple diffusivity model developed through the Phase 1 activities. The measured and calculated values were in reasonable agreement except for incomplete appreciation of the boundary conditions acting on the clay and, hence, erroneous assumptions used for the calculations. Thus, it appears feasible to predict the water uptake properties of confined highly compacted bentonitic materials. In common with the results from the borehole plugging and the shaft plugging experiments, the variations in moisture content through the clay showed that the HCB was not homogenised over the period of the test. Differential stresses were sustained. An eventual equalisation of the stresses within and densities of the clay in the very long term remains questionable.

Observations made during the disassembly of the test structures showed that the expanded bentonite conformed with the faces and edges of the inside of the bulkheads and the irregular rock surfaces. This evidence supported the conclusion that the swelling HCB can be used to effectively seal the interfaces between engineered barriers and the excavated rock surfaces. Observations of the limited penetration of the swelling HCB into fractures in the rock surface echoed findings of the BMT and the borehole sealing experiments, and further supported the suggestion that physical interactions between the rock and the clay controlled water ingress into the clay and effectively sealed the interface between the clay and the rock. Although the clay had penetrated outwards from the seal into the fractures, the outward movement was clearly limited in extent and there was no evidence of erosion of the clay under the high pressure gradients imposed on the groundwater. Moreover, it could be concluded that sand and clayey materials can be designed and used to confine HCB and preserve its long-term sealing functions.

4. Phase 3 - Grouts and Grouting

4.1 Materials

Scoping studies led to the selection of high-performance cement- and bentonite-based grouts for investigation in Phase 3. Laboratory studies of the rheological and hydraulic conductivity properties of these materials were undertaken to define the conditions under which they could be applied with maximum effect as grouts.

The results confirmed that decreasing the water content tends to increase strength (shearing resistance) and decrease the hydraulic conductivity of both material types. For these and other reasons it was anticipated that decreasing the water content would improve the long-term performance of the materials. Methods allowing for the injection of the materials with low water content yet with sufficient fluidity to permit injection were developed.

Both clay and cement grouts were determined to exhibit pseudo-plastic behaviour immediately after mixing with water. Moreover, the rheological behaviour could be modified by vibrating the freshly mixed materials. The effect of vibrations of different frequencies and amplitudes on the rheology of various grout mixtures were evaluated. It was generally concluded that applying vibrations to the grouts during injection would decrease the water content needed to provide adequate fluidity to the materials. Thus, a range of operating parameters was defined for the effective functioning of a grout injection pump that could apply vibrations as well as pressure to the grouts during injection.

It was shown that reductions in the water contents of both bentonite-based and cement-based grouts could be effected by the use of admixtures. For the clay grouts, NaCl added to the mixing water decreased the water content at which the materials were sufficiently liquid to be described as grouts. In addition to changing the rheological properties of freshly mixed grouts it was inferred from the results that the addiction of salt to the grouts could improve the long-term strength and the hydraulic properties of the materials.

The water content of cement grouts could be decreased by admixing superplasticizers. In addition, the inclusion of finely divided pozzolanic materials, in the form of silica fume, eliminated bleeding from the freshly mixed cement grouts, increased long-term strength and decreased hydraulic conductivity.

Thus, clay grouts incorporating NaCl and cement grouts including superplasticizers were studied in the in situ experiments. In addition cement grouts with and without pozzolan were tested.

Cement, pozzolan and clay products exhibit inherent variability arising from natural source variability and changing production patterns. Ageing processes, post-production and prior to use, also influence the properties of cementitious materials. Effects of these factors on the particle size and rheological properties of freshly mixed grouts were determined. It can be deduced that practical grouting operations associated with repository sealing would be obliged to accommodate much of this variability and that experienced personnel will be required to effect control procedures and make field adjustment to mix composition. Ultimately, grout mixes will be selected through the application of understanding gained from observations made during the progress of the in situ grouting operations associated with repository sealing. In this latter context it is noted that the cement grouts used in the *in situ* experiments at the Stripa mine were selected on the basis of expected material variability and possible effects on the success of the grouting operations. The materials used were varied somewhat from the grouts preselected from the results of the laboratory studies. Modifications to the mix compositions were made as the *in situ* work progressed.

4.2 Grouting around deposition holes

The hydraulic conductivity of the rock in the floor of the BMT room around two BMT holes was measured before and after injection with clay-based grout and after the grouted rock had been heated to almost 100°C. The rock was grouted from the BMT holes using a large diameter packer and injection system that was specially developed for the purpose and linked to the dynamic injection device that had been developed and tested during preliminary activities.

The hydraulic conductivity of the rock was decreased by grouting. Heating caused the hydraulic conductivity of the grouted rock to increase. However the final values tended to be less than those of the ungrouted rock. The increase in hydraulic conductivity of the grouted rock caused by heating was associated with heave in the unrestrained floor of the test room. Grouting also caused the floor to heave. Thus, it was shown that the materials and techniques developed through the programme could be used successfully to decrease the hydraulic conductivity of discretely fractured rock with values before grouting as low ad 10^{-8} m/s. It was shown that the Stripa granite with hydraulic conductivities in the range $10^{-10} < k < 10^{-8}$ could be grouted if the fractures were relatively free of natural infilling materials.

The upper 0.5 m to 1.0 m of floor of the BMT rooms was found to have been significantly disturbed by the excavation process and to have an hydraulic conductivity axially along the room that was as much as 4 orders of magnitude higher than the average value measured for the undisturbed rock. However, the values varied over 3 orders of magnitude on a scale of metres. This latter finding and the observed heave in the floor of the room, which was also locally variable at the same scale, led to the suggestion that, for the purposes of repository performance assessment, the only reasonable approach may be to attempt to predict the thermo-hydro-mechanical performance of ungrouted or grouted rock at less detail than the one achieved in this experiment.

4.3 Fracture zone grouting

The natural barrier investigations provided background information on the hydrogeological characteristics of the rock mass that was to be grouted and a concept for an experiment to determine the effectiveness with which high performance grouts could seal hydraulically active fracture zones in the rock mass was developed. Effectiveness was to be evaluated by measuring the extent to which grouting was able to decrease the rate of water flow into excavations.

The background information on the hydrogeological characteristics of the rock proved to be insufficient for the level of understanding of the rock required for the grouting experiment. Additional probes to locally characterise the rock indicated more complex hydrogeological characteristics than those originally envisaged and led to a revised concept of the hydrogeological conditions. The hydrogeological characterisation activities associated with the experiment identified a subset of fracture features, providing local control (at the scale of tens of metres) on water flows. The subset had not been recognised by preceding, more globally oriented, natural barrier studies. Similar findings may be expected should grouting activities prove to be necessary to effect repository sealing. Such grouting activities will provide additional information on the detailed structural geology of the host rock and may be used to refine hydrogeological models used for site assessment.

Hydraulic testing of the rock locally around the test room provided estimates for the apparent hydraulic conductivities of the different structures through which the excavation had been made. With this information it was possible to construct a simple axisymmetric finite element model of the flow paths. Using known hydraulic pressures in the rock and making reasonable assumptions for the effects of grouting on the hydraulic conductivity of the rock, estimates were made for the effects of the grouting on the total inflow into the room. The results showed that total inflows could be expected to be decreased by less than 10 %, which is less than the natural fluctuations measured before grouting. While some measure of this decrease may have been observed in the test results, alone, the measurements of total inflow into the room were not sufficient to evaluate the effects of grouting. In light of the hydrogeological conditions that will exist in a sealed repository after the major disturbances to the groundwater flow caused by repository construction are removed, the rate at which water flows into a room may not be the appropriate criterion by which to evaluate the effectiveness of grouting.

The physical presence of grout in the fractures, the results of the tracer experiments and the rates of evaporation from the rock surfaces before and after grouting, the measured rates of inflow from holes drilled into the ungrouted and grouted hydraulically active volumes of rock, along with hydraulic testing of ungrouted and grouted rock, all indicated that the grouting activities changed the dominant water flow paths in the rock.

From these observations it could be concluded that, using the techniques and cement-based grouting materials developed and used through the programme, it would be possible to decrease the apparent hydraulic conductivity of fracture zones, such as that exemplified by the J zone in the Stripa rock mass, from $10^{-8} > k > 10^{-9}$ m/s to $5 \cdot 10^{-10} > k > 10^{-10}$ m/s. To achieve this effect, borehole spacing would have to be closer than the ones used in the grouting experiment. Further improvements in grouting equipment techniques and processes would add further confidence in an ability to achieve the result.

4.4 Grouting the excavation disturbed zone

In situ investigations were carried out in 3 500 m³ of rock around the enclosed section of the tunnel used for the BMT. The hydraulic properties of the excavation disturbed zone (EDZ) around tunnels excavated in the Stripa granite and the ability to grout and decrease the hydraulic conductivity of the zone were investigated. The experimental setup is shown in Figure 5.

The EDZ was considered to consist of two parts: (i) a blast damaged zone (BDZ), which extended around the periphery of the room to a maximum depth of approximately 1 m from the surface of the excavation; and (ii) a stress disturbed zone, which theoretically extended to a depth of about 12 m into the rock and was investigated experimentally to a depth of 7 m. The hydraulic conductivity of the BDZ was measured locally, at the scale of metres, by water pressure injection tests. Attempts were made to measure the hydraulic conductivity of the stress disturbed zone over a distance of 13 metres parallel with the longitudinal axis of the test tunnel. These measurements were supplemented by assessments of the hydraulic conductivity of the rock made through evaluations of the water inflows under the natural hydraulic gradients into openings in the rock. Hydraulic conductivity measurements were made before and after the BDZ was grouted using both static and dynamic injection techniques with Alofix cement-based grout. Practical difficulties, associated with sealing the surface of the test chamber, led to delays in the programme and cost concerns. Thus, a programme to test the effectiveness of grouting the stress disturbed zone was not completed.

Evaluation of the results from hydraulic tests required the development of conceptual models for the fracture features and the hydraulic characteristics of the undisturbed and disturbed Stripa granite. These conceptual models were incorporated into available computer codes which were used to appraise the effects of excavation on the hydraulic properties of the granite. At the level of detail needed for the experiment, the conceptual models and computer codes, alone, were not sufficiently complete to describe the properties of the excavation disturbed rock mass and account for the test results. Combined with observations made in other parts of the Stripa mine throughout the SAC programme and Phases 1, 2 and 3 of the Stripa Project, the results of the modelling methods and the hydraulic testing led to the following conclusions with regard to the hydraulic properties of the rock around the BMT excavation. A blast damaged zone, which could be generally ascribed an hydraulic conductivity of approximately 10^{-8} m/s existed in the rock within 1 m of the surface of the excavation. The hydraulic conduction properties of the stress disturbed zone were generally anisotropic with axial and radial hydraulic conductivities of $5 \cdot 10^{-10}$ m/s and $5 \cdot 10^{-11}$ m/s, respectively. The undisturbed rock could be considered as isotropic with hydraulic conductivities in the range $3 \cdot 10^{-11}$ to $9 \cdot 10^{-11}$ m/s. The values given vary spatially within defined ranges.

With the techniques and grouting materials used for the experiment, despite clear evidence that grout had been injected into the EDZ, the hydraulic conductivity of the zone was not measurably changed. Some possible explanations for this result include insufficiently close spacing of the holes through which the grout was injected and the effects of infilling materials in the natural fractures. Alternative grouting procedures may have given different results. The lack of opportunity to conduct a second grouting phase or to grout the stress disturbed zone left these issues unresolved.

5. Longevity of Sealants

The longevity of clay- and cement-based sealants was investigated by: investigation of natural analogues; laboratory studies of material properties; and numerical modelling of thermodynamic processes.

For bentonite-based materials, attention focussed on developing detailed understanding of hydrothermal alteration of minerals; particularly, reactions causing transformation of smectite clays to hydrous clay-micas or causing silicification of the clay mass were studied. Both of these processes could decrease the swelling capacity of the bentonite and lead to loss in long-term function. For the cement-based sealants, mechanisms causing dissolution of cement in groundwater and, thereby, increasing the hydraulic conductivity of grouted rock were investigated; specifically, the leaching and hydraulic conductivity properties of the materials were studied. To allow for the development and application of numerical models of cement grout longevity, a database on the fundamental thermodynamic properties of cement grout phases was established and expanded. In addition to these basic studies, the mechanical stability of clay gels and unset cement pastes were investigated with respect to their ability to resist erosion. This information was needed to define the limiting groundwater flow conditions under which each of the materials could be applied.

The crystal structures of smectite minerals and hydrous clay-micas possess similar features. With particle sizes less than 2 μm, both mineral types consist of negatively charged lamellae of phyllosilicates comprised of covalently bonded silica and alumina layers. The lamellae in clay-mica are bonded by K^+. In smectites the lamellae are separate and discrete. Studies of the products of reaction between bentonite, finely ground silica, groundwater and rock over a wide range of temperatures and pressures showed that when K^+ is present in the groundwater, the smectite in bentonite clay will convert to hydrous clay-mica. In contrast with HCB, for which the conversion reaction will take many tens of thousands of years, conversion in clay grouts will take a few thousand years or less. At temperatures above about 130°C, the conversion reaction will be accompanied by the precipitation of silica within the clay fabric. This latter process results in cementation and embrittlement of the grout. The effects of the conversion of smectite to hydrous clay-mica on the performance of a grouted rock mass were assessed by reviewing the structure of the grouts recovered from the *in situ* tests. It was estimated that after the smectite in the grouted rock mass has undergone complete conversion to hydrous clay-mica, the grouted rock could still possess an hydraulic conductivity that is significantly less than that of ungrouted rock. The conversion process was shown to be largely controlled by the diffusion coefficient of K^+ in the clay and the concentration of K^+ in the groundwater.

To evaluate the longevity of high-performance cement-based sealants, laboratory studies were coupled with geochemical modelling of the changes that may occur within the fabric of the material and a numerical assessment of the effects of these changes on performance. Initially, modelling studies were based on contemporary understanding of normal cements and concretes and were subsequently adjusted for laboratory findings on high-performance grouts.

The studies indicated that the grouts could endure between 100 000 and 1 000 000 years in a repository environment. These predictions did not account for unhydrated materials that were found in both ancient cements and modern high-performance materials. The long endurance was related to the low dynamic porosity of the high-performance materials.

The porosity-hydraulic conductivity relationship for conventional portland cements was found not to apply to high-performance materials, laboratory specimens of which were shown to be virtually impermeable at hydraulic gradients less than approximately 15 000. In this context, it is noted that the maximum hydraulic gradient in the Stripa facility was approximately 2 000; much lower gradients are expected in a sealed repository. Initially, based on conventional wisdom, it was assumed that cements would degrade by water percolating through the grouts; as the water passed through, the cement solids would dissolve and the consequent porosity increase would increase hydraulic conductivity. Given the very low hydraulic conductivity of the high-performance grouts, substantial flow through the body of the cement did not appear to be likely. While flow will occur around the grout, diffusion processes will operate within the grout to alter the mineralogical composition and chemistry of the grout. The slow diffusional processes virtually assure an approach to chemical equilibrium and it can be inferred, as shown in laboratory tests, that void spaces represented by micro-cracks or other microporosity will be filled by the precipitation of secondary reaction products. The chemical reaction of groundwater and cement yields secondary products which occupy more space than that occupied by the original solid cement phases.

The consequence of this new understanding is that during early repository evolution, cement grout performance will be dominated by surface-controlled mechanisms. Because these mechanisms are less efficient at mass removal than the assumed processes of percolation and dissolution, the reported estimated persistence times are considered to be a lower bound on the longevity of intact cement-based sealing materials. Several uncertainties rest with this judgement. Particularly, examination of the microstructure of grouts injected in the Stripa mine revealed an inhomogeneous structure that was not present in laboratory prepared and tested materials. The full implication of the Stripa findings need to be appraised after further *in situ* tests have established whether the results are specific to the Stripa site and to the injection technique applied in the Stripa field tests.

6. Conclusions

-   The hydraulic conductivity of boreholes and excavated openings could be returned to values similar to those of intact granite by the judicious use of HCB.

-   Models are now available to predict the response of HCB to changes in stress, thermal and hydraulic gradients. The models for water transfer are not rigourously precise. In contrast, thermal properties are reasonably well understood and heat fluxes through the material can be described reliably.

-   The properties of advanced, high-performance bentonite- and cement-based materials pertinent to their successful injection as grouts in fractured rock have been well defined. Equipment and procedures for injection of the grouts have been developed and are available for use in repository development. The limits of the application of the selected materials and methodologies were defined for the Stripa granite.

- Data at the level of detail derived for the application and qualification of general groundwater flow models such as those examined in the natural barrier studies of the Stripa Project are unlikely to provide sufficient information for grouting activities intended for repository sealing. As at Stripa, the information gained from the grouting activities at repository sites will likely lead to revisions in understanding of the rock mass.

- An excavation disturbed zone (EDZ) consisting of a blast disturbed zone enveloped by a stress disturbed zone existed in the rock surrounding tunnels in the Stripa granite. In the absence of alternative, preferably site specific information, the blast disturbed zone in granitic rocks similar to those at Stripa can be taken to have an hydraulic conductivity of about 10^{-9} to 10^{-8} m/s. At the locations in the Stripa mine studied by the engineered barriers research group, the stress disturbed zone appeared to be more conductive parallel to the axis of the excavations ($3 \cdot 10^{-10} \leq k \leq 9 \cdot 10^{-10}$ m/s) than normal to it ($7.5 \cdot 10^{-12} \leq k \leq 2.3 \cdot 10^{-11}$ m/s).

- Under the low hydraulic gradients expected in the groundwater in a sealed repository site, chemical transformation of the minerals in clay- and cement-based sealants can be predicted to extend over tens of thousands to millions of years. The predicted period depends on the porosity of the as-placed materials and the ionic concentrations in the groundwater.

- The sealing properties of both clay- and cement-based sealants are most susceptible to change under high hydraulic gradients. Thus, both materials will be most vulnerable to adverse change during seal construction and the period over which the repository is open for the deposition of waste.

- The engineered barriers studies for the Stripa Project have increased confidence in the ability to engineer geological repositories for heat-generating radioactive wastes.

- Due to the perturbations caused by the presence of excavations it may never be possible to physically examine in underground laboratories, or from excavations at repository sites, the conditions under which radionuclides will migrate within either the engineered barriers or the host rocks that form the waste isolation system.

A number of unresolved items have been identified. These could be considered for resolution in future collaborative studies or national programmes of research and development related to repository design and construction.

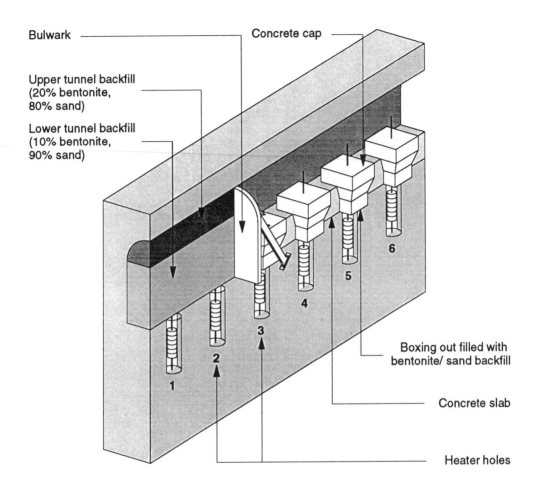

Figure 1   The layout of the Buffer Mass Test.

Six 160 mm diameter holes were drilled in line in the floor of the room. The holes were equipped with electrical heaters surrounded by highly compacted bentonite and overpacked with sand-bentonite mixtures. Temperatures, pressures and displacements were monitored over 4 years.

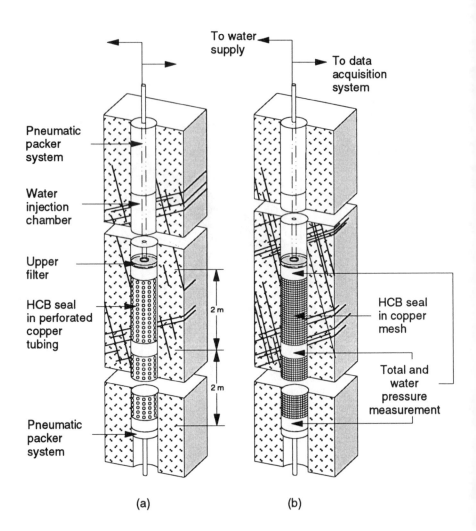

Figure 2   The layout of the vertical borehole plugging tests.

Two types of plug were tested. One [(a) above], similar to the system used in the horizontal borehole plugging test, was encased in a perforated copper tube. The other [(b) above] was encased in a copper wire mesh. Both systems effectively sealed the boreholes. System (a) was more rigid and the investigators considered it easier to emplace than system (b).

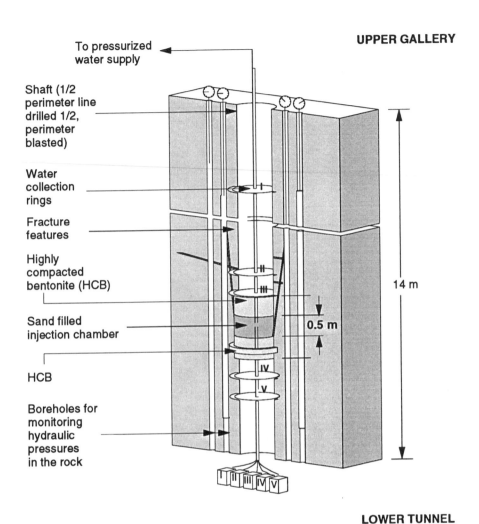

Figure 3   The layout of the shaft plug test.

Concrete bulkhead performance was first tested. A plug consisting of HCB confined by tied steel plates was then tested. The second test showed that the HCB plug had a lower hydraulic conductivity than the concrete alone. Water was lost from the test cell through the neighbouring rock.

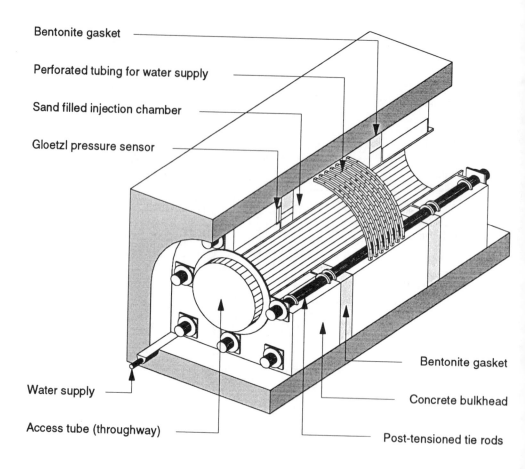

Bentonite gasket

Perforated tubing for water supply

Sand filled injection chamber

Gloetzl pressure sensor

Water supply

Access tube (throughway)

Bentonite gasket

Concrete bulkhead

Post-tensioned tie rods

Figure 4   The layout of the tunnel plug test.

HCB was placed as blocks on the inner faces of two tied concrete bulkheads. The inner steel tube (ϕ = 1.5 m) provided access to the inner bulkhead. Such a structure may be used to seal off fracture zones during repository operation and provide access for equipment and manpower.

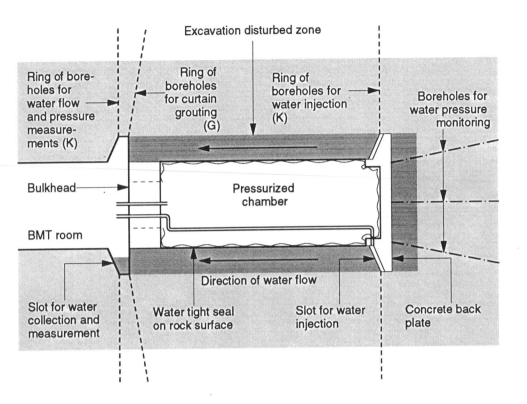

Figure 5   Investigating the excavation disturbed zone (EDZ).

The inner section of the BMT tunnel was to be sealed with a water tight lining and pressurized. Two rings of monitoring boreholes(K) were packed off in sections. This allowed for cross-hole hydraulic testing and, in combination with the end slots, for evaluation of the axial hydraulic conductivity of the EDZ, which was inferred from numerical simulations and testing to consist of a blast damaged zone and a zone influenced by stress relief. The effectiveness of cement grouts in sealing these two zones was examined.

A Note on Modelling of the Groundwater Flow and Pressure Behaviour Observed during Excavation of the SCV Drift in the Stripa Project

B. Damjanac and C. Fairhurst

University of Minnesota
Minneapolis, Minnesota, USA

Abstract

A simple back-analysis of groundwater results related to excavation of the SCV drift in the Stripa Project has been conducted in order to assess the applicability of a porous medium conceptual model to the SCV site. It has been found that a porous medium model can not explain all of the observed measurements. In particular, the predicted groundwater pressure drawdown and recovery are not consistent with the behavior observed. The authors have no simple explanation for this inconsistency.

1 INTRODUCTION

A main objective of the Site Characterization and Validation (SCV) Project conducted at the Stripa Mine, Sweden, was to verify and validate different numerical methods for modeling and prediction of groundwater flow in a fissured rock mass. Six, 100 m long boreholes were drilled subhorizontally into the central portion of the rock mass chosen as the site for the SCV studies, as seen in Figure 1(a), i.e. one central hole (D1) along the axis of the proposed 2.4 m diameter SCV drift, surrounded by five 100 m long holes (D2 – D6), each located on a radius of 1.2 m from D1, defining the periphery of the future drift.

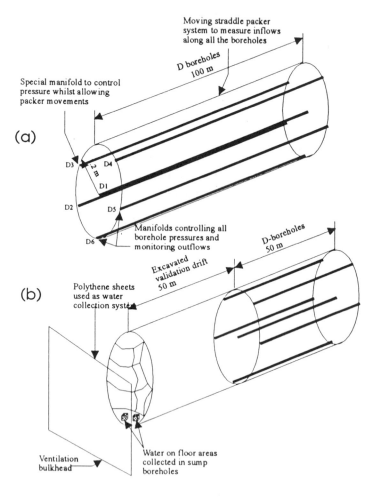

Figure 1: **Array of D boreholes in the simulated Drift Experiment (after Olsson(1992))**

Groundwater inflow was measured (i) over the full 100 m of each borehole prior to excavation (by drilling and blasting) of the 50 m long SCV drift and (ii) over the remaining 50 m long boreholes after excavation of the drift (see Figure 1(b)). The modelers were given the information on the groundwater inflow into the D boreholes prior to excavation of the SCV drift, and asked to predict, from their models, the inflow into the drift after excavation.

The numerical models used for the predictions were based on several diverse conceptual models, including equivalent porous medium and discrete fracture flow models. All models predicted an increase in total inflow into the drift compared to the total inflow into the boreholes. In fact, the inflow after excavation of the Validation drift was reduced to approximately 12% only of pre-excavation inflow into the D boreholes[1]. Possible reasons for the reduction that have been suggested include:

1. release of gas initially dissolved in the groundwater with the drop in pressure as the water moved towards the excavation. This release, it is surmised, could produce a reduction of conductivity under the conditions of two phase flow.

2. fluid flow–solid deformation coupling, i.e. significant deformation of joints due to stress concentrations around the excavation, producing a reduction in joint conductivity. This explanation is weakened by the fact that, while stress increases would tend to close joints over part of the periphery, stress decreases over the remainder of the periphery would tend to open joints, hence increasing conductivity. It should also be noted that the dominant joint systems in the vicinity of the SCV site were normal to the axis of the SCV drift, and hence would be relatively unaffected by changes in the tangential stress distribution around the excavation — i.e. as computed according to the calculated (two dimensional, plane strain) stress changes.

Without trying to pass judgment on which of these (or other[2]) hypotheses appears to be the more reasonable, the authors have attempted to determine, by back-analysis, whether or not the observed groundwater flow and pressure distribution can be made to agree with values calculated from an analysis in which the SCV site is assumed to behave as an equivalent porous medium. Hydraulic conductivities (i.e. the equivalent porous medium values) for different regions around the drift were used as arguments in the back analysis in order to attempt to reproduce the measured values.

---

[1] A similar effect, in this case a reduction to approximately 25%, was observed during sinking of the shaft for the Underground Research Laboratory (URL) in Canada.

[2] The possibilities of a reduction in permeability local to the excavation, due to effects at the drift face as it is being excavated e.g. shear stress changes produced slip on fissures, or blast induced permeability decrease, both seemingly plausible, have not been investigated.

Both of the suggestions above effectively produce an annular "skin" or zone of low permeability around the tunnel. In out analysis this skin was assumed, without specific justification, to be created by excavation of the tunnel. The reasons for the variations of hydraulic conductivity in the various zones in the SCV site were not investigated i.e. no attempt was made to model fluid flow–solid deformation coupling or two-phase flow with exsolution of dissolved gas from the groundwater. The motivation for the analysis was simply to try, using an equivalent porous medium model, to reproduce the measured results by varying the hydraulic conductivities of the different zones in the rock mass.

2   DESCRIPTION OF EXPERIMENT

In addition to monitoring of groundwater flows into the boreholes and into the drift, the change in pore pressure at given locations around the drift was also registered as the drift was excavated. One distinct fracture zone of relatively high permeability, defined as the "H zone", intersects both the D boreholes and the Validation drift approximately at 90° and has a thickness of approximately 10 m (Figure 2). The drilling of the D boreholes created a sink, draining the surrounding rock mass as observed at a piezometer in the vicinity of the drift (piezometer W2-5, Olsson (1992)). Observations in the D boreholes indicated that 85% of the inflow into the boreholes came from the H-zone.

With excavation of the Validation drift the pore pressure in W2-5 and other holes was observed to start to increase, while the total inflow into the Validation drift and boreholes was observed to decline. As noted above, the Validation drift was excavated over the first 50 m of the D boreholes only. The plot of head response to Validation drift excavation at piezometer W2-5 (located approximately at point 2 in Figure 2, i.e. approximately 65 m from the drift) showed almost complete recovery of the initial (pre borehole) pressure head after passage of the Validation drift (see Figure 3). Water flow measurements show that the inflow into the Validation drift was 12% of the inflow into the corresponding 50 m portion of the D boreholes. Also, the percentage of the total inflow coming from the H-zone increased from 85% to 97%.

3   CONCEPTUAL AND NUMERICAL MODEL

Since a fully three-dimensional porous medium numerical model was not available to the authors, the analysis was carried out using an axisymmetrical model in which the axis of symmetry was taken to be along the axis of the Validation drift. This model is not totally appropriate, because gravity pressure changes in the fluid over the vertical region around the drift were neglected, but since the value of the

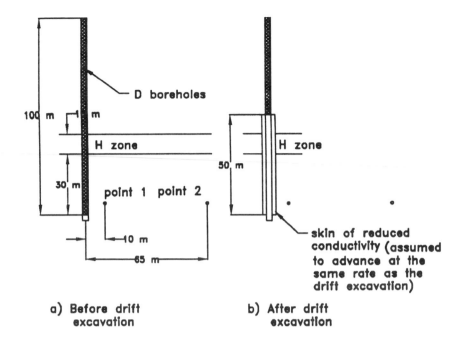

Figure 2: Layout of the model

initial pore pressure (approximatelly 200 m) was high relative to the height of the Validation drift the error due to the assumption of constant vertical pressure is considered acceptable.

As mentioned, groundwater flow in the rock mass was modeled as flow in an equivalent porous medium. The conductivities of the different zones were varied in order to more accurately take into account processes which may influence the conductivity of rock mass. Different conductivities were assigned to the various zones of the rock mass at the SCV site as follows:

1. the general rock mass at the SCV site. Since we were interested essentially in *ratios* of inflow between different zones, and in establishing the qualitative behavior of the system, the conductivity of this part of the rock mass was assumed to be unity,

2. the H zone, which exhibits an increased conductivity relative to the general rock mass. The ratio of the conductivity of H zone relative to the conductivity of the general rock mass defines the ratio of inflows into the D boreholes from the H zone and the rest of the rock mass;

441

3. the D boreholes; these boreholes were simulated as a single circular zone, of diameter 2.4 m (see Figure 2(a)) within which the conductivity was increased relative to the general rock mass. The ratio of conductivity of the D borehole zone to the conductivity of the general rock mass defines the drawdown in hydraulic head (piezometer W2-5, and point 2 in Figure 2) in the surrounding rock mass,

4. a "skin" around the Validation drift representing a zone of reduced conductivity caused by excavation of the Validation drift (as noted above, the reason for this reduction was not analyzed nor is it relevant to this analysis). The ratio of conductivity of this zone relative to the conductivity of the general rock mass defines the recovery of the hydraulic head (piezometer W2-5, and point 2 Figure 2) in the surrounding rock mass,

5. a "skin" around the Validation drift in the H zone. The ratio of conductivity of this zone relative to conductivity of the rock mass, and the conductivity of the "skin" in the rest of the drift defines the ratio of inflow from the H zone relative to the total inflow after excavation of the Validation drift.

The back-analysis used the following measurements as constraints in fitting the arguments (i.e. the ratios of hydraulic conductivities) into the numerical analysis:

1. Before excavation of the Validation drift, inflow from the H zone was 85% of the total inflow into the D boreholes.

2. After excavation of the drift, inflow from the H zone was 97% of the total inflow into the Validation drift.

3. After excavation of the drift inflow into the Validation drift was 12% of the total inflow into the D boreholes before excavation.

4. The hydraulic head drawdown and recovery curve for piezometer W2-5 — as reported by Olsson (1992) (see Figure 3).

These particular results were chosen as the most characteristic for the experiment. To simplify analysis, transient flow behavior was not considered , i.e. values of hydraulic head drawdown and recovery observed in the piezometer records were assumed to correspond to steady state values.

Thus, assuming that the piezometer measurements do not contain any severe measurement errors, we have a well defined problem — with 4 unknowns and 4 available conditions.

The numerical analysis were conducted using the two-dimensional explicit finite difference code FLAC, Itasca (1993), under axisymmetric flow conditions.

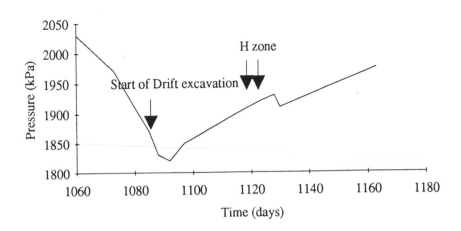

Figure 3: **Pressure response to Validation drift excavation in monitoring interval W2-5**

4 RESULTS OF CALCULATIONS

The cases analyzed are summarized in Table 1. All conductivity values indicate ratios of conductivity relative to the general rock mass at the SCV site, which is assumed to be homogeneous and isotropic. The primary concern of the analysis is the relative values of inflow from the different zones along the D boreholes and along the Validation drift, hence only relative values of conductivities are sufficient. The parameters were not varied further when the predicted values were found to be reasonably close to those measured.

The modeling results, which correspond to steady state values before and after excavation, are summarized in Table 2. It is seen that case 4 in Table 2 agrees well with constraints 1, 2, and 3 listed above. Constraint 4 was not well matched for the values in case 4, as is discussed below.

In order to obtain a hydraulic head drawdown due to drilling of the D boreholes, similar to that observed at the location of piezometer W2-5, it was necessary to increase the hydraulic conductivity of the numerical grid representing the D boreholes to an effectively infinite value. Practically, this assumes that the D boreholes are equivalent to an open drift (without a low permeability skin) of 2.4 m diameter along their full 100 m length. However, under such a condition, the numerical

case	Hydraulic conductivities			
	major part	H zone	"skin"	H zone "skin"
1	1.0	5.0	0.01	0.1
2	1.0	10.0	0.01	0.1
3	1.0	100.0	0.01	1.0
4	1.0	50.0	0.01	1.0

Table 1: **Hydraulic conductivity ratios assumed in the groundwater flow models analyzed**

model does not indicate complete recovery of the head when the Validation drift is excavated and the low permeability skin is developed. Thus, this model does not agree with the observed behavior. It should be noted that maximum drawdown in the model is reached almost at steady state (see Figure 4, point 2) while in actual record the pore pressure has not reached steady state but declines further after the start of excavation (see Figure 3). Also, measurements indicate that recovery of hydraulic head at piezometer W2-5 started almost immediately with the start of drift excavation. This behavior could not be reproduced in the numerical model. Piezometer W2-5 (point 2 Figure 2) is located approximatelly 65 m from the Validation drift which is 50 m long in total, so excavation of the section of the drift closest to W2-5 does not have a significant, distinct early influence on the piezometer. On the other hand, a point in the model some 20 m distant from the drift (i.e. point 1 Figure 2) is very much influenced by local effects of excavation (e.g. a change of hydraulic conductivity of that part of the drift closest to the point). The measured drawdown–recovery curve for W2-5 was reproduced for point 1 i.e. 20 m from the drift, by increasing the hydraulic conductivity of the zones representing the D boreholes by a factor of 100. The numerical result for point 1 then agree completely with measurements for piezometer W2-5 (other constraints are also fulfilled). Figure 4 shows the change in hydraulic head at points 1 and 2 for case 4 in the numerical experiment, with infinite hydraulic conductivity along the D boreholes. It is seen that excavation of the Validation drift has a delayed influence

case	% of inflow from H zone		% reduction of inflow
	D boreholes	Validation drift	into the Validation drift
1	55.21	71.03	10.28
2	70.82	78.01	11.24
3	95.66	95.61	12.47
4	92.02	94.75	7.19

Table 2: **Results of calculation**

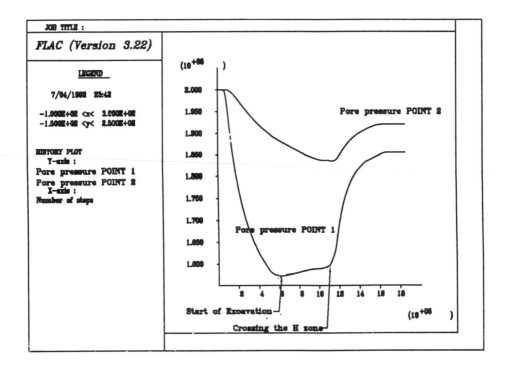

Figure 4: **History of hydraulic head (MPa) at locations 1 and 2 predicted by the numerical modeling**

on point 2 while the influence is almost immediate at point 1. Thus, the pressure behavior of point 1 agrees much better with the behavior of piezometer W2-5 than does the behavior of point 2.

5   SUMMARY

An equivalent porous medium model of the SCV site is capable of matching all of the observed flow and pressure behavior except that of the hydraulic head drawdown–recovery curve at the location corresponding to piezometer W2-5, i.e. 65 m from the SCV drift. Very high hydraulic conductivity in the area of the D boreholes can produce the measured drawdown, but in this case the pressure recovery is not complete. The model results show smaller sensitivity due to "skin" of reduced permeability around excavation than is observed in the piezometer record.

Piezometer W2-5 behaves similarly to a point placed 20 m from the drift in the numerical model. This suggests that there may be a "short circuit" between W2-5 and a point much closer to the drift, so that the influence of the Validation drift on piezometer W2-5 is amplified considerably over what could otherwise be expected. The short circuit could be a highly permeable zone or a distinct fracture (or connected fractures) from the drift to the borehole containing piezometer W2-5. However, this explanation is not consistent with the fact that the recovery curve that results from excavation of the Validation drift in general, exhibits a uniform smooth increase. In the case of a distinct fracture zone crossing the Validation drift, the recovery curve would be more rapid, as in the case of Point 1.

The main conclusion from this analysis is that a porous medium model does not appear to be sufficient to describe all of the groundwater flow and pressure phenomena observed around the SCV drift. However, the reason for the observed anomaly in pressure drawdown and recovery of piezometer W2-5 is not clear to the authors. Possible explanations may be:

- error in measurement or

- some local effect, in the vicinity of W2-5, i.e. something (other than that of a connection between W2-5 and a drift, or a point close to the drift, via a high conductivity path to the drift) which influenced record at piezometer W2-5. The existence of such local influences is not consistent with the assumption of a porous medium model.

References

[1] Ollsen O.(Editor), 1992. *Site characterization and validation — final report.* Stripa Project TR 92-22 SKB Stockholm, Sweden

[2] FLAC 3.22, 1992. Itasca Consulting Group Inc., Suit 310, 708 South Third Street, Minneapolis, Minnesota 55415, USA, Fax No (1) 612-371-4717.

SESSION V

PANEL DISCUSSION

SÉANCE V

TABLE RONDE

Chairman - Président
J. Hunter (United States)

Waste Management Research
The past, the present and the future

Per-Eric Ahlström
SKB (Sweden)

INTRODUCTION

After completion of fifteen years of experimental work at the Stripa mine and at the end of the final Stripa symposium it seems appropriate to look backwards and try to evaluate what has been achieved. More important is however to look forward and try to give some thoughts on where to go. Several speakers at this symposium have already done both, so you must forgive me if some comments I will make will look as repetitions of what has been said previously. The echoes may be an expression for some consensus. My reflections today are however my own and not necessarily shared by all persons at SKB or participants in the Stripa project. They are based on my experiences obtained by participating in the KBS-studies, by attending and sometimes chairing numerous meetings within the Stripa project since late 1976 and also as director for the research and development division for SKB since 1984.

THE EARLY YEARS

Research on management and disposal of radioactive wastes started already in the 1950s, when the first plans for nuclear power plants were announced. By that time the research was however limited in scope and mainly concerned with the treatment of low and intermediate level wastes from the operation of the plants to be built. Disposal of such wastes were generally considered to be made either by shallow land burial or by dumping in deep sea waters. The treatment and disposal of high level wastes and other longlived wastes were given only occasional attention by the civilian nuclear power industry. It was thought that this problem was the duty of the reprocessing industry, which by that time mainly operated for the military sector. It was recognized that the volumes of such wastes were small and that the reprocessing plants would be large units serving many reactors in several countries. The services offered by the reprocessors also included the acceptance of all the wastes arising from the reprocessing. This state of affairs prevailed until the early 1970s.

In Sweden the first sign of a change was the appointment of a government special committee - the Aka-committee - in 1972 to review the management of radioactive wastes from sources in Sweden. At first also this committee was mainly concerned with low and intermediate level wastes. Soon enough it became however obvious that also the high level wastes would be an issue for all countries with nuclear power. The mandate of the committee was expanded to include also high level wastes. In the mid-1970s the first R&D-programme covering all aspects of waste management was launched by the committee.

By that time the radioactive waste had become a main issue in the political arena. In the autumn of 1976 a new Swedish government announced new very restrictive legislation concerning high level waste management to be put on the operators of nuclear power plants. This shifted in a couple of weeks the main responsibility for research in this area from the government to the nuclear power industry. A "crash programme" was started and I recall that the very first international discussion on experiments to be made in Stripa was held in Stockholm only some five to six weeks after the new rules were announced by the government. This happened 16 years ago.

SIXTEEN YEARS SINCE STRIPA STARTED

Where have we come during these sixteen years? Well, I don't think I have time enough to review the developments comprehensively or in a way that would give justice to all the scientist, engineers, technicians, managers and decisionmakers involved even if I concentrated on only the Swedish developments. Instead I have chosen to give some more or less random examples which I personally think illustrate the progress that has been made. Of course as you already have noticed I have a Swedish bias or perspective.

Research has been conducted in all areas of importance for the safe disposal of radioactive wastes. Before the mid-1970s the research was concentrated on the waste forms - for HLW solidified, mostly vitrified, waste - and on the (deep) geologic isolation of the waste forms. Disposal of unreprocessed spent nuclear fuel was not even considered. The Aka-committee and the early KBS-studies in Sweden introduced two major new aspects into the waste management research. Direct disposal of unreprocessed spent fuel was investigated and shown to be feasible in the KBS-2 report. The use of long-lived engineered barriers was - I believe - for the first time shown to be able to give a major contribution to long-term safety in both the KBS-1 report for vitrified high-level waste and the KBS-2 report. The multibarrier concept has stayed as one of the cornerstones for the safe isolation of nuclear waste. No major new principle ideas on how to dispose of the wastes have been suggested since then.

The research on waste forms have continued throughout the sixteen years both for vitrified high-level waste and for spent fuel. An indepth understanding of some basic phenomena in the interaction between groundwater, the waste form and other species in the repository environment has contributed to the development of models for spent fuel corrosion and for vitrified high-level waste corrosion in such environment. These models are still conservative but show that e g the alteration of spent fuel in the ground waters found in deep granitic bedrocks will take at least millions of years. This is a very important result because it means that from a health point of view the most hazardous long-lived radionuclides e g Pu-239 will decay long before any significant amount would go in solution. Other more easily soluble nuclides like I-129 are very much less hazardous and should thus be of much less concern in safety assessments.

Studies of waste canisters have confirmed that it is possible to design and construct a canister for high level waste with a very long lifetime if it is disposed of in a deep repository in granitic bedrock. A service life of the order of a thousand years - i e well beyond the decay time of the most dangerous fission products - can be obtained with several different materials - e g steel and titanium. A copper canister would have a service life of again millions of years. This means that the waste can be isolated from the groundwater until the radiotoxicity has decayed to the level found in rich uranium ores for spent fuel and to a much lower level for vitrified HLW.

Some of the developments on buffer materials have been extensively reported during this symposium. It would be presumptuous of me to try to add something in that area. A general conclusion concerning the waste form, the engineered barriers and the near field is that we have now reached a rather mature state of knowledge. We are able to develop models for a reasonably cautious safety assessment for almost any significant scenario to be included in a comprehensive safety analysis. However we still lack the ability to quantify the safety margins that we know exist with respect to spent fuel corrosion in repository environment. This is a challenge for future research.

In the conceptualisation of the engineered barriers for a deep repository in granitic bedrock the emphasis was on looking for materials that were known to be stable in the geochemical environment that exists in such rocks. Thus it was possible already from the beginning to have a high confidence both qualitatively and quantitatively in the isolating properties of the engineered barriers. For the geologic barrier the situation was quite different, only some qualitative understanding existed of the transport properties of the deep groundwaters. Safety assessments had to be based on few data and on very simplified models which were chosen to give bounding estimates of the barrier properties.

Thus it is no surprise that the main advancements during the sixteen year period have been made in the understanding of the geologic barrier. The improvements touch all areas including measurement methodology, techniques and data, interpretation of measurements, conceptual modelling, quantative modelling and also assessment of the barrier isolation and retardation properties. The measurement methods have been vastly improved in many respects. Within the Stripa project geophysical techniques like radar and seismics have been developed to useful tools for site characterization. Outside the project probes for accurate in situ registration of important geochemical parameters as pH and redoxpotential - Eh - in deep boreholes have been developed. This has contributed to the development of a reasonably consistent understanding of the geochemical properties and behaviour of deep groundwaters.

Geohydrological, geochemical and geophysical data have been collected extensively during the past sixteen years in the Stripa project and in many other projects. The SKB GEOTAB database now contains some 140 Megabyte data from 562 boreholes with a combined length of 85 km at about a dozen sites in Sweden. One hundred different methods of measurement were employed. This very large amount of geological data gives us a solid confidence that there exists many sites in Sweden which have geological conditions pertinent for the construction of a safe repository.

The improvements in data collection methodology and quality have also contributed to the improved conceptual understanding of e g groundwater movements in fractured rock. This and the databases for real sites and the remarkable improvements in computer technology are prerequisites for the big steps towards more realistic quantitative modelling of groundwater flow and transport that have been taken in the latter part of the 1980s and in the early 1990s. In this area we still have a long way to go, however, until we can make a good realistic quantitative assessment of the safety margins provided by the geological barrier. One important goal for the future is thus improved quantitative understanding of groundwater flow and solute transport in deep, fractured rocks.

WHERE ARE WE HEADING?

Thus there are two areas of interest for repository safety where I see considerable room for improvements in future research - spent fuel corrosion in repository environment and transport of dissolved radionuclides in fractured crystalline rock. I would however in this context like to emphasize two things - first: these improvements are not necessary for the assurance of the safety of a deep repository - and second: the perspective is of course the present Swedish plans for direct disposal of unreprocessed spent fuel in the deep bedrock.

In general I would say that the time of generic research in the radioactive waste management area is approaching the end of the road. The variation in geologic settings is so large that you have to go to specific sites in order to come further in the development of repositories. This does not mean that experiences from one site could not be transferred to another. It must however be kept in mind that each site has its own characteristics for good and for worse. When working with generic sites you will tend to do the impossible - build a repository on a site with all the bad properties that exist on any site combined on one site. Fortunately it is unlikely that such a site exists and if it exists you will not select that site for your repository. Of course you also may run the opposite risk that you prescribe the ideal site with no bad property which you never will be able to find. These are a couple of reasons why you have to be site specific in order to make a realistic design and realistic safety assessments for a deep repository. They are not new and they have been pointed out many times by many persons.

I think I pointed out already at the previous Stripa symposium three years ago that we are moving from *research and development* towards *development and demonstration*. The tendencies I just have mentioned are clearly part of this process. In Sweden we are now well on way constructing the Äspö Hard Rock Laboratory. This is of course a facility for research but even more for development and demonstration. Important stage goals are

- demonstration and verification of methodology for preliminary characterization of a site from the surface and by borehole measurements,

- further development and improvement and demonstration of methodology for detailed characterization of a repository site,

- further development and improvement of modelling of groundwater flow and solute transport,

- demonstration of methodology for repository design and construction - application of "design as you go"-philosophy,

- demonstration and (inactive) testing of important equipments or components for the disposal of spent fuel in a repository.

I have said several times that the Äspö HRL is the "dress rehearsal" for the repository.

SKBs new "Research, Development and Demonstration" - programme, which was submitted to the Nuclear Power Inspectorate two weeks ago, emphasizes the ongoing transition even further. Our plans are to complete the R&D-work started some sixteen years ago by a demonstration of the disposal of real spent nuclear fuel in a deep repository. This could be the first stage of a repository for all spent fuel from the present Swedish nuclear programme. That is our belief and intention. It will however leave the option open for retrieving the fuel that is deposited as part of the demonstration and bring it back to an interim storage if those who are deciding in some twenty years from now should want to do that.

In order to make the demonstration disposal we will need an encapsulation plant and a deep repository. Our plans are to build an extension or annex to CLAB - our interim storage facility for spent nuclear fuel - for the encapsulation. The site for the deep repository has not yet been selected.

What can we really demonstrate in such an exercise? Well, there is one thing that you really cannot demonstrate experimentally and that is the long term safety. But a lot of other important things some of which may seem trivial, but have turned out to be the most difficult can be demonstrated:

- we can demonstrate the encapsulation of very radioactive spent nuclear fuel,

- we can demonstrate the physical transportation, handling and disposal of the canisters containing radioactive spent nuclear fuel from the encapsulation plant to the disposal positions deep in the bedrock,

- we can demonstrate all aspects of the licensing which has to be made in several steps,

- we can demonstrate the multistep decision process,

- we can demonstrate the necessary quality assurance programme,

- we can demonstrate the operation aspects of a repository,

- we may be able to demonstrate the retrieval of disposed canisters with spent fuel.

The basic philosophy is that we will do everything necessary to be able to continue with disposal of the remaining spent nuclear fuel once the demonstration phase is completed in some twenty years if so will be decided. A prerequisite will however be a new licensing and decision process. This means that any decision taken throughout the demonstration phase will not be forever, but only for the time needed to do the demonstration. Our interpretation is that the main concern for political decisionmakers in a situation where they feel uncertain about the risk is to take decisions which in some sense are not possible to take back. The option of retrieval after demonstration will hopefully diminish the *"eternity"* aspects of the necessary political siting decisions in favor of more beneficial aspects like e g *advanced clean industry with many secure and attractive jobs*.

This year it is fifty years since the first nuclear chain reaction was demonstrated. It took about fifteen years from then until we had the first electricity producing reactors in operation. After another fifteen years nuclear power became a political issue mainly due to lack of demonstration of final disposal of high level waste. Twenty years of research and development under the pressure of the debate caused by the political issue has convinced the vast majority of the informed scientific community that the waste can be disposed of in a safe way by use of existing technology. We will probably need another fifteen to twenty years to physically demonstrate all practical aspects of the technology and the associated administrative process. I am convinced that once we have completed such a demonstration the nuclear waste will since long have ceased to be a political issue perhaps in favor of other more pressing problems for mankind.

I think the Stripa programme has brought an important contribution towards not *the* final but *a* final solution of the waste disposal issue. I think it is very important that we continue to explore the route!

LIST OF PARTICIPANTS   LISTE DES PARTICIPANTS

AUSTRALIA - AUSTRALIE

GOLIAN, C., Australian Nuclear Science & Technology, PMB 1, PO Menai
NSW 2234.

AUSTRIA - AUTRICHE

ITO, F., University of Innsbrück, Taisei Corporation, Technikerstrasse 13,
A-6020 Innsbrück.

BELGIUM - BELGIQUE

NEERDAEL, B., CEN/SCK, Boeretang 200, B-2400 Mol.

CANADA

DORMUTH, K. W., Atomic Energy of Canada Limited, Whiteshell Laboratories,
Pinawa, Manitoba R0E 1L0.

FLAVELLE, P., Atomic Energy Control Board, P.O. Box 1046 Station B
Ottawa, Ontario K1P 5S9.

GALE, J., Fracflow Consultants, 36 Pearson Street, St. John's,
Newfoundland A1B 3R1.

GRAY, M., AECL Research, Whiteshell Laboratories, Pinawa,
Manitoba R0E 1L0.

JOHNSON, L., AECL Research, Whiteshell Laboratories, Pinawa,
Manitoba R0E 1L0.

KUZYK, G. W., AECL Research, Whiteshell Laboratories, Pinawa,
Manitoba R0E 1L0.

LOGHA, G. S., Atomic Energy of Canada Limited, Whiteshell Laboratories,
Pinawa, Manitoba R0E 1L0.

METCALFE, D., Atomic Energy Control Board, P.O. Box 1046, Station B, 270 Albert Street, Ottawa, Ontario K1P 5S9.

ONOFREI, M., AECL Research, Whiteshell Laboratories, Pinawa, Manitoba R0E 1L0.

WHITAKER, S. H., Atomic Energy of Canada Limited, Whiteshell Laboratories, Pinawa, Manitoba R0E 1L0.

FINLAND - FINLANDE

AHOKAS, H., Fintact, Hopratie 1 B, FIN-00440 Helsinki.

ANTTILA, P., Imatran Voima Oy, P.O. Box 112, FIN-01601 Vantaa.

COSMA, C., Vibrometric Oy, Taipaleentie 117, FIN-01860 Perttula.

ELORANTA, E., Finnish Centre for Radiation and Nuclear Safety, P.O. Box 268, FIN- 00101 Helsinki.

GARDEMEISTER, R., Imatran Voima Oy, P.O. Box 112, FIN-06101 Vantaa.

HAUTOJÄRVI, A., VTT/Nuclear Engineering Laboratory, P.O. Box 208, FIN-02151 Espoo.

KALLIO, T., Imatran Voima Oy, P.O. Box 112, FIN-01601 Vantaa.

KORKEALAAKSO, J., Technical Research Centre of Finland, Traffic and Geotechnical Laboratory, P.O. Box 108, FIN-02151 Espoo.

KOSKINEN, L., Technical Research Center of Finland, Nuclear Engineering Laboratory, P.O. Box 208, FIN-02151 Espoo.

LEINO-FORSMAN, H., Technical Research Centre of Finland, Reactor Laboratory, P.O. Box 200, FIN-02151 Espoo.

NYKYRI, M., Teollisuuden Voima Oy, Annankatu 42 C, FIN-00100 Helsinki.

PALMU, J., Imatran Voima Oy, P.O. Box 112, FIN-01601 Vantaa.

PITKÄNEN, P., Technical Research Center of Finland, P.O. Box 108, FIN-02231 Espoo.

POLLA, J., VTT,Technical Research Centre of Finland, P.O. Box 108,
FIN-02151 Espoo.

RYHÄNEN, V., Teollisuuden Voima Oy, Annankatu 42 C,
FIN-00100 Helsinki.

SALO, J.-P., Teollisuuden Voima Oy, Annakatu 42 C, FIN-00100 Helsinki.

TAIVASSALO, V., VTT, Technical Research Centre of Finland, Nuclear
Engineering Laboratory, P.O. Box 208, FIN-02151 Espoo.

VIRA, J., Teollisuuden Voima Oy, Annankatu 42 C, FIN-00100 Helsinki.

FRANCE

DEWIERE, L., ANDRA, Route du Panorama Robert Schumann, B.P. 38,
F-92266 Fontenay-aux-Roses Cedex.

LEMEILLE, F., Institut de Protection et de Sûreté Nucléaire, DES/SESID,
B.P. 6, F-92265 Fontenay-aux-Roses Cedex.

MASSAL, P., BRGM - 4S/STO, Avenue de Concyr, B.P. 6009,
F-45060 Orléans Cedex.

PEAUDECERF, P., BRGM - 4S/STO, Avenue de Concyr, B.P. 6009,
F-45060 Orléans Cedex

PLAS, F., ANDRA, Route du Panorama Robert Schumann, B.P. 38,
F-92266 Fontenay-aux-Roses Cedex.

VIGNAL, B., ANDRA, Route du Panorama Robert Schumann, B.P. 38,
F-92266 Fontenay-aux-Roses Cedex.

GERMANY - ALLEMAGNE

BREWITZ, W., GSF Forschungszentrum für Umwelt und Gesundheit,
Theodor-Heuss-Strasse 4, D-3300 Braunschweig.

FUCHS, H., GNS Gesellschaft fur Nuklear-Service mbH, Lange Laube 7,
D-3000 Hannover 1.

HOFFMANN, U., Bavarian Geological Survey, Hessstrasse 128,
D-8000 München 40.

JANBERG, K., GNS Gesellschaft fur Nuklear-Service mbH,
Zweigertstrasse 28-30, D-4300 Essen 1.

LIDTKE, L., c/o Federal Institute for Geosciences and Natural Resources
Stilleweg 2, D- Hannover 3.

MIRSCHINKA, V., RWE Engergie AG, Kruppstrasse 5, D-4300 Essen 1.

ITALY - ITALIE

GERA, F., ISMES S.p.A., Via dei Crociferi 44, I-00187 Roma.

JAPAN - JAPON

FUJIWARA, A., Radioactive Waste Management Center, No 15 Mori Building,
2-8-10 Toranomon, Minato-ku, Tokyo.

GOTO, K., JGC, c/o 2-1-1 Uchisaiwai-cho, Chiyoda-ku, Tokyo.

HIRATA, Y., Taisei Kiso Sekkei Co. Ltd, c/o 2-1-1 Uchisaiwai-cho, Chiyoda-ku
Tokyo.

KAMEMURA, K, Shinozuka Research Institute, 5 F Maguna Kogyo Building,
1-31-13 Yoyogi, Shibuya-ku, Tokyo.

KOJIMA, K., University of Tokyo, c/o 2-1-1 Uchisaiwai-cho, Chiyoda-ku,
Tokyo.

MINAGAWA, J., Daia Consultants Co. Ltd, c/o 2-1-1 Uchisaiwai-cho,
Chiyoda-ku, Tokyo.

MORITA, A., Nuclear Safety Research Association, c/o 2-1-1 Uchisaiwai-cho,
Chiyoda-ku, Tokyo.

MURANO, T., The Institute of Applied Energy, Shinbashi SY Bldg.,
14-2 Nishishinbashi, 1-Chome, Minato-ku, Tokyo 105.

NAKAGOSHI, A., Hazama Corporation, c/o 2-1-1 Uchisaiwai-cho, Chiyoda-ku
Tokyo.

NAKAMURA, H., Radioactive Waste Management Center, No. 15 Mori Bldg, 2-8-10 Toranomon, Minato-ku, Tokyo 105.

OKAMURA, J., Kojima Corporation, c/o 2-1-1 Uchisaiwai-cho, Chiyoda-ku, Tokyo.

SAKUMA, H., PNC, 1-9-13 Akasaka, Minato-ku, Tokyo 107.

SATO, H., Nuclear Safety Research Association, c/o 2-1-1 Uchisaiwai-cho, Chiyoda-ku, Tokyo.

SHIGENO, Y., Takenaka Corporation, c/o 2-1-1 Uchisaiwai-cho, Chiyoda-ku, Tokyo.

TANAKA, M., Kyoto University, c/o 2-1-1 Uchisaiwai-cho, Chiyoda-ku, Tokyo.

UOKAWA, N., Maeda Corporation, c/o 2-1-1 Uchisaiwai-cho, Chiyoda-ku, Tokyo.

UTSUGIDA, Y., Shimizu Corporation, c/o 2-1- Uchisaiwai-cho, Chiyoda-ku, Tokyo.

WATANABE, K., Saitama University, Faculty Engineering, 255 Shimo Okubo, Urawa Saitama 338.

YAMAMOTO, M., Sato Kogyo Co. Ltd, c/o 2-1-1 Uchisaiwai-cho, Chiyoda-ku, Tokyo.

KOREA (REP. OF) - COREE (REP. DE)

HWANG, Y.S., Nemac/Kaeri, Safety Analysis Department, Daeduk-danji, P.O. Box 7, Taejeon.

LEE, S. K., Korea Institute of Geology, Mining and Materials, 30 Kajungdong, Yusongku, Taejeon, 305-350.

RHEE, C. G., Korea Atomic Energy Research Institute, 305-606 P.O. Box 7, Daeduk-Danji Taejeon.

NORWAY - NORVEGE

BARTON, N., Norwegian Geotechnical Institute, P.O. Box 40 Tasen, N-0801 Oslo 8.

RUSSIAN FEDERATION - FEDERATION DE RUSSIE

PEVZNER, L. Abramovich, Company of Superdeep Drilling and Comprehensive Study of the Earth "Nedra", Volkushi 28, 150003 Yaroslavl.

SPAIN - ESPAGNE

ASTUDILLO, J., ENRESA, Emilio Vargas 7, 28043 Madrid.

BAJOS PARADA, C., ENRESA, Emilio Vargas 7, 28043 Madrid.

MAYOR, J. C., ENRESA, 7 Emilio Vargas, 28043 Madrid.

SWEDEN - SUEDE

AGREN, T., KEMAKTA, Box 12655, 112 93 Stockholm.

AHAGEN, H., SINTHIS, Skaldevägen 62-64, 161 42 Bromma.

AHLSTROM, P.-E., SKB, Box 5864, 102 48 Stockholm.

ALMEN, K. E., SKB, Box 5864, 102 48 Stockholm.

ANDERSSON, J., Swedish Nuclear Power Inspectorate, Box 27106, 102 52 Stockholm.

ANDERSSON, P., BeFo, Swedish Rock Engineering Research Foundation, Box 5501, 114 85 Stockholm.

ANDERSSON, O., IPA-Konsult AB, Box 622, 572 27 Oskarshamn.

ANNERTZ, K., SKB/Äspö Hard Rock Laboratory, PL 300, 570 93 Figeholm.

APELQVIST, G., Vattenfall, Koncernutveckling, 162 87 Vällingby.

BÄCKBLOM, G., SKB, Box 5864, 102 48 Stockholm.

BARRDAHL, R., SSI, P.O. Box 60204, 104 01 Stockholm.

BIRGERSSON, L., Kemakta Konsult AB, Box 12655, 112 93 Stockholm.

BJURSTRÖM, S., SKB, Box 5864, 102 48, Stockholm.

BÖRGESSON, L., Clay Technology, IDEON, 223 70 Lund.

CARLSSON, H., SGAB International AB, Box 4607, 116 91 Stockholm.

CARLSSON, J., SKB, Box 5864, 102 48 Stockholm.

CARLSTEN, S., Geosigma AB, Box 894, 751 08 Uppsala.

DE C. PEREIRA, A., Stockholm University, Fysikum, Vanadisvägen 9, 113 46 Stockholm.

DVERSTORP, B., SKI, Box 27106, 102 52 Stockholm.

ENG, T., SKB, Box 5864, 102 48 Stockholm.

ERICSSON, L. O., SKB, Box 5864, 102 48 Stockholm.

FRANZEN, T., Swedish Rock Engineering Research Foundation, BeFo, Box 5501, 114 85 Stockholm.

GEIER, J., Golder Geosystem AB, Björkgatan 73, 753 23 Uppsala.

GRUNDFELT, B., Kemakta Konsult AB, Box 12655, 112 93 Stockholm.

GUDOWSKI, W., Royal Institute of Technology, Department of Neutron and Reactor Physics, 100 44 Stockholm.

GUSTAFSON, G., Chalmers, Institute of Geology, 412 96 Göteborg.

GUSTAFSSON, B., SKB, Box 5864, 102 48 Stockholm.

HAKAMI, E., KTH, Teknisk Geologi, 100 44 Stockholm.

ISANDER, A., Sydkraft Konsult AB, 205 09 Malmö.

JENSEN, M., SSI, Box 60204, 104 01 Stockholm.

JING, L., KTH, Engineering Geology, 100 44 Stockholm.

KARLSSON, F., SKB, Box 5864, 102 48 Stockholm.

KAUTSKY, F., SKI, Swedish Nuclear Power Inspectorate, Box 27106, 102 52 Stockholm.

KUNG, C. S., Kungliga Tekniska Högskolan, Vattenvardsteknik, 100 44 Stockholm.

LARSON TULLBORG, E.-L., Terralogica, Box 140, 440 06 Grabo.

MARKSTRÖM, I., Sydkraft Konsult AB, 205 09 Malmö.

MIYAKAWA , K., Criepi, c/o SKB, Äspö Hard Rock Laboratory, Pl300, 570 93 Figeholm.

MUNIER, R., Uppsala University, Love Almquist väg 12, 112 53 Stockholm.

NERETNIEKS, I., KTH, Department of Chemical Engineering, 100 44 Stockholm.

NILSSON, L. B., SKB, Box 5864, 102 48 Stockholm.

NORDIN, S., Vattenfall AB, 162 87 Vällingby.

NORDLUND, E., Lulea University of Technology, 941 87 Lulea.

NORRBY, S., SKI, Box 27106, 102 52 Stockholm.

OLSSON, O., Conterra AB, Box 493, 751 06 Uppsala.

OLSSON, T., Golder Geosystem AB, Björkgatan 73, 753 23 Uppsala.

OSAWA, H., Power Reactor and Nuclear Fuel Develop., SKB, c/o Äspö Hard Rock Laboratory, Pl300, 572 93 Figeholm.

PAPP, T., SKB, Box 5864, 102 48 Stockholm.

PUSCH, R., Clay Technology AB, IDEON, 223 70 Lund

RAMQVIST, G., Elterkno AB, Gruvvägen 1 Ställberg, 714 00 Kopparberg.

REHN, I., VBB V lak, Box 2203, 403 14 Göteborg 2.

RYDELL, N., KASAM, Miljödepartementet, 103 33 Stockholm.

SHAIKH, N.-A., Geological Survey of Sweden, Box 670, 751 28 Uppsala.

SÖDERBERG, O., KASAM, Ellagstiftningen c/o SKN, Box 27824,
115 93 Stockholm.

STANFORS, R., Kiliamsgatan 10, 223 50 Lund.

STENBERG, L., SKB/Äspö Hard Rock Laboratory, PL300, 570 93 Figeholm.

STEPHANSSON, O., KTH, Teknikringen 72, 100 44 Stockholm.

STIGH, J., KASAM, Geologiska Institutionen, Chalmers, Göteborgs Universitet,
Göteborg.

STILLBORG, B., SKB, BOX 5864, 102 48 Stockholm.

STRÖM, A., SKB, BOX 5864, 102 48 Stockholm.

SVEMAR, C., SKB, Box 5864, 102 48 Stockholm.

TALBOT, C., Uppsala University, Institute of Geology, Box 555,
751 22 Uppsala.

TOVERUD, Ö., SKI, Box 27106, 102 52 Stockholm.

WIKBERG, P., SKB, Box 5864, 102 48 Stockholm.

ZELLMAN, O., SKB/Äspö Hard Rock Laboratory, PL300, 570 93 Figeholm.

SWITZERLAND - SUISSE

ALBERT, W., NAGRA, Att: R. Lieb, Hardstrasse 73, 5430 Wettingen.

BECK, R., Sagestrasse 4 F, 8371 Oberwangen TG.

CORREA, N., NAGRA, Att: R. Lieb, Hardstrasse 73, 5430 Wettingen.

LIEB, R., NAGRA, Hardstrasse 73, 5430 Wettingen.

MARSCHALL, P, NAGRA, Att. R. Lieb, Hardstrasse 73, 5430 Wettingen.

TAKEUCHI, K., Obayashi Corporation (Currently NAGRA), NAGRA, Engineering Department, Hardstrasse 73, 5430 Wettingen.

TAIWAN

CHYEN, P.-C., Taiwan Power Company, 4F. 2, Alley 15, Lane 196, Sec. 4, Roosevelt Road, Taipei.

CHYR, G.-T., Taiwan Power Company, 4F. 2, Alley 15, Lane 196, sec. 4, Roosevelt Road, Taipei.

LIU, W.-C., Radwaste Administration, Atomic Energy Council, 65, Lane 144, Keelung Road, Section 4, Taipei.

UNITED KINGDOM - ROYAUME-UNI

BLACK, J., Golder Associates UK Ltd, Wheatcroft Buildings, Landmere Lane, Edwalton, Nottingham NG12 4DE.

BRIGHTMAN, M., Golder Associates UK Ltd, Wheatcroft Buildings, Landmere Lane, Edwalton, Nottingham NG12 4DE.

CHAPMAN, N., Intera Information Technologies, 14B, Burton Street, Melton Mowbray.

DERSHOWITZ, W., Golder Associates UK Ltd, Wheatcroft Buildings, Landmere Lane, Edwalton, Nottingham NG12 4DE.

FRANCIS, A., United Kingdom Nirex Limited, Curie Avenue, Harwell, Didcot, Oxfordshire OX1 1 ORH.

HERBERT, A., AEA Decommisioning + Radwaste, Hydrogeology Department, Harwell Laboratory, Didcot, Oxfordshire OX11 ORH.

HODGKINSON, D., Intera Information Technologies, Chiltern House, 45 Station Road, Henley-on-Thames, Oxfordshire.

JACKSON, R., Department of Environment, Radioactive Substances Division, Romney House, 43 Marsham St., London SW1 P 3PY.

LANYON, G., Geoscience Ltd, Falmouth Business Park, Falmouth, Cornwall.

LEWIS, H., BNFL, R 410, Risley, Warrington, Cheshire.

LITTLEBOY, A., United Kingdom Nirex Limited, Curie Avenue, Harwell, Didcot, Oxfordshire OX11 ORH.

LLOYD, J., BNFL, R 410, Risley, Warrington, Cheshire.

STEADMAN, J., Building Research Establishment, Garston, Watford WD2 7 JR.

YEARSLEY, R., Her Majesty's Inspectorate of Pollution, Technical Policy Division, Romney House, 43 Marsham Street, London SW1 P 3PY.

UNITED STATES - ETATS-UNIS D'AMERIQUE

ALCORN, S. R., RE/SPEC Inc., 4775 Indian School Road NE, Albuquerque, NM 87110.

BARR, G. E., Sandia National Laboratories, P.O. Box 5800, Department 6312, Albuquerque, NM 87185-5800.

COONS, W. E., RE/Spec Inc., 4775 Indian School Road, N.E. Suite 300, Albuquerque, NM.

DANKER, W., US Department of Energy, 1000 Independence Avenue, SW, Washington. DC 20585.

DAVIS, S., University of Arizona, Department of Hydrology & Water Resource, Tucson, AZ 85721.

DOCKERY, H. A., Sandia National Laboratories, P.O. Box 5800, Department 6312, Albuquerque, NM 87185-5800.

DOE, T., Golder Associates, 4104 148th Avenue NE, Redmond, WA.

FAIRHURST, C., University of Minnesota, Department of Civil and Mineral Engin., 500 Pillsbury Drive S.E, Minneapolis, MN 55455-0220.

GNIRK, P., Table Top Consultants, HCR 33 -- Box 1202, Rapid City, SD 57701.

GODMAN, R, TRW Environmental Saftey Systems Inc., 2650 Park Tower Drive, Suite 800, Vienna, VA 22180.

HUNTER, T., Sandia National Laboratories, P.O. Box 5800, Albuquerque, NM 87185-5800.

ISAACS, T., US Department of Energy, 1000 Independence Avenue, SW, Washington, DC 20585.

LEVICH, B., Yucca Mountain Site Characterisation Proj. Of., 101 Convention Center Drive, P.O. Box 98608, Las Vegas, Nevada 89193.

LONG, J., Lawrence Berkeley Laboratory, One Cyclotron Road, Building 50 E, Berkeley, CA 94720.

OSTENSEN, R. W., Sandia National Laboratories, Department 6119, Albuquerque, NM 87185-5800.

WEBB, S. W., Sandia National Laboratories, P.O. Box 5800, Department 6119, Albuquerque, NM 87111.

INTERNATIONAL ATOMIC ENERGY AGENCY/AGENCE DE L'ENERGIE ATOMIQUE

BELL, M., International Atomic Energy Agency, P.O. Box 100, Wagramerstrasse 5, A-1400 Vienna.

OECD NUCLEAR ENERGY AGENCY/AGENCE DE L'OCDE POUR L'ENERGIE NUCLEAIRE

OLIVIER, J.-P., OECD/NEA, Le Seine-St-Germain, 12, boulevard des Iles, F-92130 Issy-les-Moulineaux.

PATERA, E.S., OECD/NEA, Le Seine-St-Germain, 12, boulevard des Iles, F-92130 Issy-les-Moulineaux.

MAIN SALES OUTLETS OF OECD PUBLICATIONS
PRINCIPAUX POINTS DE VENTE DES PUBLICATIONS DE L'OCDE

ARGENTINA – ARGENTINE
Carlos Hirsch S.R.L.
Galería Güemes, Florida 165, 4° Piso
1333 Buenos Aires   Tel. (1) 331.1787 y 331.2391
Telefax: (1) 331.1787

AUSTRALIA – AUSTRALIE
D.A. Information Services
648 Whitehorse Road, P.O.B 163
Mitcham, Victoria 3132   Tel. (03) 873.4411
Telefax: (03) 873.5679

AUSTRIA – AUTRICHE
Gerold & Co.
Graben 31
Wien I   Tel. (0222) 533.50.14

BELGIUM – BELGIQUE
Jean De Lannoy
Avenue du Roi 202
B-1060 Bruxelles   Tel. (02) 538.51.69/538.08.41
Telefax: (02) 538.08.41

CANADA
Renouf Publishing Company Ltd.
1294 Algoma Road
Ottawa, ON K1B 3W8   Tel. (613) 741.4333
Telefax: (613) 741.5439
Stores:
61 Sparks Street
Ottawa, ON K1P 5R1   Tel. (613) 238.8985
211 Yonge Street
Toronto, ON M5B 1M4   Tel. (416) 363.3171
Telefax: (416)363.59.63
Les Éditions La Liberté Inc.
3020 Chemin Sainte-Foy
Sainte-Foy, PQ G1X 3V6   Tel. (418) 658.3763
Telefax: (418) 658.3763

Federal Publications Inc.
165 University Avenue, Suite 701
Toronto, ON M5H 3B8   Tel. (416) 860.1611
Telefax: (416) 860.1608

Les Publications Fédérales
1185 Université
Montréal, QC H3B 3A7   Tel. (514) 954.1633
Telefax : (514) 954.1635

CHINA – CHINE
China National Publications Import
Export Corporation (CNPIEC)
16 Gongti E. Road, Chaoyang District
P.O. Box 88 or 50
Beijing 100704 PR   Tel. (01) 506.6688
Telefax: (01) 506.3101

DENMARK – DANEMARK
Munksgaard Book and Subscription Service
35, Nørre Søgade, P.O. Box 2148
DK-1016 København K   Tel. (33) 12.85.70
Telefax: (33) 12.93.87

FINLAND – FINLANDE
Akateeminen Kirjakauppa
Keskuskatu 1, P.O. Box 128
00100 Helsinki
Subscription Services/Agence d'abonnements :
P.O. Box 23
00371 Helsinki   Tel. (358 0) 12141
Telefax: (358 0) 121.4450

FRANCE
OECD/OCDE
Mail Orders/Commandes par correspondance:
2, rue André-Pascal
75775 Paris Cedex 16   Tel. (33-1) 45.24.82.00
Telefax: (33-1) 49.10.42.76
Telex: 640048 OCDE

OECD Bookshop/Librairie de l'OCDE :
33, rue Octave-Feuillet
75016 Paris   Tel. (33-1) 45.24.81.67
(33-1) 45.24.81.81
Documentation Française
29, quai Voltaire
75007 Paris   Tel. 40.15.70.00
Gibert Jeune (Droit-Économie)
6, place Saint-Michel
75006 Paris   Tel. 43.25.91.19
Librairie du Commerce International
10, avenue d'Iéna
75016 Paris   Tel. 40.73.34.60
Librairie Dunod
Université Paris-Dauphine
Place du Maréchal de Lattre de Tassigny
75016 Paris   Tel. (1) 44.05.40.13
Librairie Lavoisier
11, rue Lavoisier
75008 Paris   Tel. 42.65.39.95
Librairie L.G.D.J. - Montchrestien
20, rue Soufflot
75005 Paris   Tel. 46.33.89.85
Librairie des Sciences Politiques
30, rue Saint-Guillaume
75007 Paris   Tel. 45.48.36.02
P.U.F.
49, boulevard Saint-Michel
75005 Paris   Tel. 43.25.83.40
Librairie de l'Université
12a, rue Nazareth
13100 Aix-en-Provence   Tel. (16) 42.26.18.08
Documentation Française
165, rue Garibaldi
69003 Lyon   Tel. (16) 78.63.32.23
Librairie Decitre
29, place Bellecour
69002 Lyon   Tel. (16) 72.40.54.54

GERMANY – ALLEMAGNE
OECD Publications and Information Centre
August-Bebel-Allee 6
D-53175 Bonn   Tel. (0228) 959.120
Telefax: (0228) 959.12.17

GREECE – GRÈCE
Librairie Kauffmann
Mavrokordatou 9
106 78 Athens   Tel. (01) 32.55.321
Telefax: (01) 36.33.967

HONG-KONG
Swindon Book Co. Ltd.
13–15 Lock Road
Kowloon, Hong Kong   Tel. 366.80.31
Telefax: 739.49.75

HUNGARY – HONGRIE
Euro Info Service
Margitsziget, Európa Ház
1138 Budapest   Tel. (1) 111.62.16
Telefax : (1) 111.60.61

ICELAND – ISLANDE
Mál Mog Menning
Laugavegi 18, Pósthólf 392
121 Reykjavik   Tel. 162.35.23

INDIA – INDE
Oxford Book and Stationery Co.
Scindia House
New Delhi 110001   Tel.(11) 331.5896/5308
Telefax: (11) 332.5993
17 Park Street
Calcutta 700016   Tel. 240832

INDONESIA – INDONÉSIE
Pdii-Lipi
P.O. Box 269/JKSMG/88
Jakarta 12790   Tel. 583467
Telex: 62 875

IRELAND – IRLANDE
TDC Publishers – Library Suppliers
12 North Frederick Street
Dublin 1   Tel. (01) 874.48.35
Telefax: (01) 874.84.16

ISRAEL
Praedicta
5 Shatner Street
P.O. Box 34030
Jerusalem 91430   Tel. (2) 52.84.90/1/2
Telefax: (2) 52.84.93

ITALY – ITALIE
Libreria Commissionaria Sansoni
Via Duca di Calabria 1/1
50125 Firenze   Tel. (055) 64.54.15
Telefax: (055) 64.12.57
Via Bartolini 29
20155 Milano   Tel. (02) 36.50.83
Editrice e Libreria Herder
Piazza Montecitorio 120
00186 Roma   Tel. 679.46.28
Telefax: 678.47.51
Libreria Hoepli
Via Hoepli 5
20121 Milano   Tel. (02) 86.54.46
Telefax: (02) 805.28.86
Libreria Scientifica
Dott. Lucio de Biasio 'Aeiou'
Via Coronelli, 6
20146 Milano   Tel. (02) 48.95.45.52
Telefax: (02) 48.95.45.48

JAPAN – JAPON
OECD Publications and Information Centre
Landic Akasaka Building
2-3-4 Akasaka, Minato-ku
Tokyo 107   Tel. (81.3) 3586.2016
Telefax: (81.3) 3584.7929

KOREA – CORÉE
Kyobo Book Centre Co. Ltd.
P.O. Box 1658, Kwang Hwa Moon
Seoul   Tel. 730.78.91
Telefax: 735.00.30

MALAYSIA – MALAISIE
Co-operative Bookshop Ltd.
University of Malaya
P.O. Box 1127, Jalan Pantai Baru
59700 Kuala Lumpur
Malaysia   Tel. 756.5000/756.5425
Telefax: 757.3661

MEXICO – MEXIQUE
Revistas y Periodicos Internacionales S.A. de C.V.
Florencia 57 - 1004
Mexico, D.F. 06600   Tel. 207.81.00
Telefax : 208.39.79

NETHERLANDS – PAYS-BAS
SDU Uitgeverij Plantijnstraat
Externe Fondsen
Postbus 20014
2500 EA's-Gravenhage   Tel. (070) 37.89.880
Voor bestellingen: Telefax: (070) 34.75.778

**NEW ZEALAND
NOUVELLE-ZÉLANDE**
Legislation Services
P.O. Box 12418
Thorndon, Wellington   Tel. (04) 496.5652
Telefax: (04) 496.5698

NORWAY – NORVÈGE
Narvesen Info Center – NIC
Bertrand Narvesens vei 2
P.O. Box 6125 Etterstad
0602 Oslo 6   Tel. (022) 57.33.00
                                     Telefax: (022) 68.19.01

PAKISTAN
Mirza Book Agency
65 Shahrah Quaid-E-Azam
Lahore 54000   Tel. (42) 353.601
                                     Telefax: (42) 231.730

PHILIPPINE – PHILIPPINES
International Book Center
5th Floor, Filipinas Life Bldg.
Ayala Avenue
Metro Manila   Tel. 81.96.76
                                     Telex 23312 RHP PH

PORTUGAL
Livraria Portugal
Rua do Carmo 70-74
Apart. 2681
1200 Lisboa   Tel.: (01) 347.49.82/5
                                     Telefax: (01) 347.02.64

SINGAPORE – SINGAPOUR
Gower Asia Pacific Pte Ltd.
Golden Wheel Building
41, Kallang Pudding Road, No. 04-03
Singapore 1334   Tel. 741.5166
                                     Telefax: 742.9356

SPAIN – ESPAGNE
Mundi-Prensa Libros S.A.
Castelló 37, Apartado 1223
Madrid 28001   Tel. (91) 431.33.99
                                     Telefax: (91) 575.39.98

Libreria Internacional AEDOS
Consejo de Ciento 391
08009 – Barcelona   Tel. (93) 488.30.09
                                     Telefax: (93) 487.76.59

Llibreria de la Generalitat
Palau Moja
Rambla dels Estudis, 118
08002 – Barcelona
                   (Subscripcions) Tel. (93) 318.80.12
                   (Publicacions) Tel. (93) 302.67.23
                                     Telefax: (93) 412.18.54

SRI LANKA
Centre for Policy Research
c/o Colombo Agencies Ltd.
No. 300-304, Galle Road
Colombo 3   Tel. (1) 574240, 573551-2
                       Telefax: (1) 575394, 510711

SWEDEN – SUÈDE
Fritzes Information Center
Box 16356
Regeringsgatan 12
106 47 Stockholm   Tel. (08) 690.90.90
                                     Telefax: (08) 20.50.21

Subscription Agency/Agence d'abonnements :
Wennergren-Williams Info AB
P.O. Box 1305
171 25 Solna   Tel. (08) 705.97.50
                                     Téléfax : (08) 27.00.71

SWITZERLAND – SUISSE
Maditec S.A. (Books and Periodicals - Livres
et périodiques)
Chemin des Palettes 4
Case postale 266
1020 Renens   Tel. (021) 635.08.65
                                     Telefax: (021) 635.07.80

Librairie Payot S.A.
4, place Pépinet
CP 3212
1002 Lausanne   Tel. (021) 341.33.48
                                     Telefax: (021) 341.33.45

Librairie Unilivres
6, rue de Candolle
1205 Genève   Tel. (022) 320.26.23
                                     Telefax: (022) 329.73.18

Subscription Agency/Agence d'abonnements :
Dynapresse Marketing S.A.
38 avenue Vibert
1227 Carouge   Tel.: (022) 308.07.89
                                     Telefax : (022) 308.07.99

See also – Voir aussi :
OECD Publications and Information Centre
August-Bebel-Allee 6
D-53175 Bonn (Germany)   Tel. (0228) 959.120
                            Telefax: (0228) 959.12.17

TAIWAN – FORMOSE
Good Faith Worldwide Int'l. Co. Ltd.
9th Floor, No. 118, Sec. 2
Chung Hsiao E. Road
Taipei   Tel. (02) 391.7396/391.7397
                           Telefax: (02) 394.9176

THAILAND – THAÏLANDE
Suksit Siam Co. Ltd.
113, 115 Fuang Nakhon Rd.
Opp. Wat Rajbopith
Bangkok 10200   Tel. (662) 225.9531/2
                                     Telefax: (662) 222.5188

TURKEY – TURQUIE
Kültür Yayinlari Is-Türk Ltd. Sti.
Atatürk Bulvari No. 191/Kat 13
Kavaklidere/Ankara   Tel. 428.11.40 Ext. 2458
Dolmabahce Cad. No. 29
Besiktas/Istanbul   Tel. 260.71.88
                                     Telex: 43482B

UNITED KINGDOM – ROYAUME-UNI
HMSO
Gen. enquiries   Tel. (071) 873 0011
Postal orders only:
P.O. Box 276, London SW8 5DT
Personal Callers HMSO Bookshop
49 High Holborn, London WC1V 6HB
                                     Telefax: (071) 873 8200
Branches at: Belfast, Birmingham, Bristol, Edin-
burgh, Manchester

UNITED STATES – ÉTATS-UNIS
OECD Publications and Information Centre
2001 L Street N.W., Suite 700
Washington, D.C. 20036-4910 Tel. (202) 785.6323
                                Telefax: (202) 785.0350

VENEZUELA
Libreria del Este
Avda F. Miranda 52, Aptdo. 60337
Edificio Galipán
Caracas 106   Tel. 951.1705/951.2307/951.1297
                           Telegram: Libreste Caracas

Subscription to OECD periodicals may also be
placed through main subscription agencies.

Les abonnements aux publications périodiques de
l'OCDE peuvent être souscrits auprès des
principales agences d'abonnement.

Orders and inquiries from countries where Distribu-
tors have not yet been appointed should be sent to:
OECD Publications Service, 2 rue André-Pascal,
75775 Paris Cedex 16, France.

Les commandes provenant de pays où l'OCDE n'a
pas encore désigné de distributeur devraient être
adressées à : OCDE, Service des Publications,
2, rue André-Pascal, 75775 Paris Cedex 16, France.

OECD PUBLICATIONS, 2 rue André-Pascal, 75775 PARIS CEDEX 16
PRINTED IN FRANCE
(66 94 02 1) ISBN 92-64-14225-8 - No. 46968 1994